WIRELESS EXPERIMENTER'S MANUAL

BY

ELMER E. BUCHER

Fredonia Books
Amsterdam, The Netherlands

The Wireless Experimenter's Manual

by
Elmer E. Bucher

ISBN: 1-4101-0803-1

Copyright © 2005 by Fredonia Books

Reprinted from the 1920 edition

Fredonia Books
Amsterdam, The Netherlands
http://www.fredoniabooks.com

In order to make original editions of historical works available to scholars at an economical price, this facsimile of the original edition of 1920 is reproduced from the best available copy and has been digitally enhanced to improve legibility, but the text remains unaltered to retain historical authenticity.

AUTHOR'S NOTE

It is an error of statement to characterize the whole army of wireless experimenters in the United States as mere "amateurs." Hundreds of men, young and old, who have been classed as wireless amateurs are in fact physicists of the highest calibre. They find experimentation in radio not only an instructive mode of recreation, but many of them engage in the work with serious intentions hoping thereby to contribute their mite towards the general progress of the art. That many have made good in that respect is a matter of historical record.

The amateur wireless experimenter is no longer considered a menace. His status is now settled. He proved his worth in the recent European conflict as Government officials have publicly acknowledged. The amateur radio expert today is recognized as a safeguard against possible future emergencies. He has the backing of America's foremost scientists, the good will of the Army and Navy, and the commercial company sees in him a potential engineer or expert operator.

Wireless Telegraphy is an all-embracing art. It covers in some degree the fields of the chemical, mechanical, and electrical sciences. No young man can engage seriously in wireless experimenting without becoming a keener and more intelligent individual of greater value to the community. The amazing growth of the amateur fraternity during the past decade is one of the outstanding developments of the twentieth century.

In the following pages the author has endeavored to acquaint the beginner with recent and past practices, laying particular stress on latter day developments. It is hoped that the brief treatment of the elementary theory of wireless transmitting and receiving apparatus will be meditorial to more intensive study of the basic principles Although many suggestive designs of radio transmitters and receivers are presented the reader should recognize the possible latitude of variation from the fundamental idea, and use his own initiative in altering the mechanical details or even in devising new methods to accord with the material he has at hand.

The author has attempted to point out the particular types of apparatus and circuits most suited to the amateur station. The vacuum tube transmitter and receivers have been given more than the usual attention. Representative circuits for long distance reception are shown and for the first time data on closed loop aerials for indoor reception is disclosed. Simple but important radio measurements, such as the amateur may carry out with a wave-meter and associated apparatus, have been explained. It is hoped that the amateur will investigate the direction finder and make use of it for both long and short distance communication.

Another field for intensive experiment is the application of recently developed static eliminators to amateur communications. Here, so far as the amateur is concerned, is an untouched sphere.

For a more comprehensive treatment of the mathematical theory of radio design many text-books are available which the reader should consult from time to time. If the experimenter by perusing this volume, gains a better understanding of the equipment he has at hand, or is enabled to construct apparatus that, heretofore, he did not understand, the book will have fulfilled the author's desire.

E. E. B.

New York City, May 1919.

CONTENTS

CHAPTER I

ADVICE TO THE AMATEUR

It is generally conceded that *amateur wireless telegraphy* had its beginning in the United States. In fact the pursuit of the hobby has become so widespread that there is hardly a village or hamlet in which there is not at least one trailing antenna to show the popularity of the art. In all places where amateur wireless telegraphy has made a niche for itself the advantages of forming a radio club are sooner or later recognized and then arises the question, "How shall we go about it?"

A *radio club* should be educational, instructive, and productive of advancement in wireless, for it is quite possible that some of its members will, in time, develop into radio engineers or become connected with allied branches of the art. In any event the members of radio clubs will never regret the training they will receive by joining wireless organizations.

It was not found feasible in preparing this book to treat separately the technical requirements of the organization and individual members; consequently the following pages will be devoted largely to instructions for the amateur at his home station as well as for the club. The formation of a radio club generally follows the rise and stimulation of individual interest. Moreover the knowledge gained at home is bound to benefit the members of the club organization as a whole.

Thus, we will first offer advice for prospective wireless club members and describe the procedure for the formation of an organization, then give a detailed account of a series of experiments with radio transmitting and receiving apparatus.

PRELIMINARY EDUCATION.—It should be kept in mind that an interchange of signals between two amateur stations will not be allowed unless the owners possess United States operators' *license certificates* and *station license certificates*. It is obvious that the prospective amateur is not qualified to take the examination for an operator's license without preliminary code and technical training.

1

FIG. 1.   Students of the Marconi Institute acquiring familiarity with measurements of wave lengths and frequency

The apparatus of the amateur who cannot interpret the signals of the *international continental telegraph code* becomes to its owner a useless toy and attempts at communication with his friends—even if allowed—will result in hopeless bungling. He should, therefore, begin immediately the practice of the telegraph code. This he cannot do alone. He requires assistance from an amateur friend who has already passed this stage of development and who will send to him from a practice buzzer for hours at a time.

It is difficult to recognize the characters of the code when sent by another and it is a process which requires considerable practice. While it is necessary for the beginner to attain a certain degree of proficiency in this respect, it is not essential that he wait until becoming a high speed telegrapher before entering the amateurs' ranks. In fact when he attains a receiving speed of fifteen words a minute he is quite eligible to engage in wireless transmission and reception.

PROFICIENCY IN THE TELEGRAPH CODES.—The laws of the United States require that amateurs, before acquiring an operator's license certificate, shall be able to transmit and receive in the *international telegraph code* at the rate of *ten words per minute.* Irrespective of this legislation it would be of little benefit for the owner to possess a radio set without the ability to operate it and he would, at the same time, be a menace to the profession as a whole.

The interrupted currents of an *electric buzzer*, adjusted to a high pitch, give, in the head telephone, a very faithful reproduction of wireless signals. In fact any device that will impulse a telephone receiver from 500 to 1000 times per second is serviceable for code practice.

FIG. 2. The laboratory of the up-to-date amateur represents a maze of intricacies to the beginner but it is ever a source of pleasure to its owner. The station here shown has performed notable long distance receiving work.

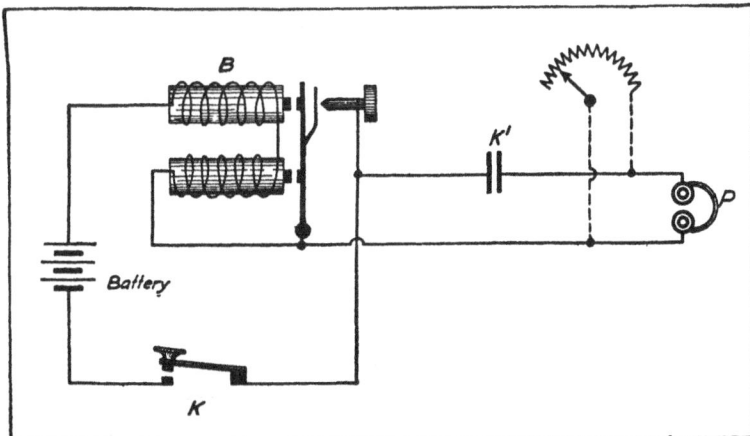

FIG. 3 Buzzer circuits for the production of artificial radio signals. With this apparatus the beginner acquires proficiency in the telegraph codes.

Fig. 3 shows a buzzer, head telephone and key connected up for code practice; Fig. 4 another system for practice by several club members simultaneously which differs from Fig. 3 in that a motor is employed to generate the telephone signal; and Fig. 5 represents a more elaborate plan where two tables, one for the instructor, and the other for the pupils, are connected to a buzzer. The complete wiring is shown.

In Fig. 3 a line for the head telephones is shunted across the buzzer contacts with a 2 microfarad condenser K' connected in series. The latter can be purchased from any telephone supply house. The buzzer B is energized by the one or two dry cells. The pupil's key (not shown) is shunted across the main key K, so that the receiving operator can interrupt or communicate with the instructor.

To imitate the note of a 500 cycle transmitter the buzzer should be adjusted to interrupt the current, say, 1000 times per second. Any buzzer may be adjusted to emit a high pitched note if the elastic spring contact on the armature is fastened directly to the iron. The buzzer itself will not make much noise, but a very clear and, perhaps, loud tone will be obtained in the telephones. Some buzzers are notoriously erratic in operation and most difficult to keep in continuous adjustment but there are several continuously operative types, sold by electrical supply houses, which are very dependable.

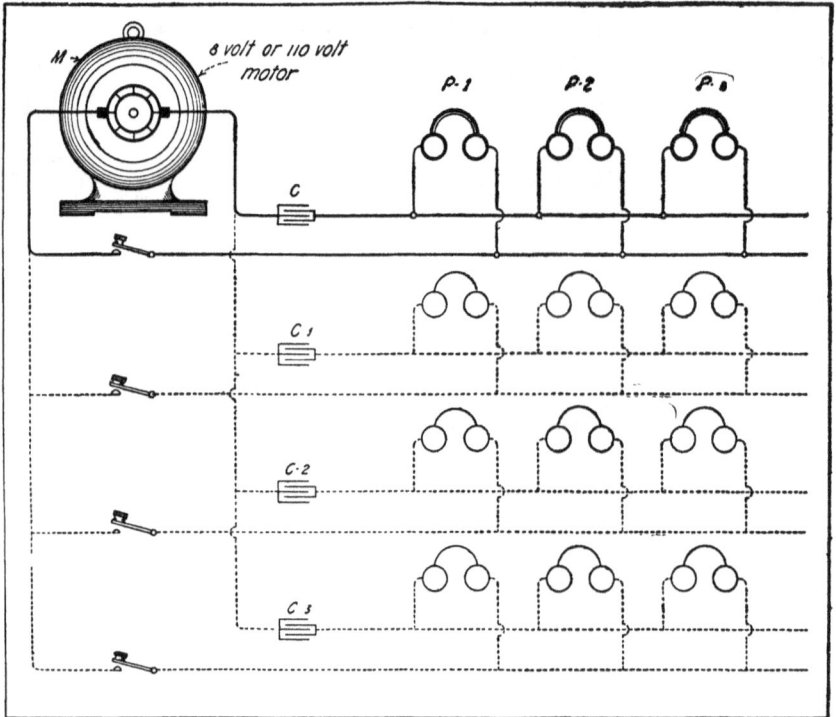

Fig 4   Code practice circuit employing a small motor to generate signal currents   The head telephones with a 1 or 2 mfd. condenser in series are shunted across the brushes of the motor.

Fig. 4 shows the preferred system for code practice. The telephones *P*-1, *P*-2, *P*-3 are connected across the commutator of a d.c. motor and, as usual, a condenser *C* of about 2 microfarads is connected in series. The voltage fluctuations across the brushes produce notes in the telephones that equal the best buzzers, with the proper type of motor. Some motors rotate at such speeds that the voltage fluctuations produced by the commutator are above the practical limits of audibility for the ordinary telephone. With such types a series rheostat should be placed in the line to reduce the speed and, consequently, the frequency of the telephone current.

Many types of motors are suitable for generating artificial radio signals, particularly small fan motors. Motors fitted with rocker arms are especially desirable as the pitch and the volume of the tone may thus be easily regulated. The voltage or the power of the motor is immaterial. A $\frac{1}{32}$ h.p. motor will energize a dozen or more head telephones.

The diagram in Fig. 4 shows several separate code practice circuits connected in parallel. They are intended for classes of different speeds. Such an arrangement is essential for a mixed class and a circuit of this type ought to be installed in the radio club headquarters for use of the members.

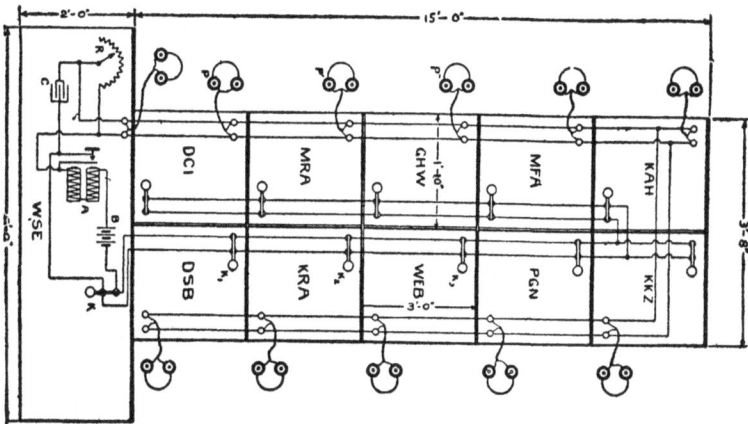

FIG. 5   Wiring diagram of a code practice table for class instruction. Call letters of well known stations are assigned to each position.

Fig. 5 requires no detailed explanation as it is an extension of Fig. 3. The reader should note the position of the shunt variable rheostat *R* which may have a maximum resistance of 100 or 200 ohms. It is used to regulate the strength of signals. Radio students should be taught to read weak signals as not all wireless telegraph transmission is conducted by strong signals. Representative call letters have been assigned to the various student positions in Fig. 5 in order to imitate conditions as they exist in actual service.

Numerous other devices and circuits for code practice have been suggested or developed. For example the telephones, with a condenser of one or two microfarads capacitance connected in series, may be connected to the terminals of a 110-volt d.c. city supply mains. The fluctuations of the line voltage are usually of sufficient intensity and of the proper order of frequency to give a note of fair pitch in the telephones. A telegraph key is connected in series with the telephones for signalling.

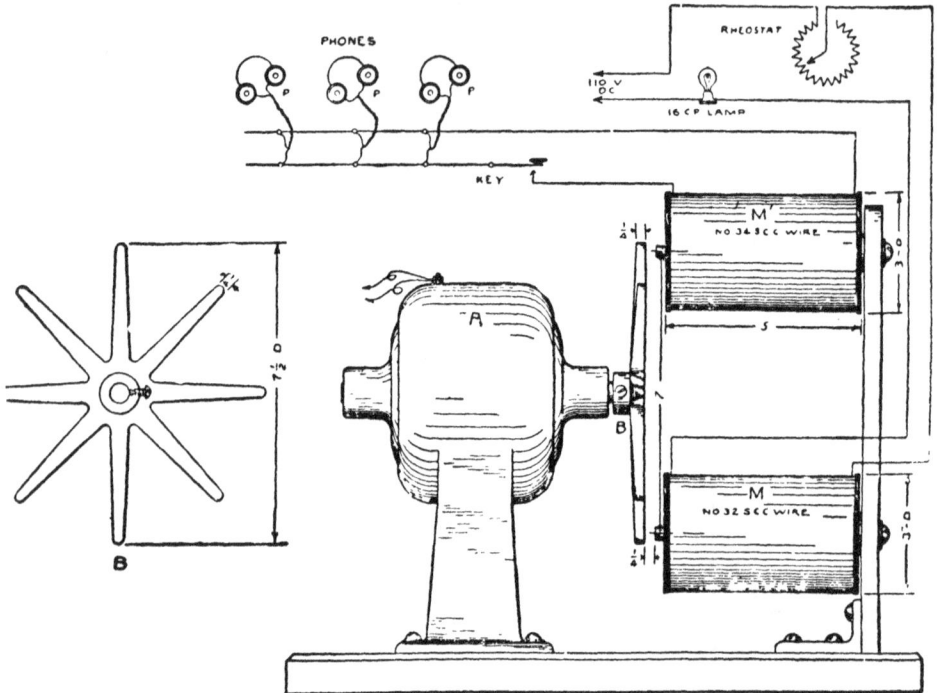

FIG. 6.   A small high frequency generator for the production of artificial radio signals.

A small high frequency generator, such as is shown in Fig. 6, has been employed.   A detailed explanation is not essential.   The lower magnet M is fed with d.c. from the city mains and the revolving iron spider closes the magnetic circuit through magnet M-1, periodically.   The motor should rotate 3600 r.p.m. to obtain a note of high pitch.

Ordinary telegraph keys are employed in all the foregoing code practice circuits.

TIMELY ADVICE.—The beginner should not attempt the more elaborate fields of wireless experimentation until he is skilled in the simple methods.   Once proficient in the code, he should purchase or construct a *receiving equipment* of elementary design.   Since a license is not required for a receiving station the receiving aerial may be within reasonable limit of any dimensions desired.   Here again, the experimenter must be guided by a sense of the fitness of things.   He requires an elementary knowledge of wireless technique.

The beginner should acquire an understanding of the elements of electricity and magnetism if he wishes to operate his instruments to good effect.   In studying the principles of electricity the author recommends that the beginner immediately learn the *difference between alternating current and direct current.* He should familiarize himself with the general conditions under which such currents are handled, particularly such knowledge *as will enable him to judge when a circuit is overloaded* and what size fuse should be installed to *carry a given amount of current.* The prospective amateur should learn the current-carrying capacity of various sizes of wire, thereby making sure that the power circuits at his station will not become overheated.

He should then make a thorough study of the underwriters' rules concerning the installation of power circuits, with particular attention to the rules for the erection of wireless telegraph apparatus. The underwriters' rules vary in different cities and a copy of them can easily be obtained for reference.

Summing up the foregoing it will be seen that the experimenter has prepared himself in two respects for amateur wireless work:

(1) He is able to telegraph at a fair speed and is therefore qualified to interpret wireless signals.

(2) He understands the elementary principles of electricity and also the fundamentals of radio telegraphy.

He is now fully qualified to embark on his initial radio experiments and should begin with a *simple receiving equipment.* The author recommends to the new-comer the simple two-slide tuner, connected as shown in Fig. 7.

A BEGINNER'S RECEIVING SET.—The set in Fig. 7 comprises a single two-wire aerial, a two-slide tuning coil *A B*, having the sliding contacts *S*-1 and *S*-2, a *silicon* or *galena* detector *D*, a small fixed condenser *C*, (0.001 mfd.) and a high resistance telephone *P*. Telephones of less than 1000 ohms resistance are not recommended for use with crystalline detectors. The tuning coil *A B* should be 8″ in length, 3″ in diameter, wound closely with a single layer of No. 26 s.s.c. wire. The wire may be bared for contact with the slider by means of a sharp pointed knife.

The condenser *C* may be constructed of 12 sheets of tin foil 3″ x 4″, separated by paraffined paper, 6 sheets being connected in parallel on each side. When properly stacked up the sheets may be compressed tightly between two boards, connections being brought from the opposing plates to appropriate binding posts.

The aerial or antenna shown in this figure, for the reception of 200 meter signals, may consist of two or four wires but it should be no more than 100′ in length.

To place this apparatus in resonance with a radio transmitting station, set slider *S*-2 at the middle of the coil *A B*. Take the sharp-pointed contact on the crystal *D* and touch lightly the crystal. Move contact *S*-1 along the coil until signals are heard. After response is obtained from some station move both sliders for louder signals and then try a new point on the crystal to see if still louder signals can be obtained.

Do not blame the apparatus if signals are not heard at once. There may be no stations within range in the act of sending. And above all do not forget the importance of a good earth connection. Attach the earth wire to the water pipes on the street side of the water meter. Make a good solid connection and examine it from time to time to see that it is not fouled.

With a receiving set of this type the student has an opportunity of becoming a keen observer of the manner in which commercial and amateur wireless telegraph traffic is handled and he obtains thus an excellent preliminary education.

It is difficult to conjecture the receiving range of an amateur's set without complete data in regard to the type of transmitting and receiving apparatus in use and the local conditions surrounding both the sending and the receiving station.

There are limits to the length of an antenna for use over a given range of wave lengths. It is preferable in all cases for the receiving aerial to have a fundamental wave length below that of the transmitting aerial of the distant station in order that a sufficient amount of inductance may be inserted at the base for coupling to the detector circuit.

Radio range is dependent upon several variable factors such as the antenna current at the transmitter, the damping decrement, local obstructions, absorption, dielectric losses, the type of receiving apparatus, etc.

An aerial comprising from two to four wires, 60' in height by 70' in length, with the wires spaced 2' apart, will have a natural wave length of about 160 meters, which is of the correct dimensions to be loaded by the insertion of the *tuning coil* (shown in Fig. 7) to a wave length of 200 meters. This aerial can be employed for the reception of longer waves say, up to 10,000 meters, with the vacuum tube detector circuits to be described in another chapter.

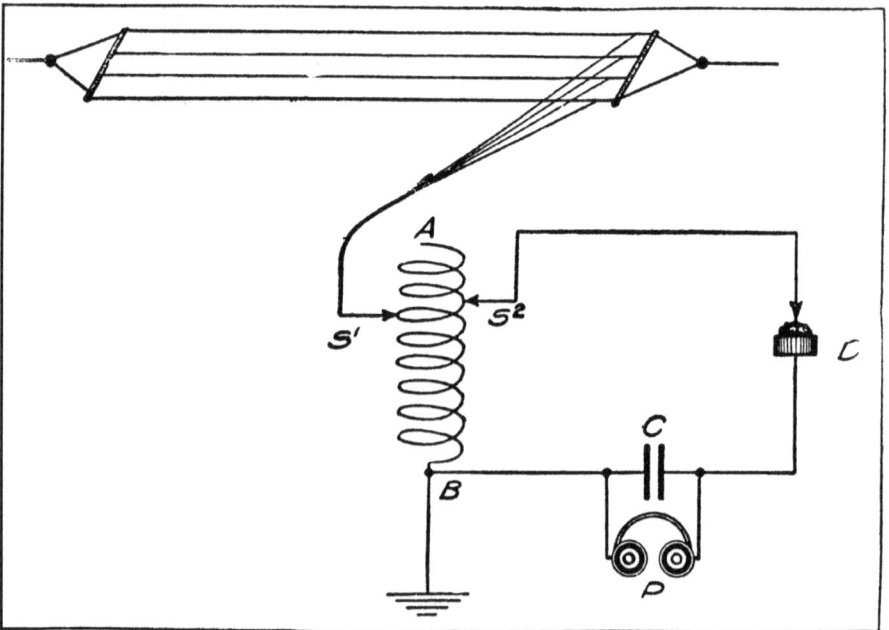

Fig. 7. A simple receiving set for the beginner. This is suitable for reception from nearby transmitters and provides excellent code practice for the beginner. All apparatus except the head telephones may be constructed by the experimenter.

However it should be borne in mind that *an aerial* having a natural wave length of *600 meters* is altogether too long for the reception of signals from amateur stations working on *200 meters*. While the wave length of such an aerial system can be reduced by a series condenser to nearly one-half its original value, it never can be cut down to a period of 200 meters.

Having progressed so far in his wireless education, the student should devote himself diligently to the use of the receiving apparatus, and familiarize himself with the methods of communication employed by amateurs and commercial stations. He will find that many amateurs are accustomed to make use of abbreviated words of the *Phillips code* and he should learn some of those commonly used in practice.

OSCILLATION DETECTORS.—It is safe to say that the amateur experimenter will employ one of four oscillation detectors, viz: the *carborundum, galena,* or *silicon rectifiers,* or the *three-electrode vacuum tube.*

Galena and silicon crystals, while fairly sensitive, are difficult to adjust and to maintain in a sensitive condition. Carborundum crystals are nearly as sensitive and tend to hold an adjustment over long periods. These crystals are in fact extremely rugged and they are preferred above all other contact rectifiers for use in stations where a nearby transmitter is apt to destroy their sensitiveness. The vacuum tube detectors are extremely sensitive, they possess marked stability and will amplify incoming radio signals enormously.

There is no reason for the beginner who has only been able to cover from fifteen to forty miles with his receiving apparatus to feel discouraged when he hears other amateurs declare that they have received 200-meter signals at a distance of 2000 miles or more. He should bear in mind that reception over extremely great distances at 200 meters is only possible at night-time, during the months of the year most favorable to long distance transmission. In the northern part of the United States the favorable period extends from about September 25th to April 15th of the following year. The question as to what results can be obtained with short waves and low power sets from the middle of April to the latter part of September is problematical.

With the great development that has taken place in vacuum tube amplifiers the amateur is now enabled to increase his range of daylight transmission and reception very considerably.

Many amateurs do not take into account the effect of local conditions upon their transmitting and receiving range. For example: if the aerial is located behind a steel building, in the tree tops, behind other structural steel work, or in valleys, signals from distant stations are not received as well or transmitted as far as with aerials which are erected in the open country, free from obstructions, which would tend to absorb the energy of passing waves.

RADIO LAWS AND REGULATIONS.—The beginner should familiarize himself with the general restrictions imposed by United States legislation by studying a booklet entitled "Radio Communication Laws of the United States and the International Radio Telegraphic Convention," which can be purchased from the Government Printing

Office, Washington, D. C. This booklet gives full information concerning the *regulations governing wireless operators* and the use of radio *telegraph apparatus at sea and ashore.*

The experimenter should first refer to the pages containing information about amateur station licenses, etc. From these he learns that, when especially qualified and after at least two years of experience the amateur may, in certain districts, secure a special license for an *exceptional station.* Such experimenters, of course, belong to the class of amateurs considerably advanced in the art   In paragraph 65 he finds that general *amateur stations* are restricted in transmission to the *wave length of 200 meters.* In the same paragraph it *also is stated that if a station is located within five miles of a naval station, the wave length for transmitting purposes is limited to 200 meters and the consumption of the power transformer to $\frac{1}{2}$ kw.* This station is said to be in the restricted class. A general or *restricted amateur station* must be in charge of an operator having an *amateur's first grade or amateur's second grade operator's certificate* who shall be responsible for its operation in accordance with the United States regulations. In fact this station must always be under the supervision of a licensed man. For a receiving station, however, no license whatsoever is required.

Provisional licenses are issued to amateurs far remote from radio inspectors. If, after actual inspection, such stations are found to comply with the law fully, the term "provisional" is struck out and the station is indicated as having been inspected. *Amateur station licenses and amateur operator's licenses hold good for a period of two years, when they can be renewed. No fees are required for either license.*

REQUIREMENTS FOR AN AMATEUR'S OPERATING LICENSE.

—In order to secure an amateur's first grade license certificate, the applicant must be familiar with the adjustment and operation of wireless telegraph apparatus. He must be familiar with the rules of the International Radio Telegraphic Convention, particularly those concerning the requirements in regard to interference. He must be able to transmit and receive at a speed sufficient to recognize distress or "keep out" signals. To qualify for an *amateur first grade license*, he must be able to take down telegraph signals in the international code at a speed of *ten words per minute.*

For an amateur second grade certificate the requirements are similar to those for a first grade certificate, except that the former license is issued to an applicant who cannot be examined. If amateurs, for valid reasons, cannot appear in person and are able to convince and satisfy the government authorities as to their knowledge of the subject, a license of this kind may be issued. If a license is secured the beginner should purchase or construct a simple transmitting set.

WHERE TO TAKE THE EXAMINATIONS.

—Operators' examinations may be taken at the following United States Naval radio stations: San Juan, Porto Rico; Colon and Darien, Canal Zone; Honolulu, Hawaiian Islands, and at the United States Army station at Fort Valdez, Alaska.

Amateurs residing in Washington and vicinity may take examinations at the Bureau of Navigation, Department of Commerce, Washington, D. C. Examinations are also held at the radio inspectors' offices in the following cities:

Custom House, Boston, Mass.
Custom House, New York, N. Y.
Custom House, Baltimore, Md.
Custom House, New Orleans, La.
Custom House, San Francisco, Calif.
Federal Building, Chicago, Ill.
Federal Building, Detroit, Mich.
205 Citizens Bank Building, Norfolk, Va.
2301 L. C. Smith Building, Seattle, Wash.

Applicants should write to the examining officer nearest to their station and secure a copy of form 756, the application blank for an operator's license, and to the radio inspector for form 757 which is an application for a license for a land station. Amateurs at points remote from examining officers and radio inspectors can obtain second grade amateur licenses without personal examination. Examinations for first grade licenses will be conducted by the radio inspector when he is in the vicinity of their stations, but special trips cannot be made for this purpose. Persons holding amateur second grade operating licenses should make every effort to take the examinations for an amateur first grade license or higher.

LAND STATION LICENSES.—To secure a land station license the applicant fills out a blank form on which he states the nature, type and character of his apparatus. The authorities use this information in making calculations to determine the probable wave length and range of the set. In their final decisions they are not guided wholly by the type of the set alone but by the local conditions surrounding the station and the probable interference that it will set up. The license once granted the beginner may communicate with other amateurs, happy in the feeling that he has moved up a round on the wireless ladder.

CHAPTER II

THE FORMATION OF A RADIO CLUB

It is not difficult for amateurs to get in touch with one another, as a rule, for as soon as a new station has been erected and the owner begins to operate his set other wireless enthusiasts communicate with him. Names and addresses, which are among the chief requisites in the preliminary plans for the organization of a radio club, can be obtained easily.

HOW TO GET TOGETHER.—In a community which is without a wireless organization the amateurs should meet and compose a letter inviting those in their neighborhood interested in the art who wish to join a radio club to correspond regarding the subject. This letter should be sent to the National Wireless Association, 233 Broadway, New York City, with a request for its publication in *The Wireless Age*. After it has been published a letter should be sent to prospective members of the organization, giving the name of the place where the first meeting of the club will be held and the date.

It is the purpose of the National Wireless Association to aid in forming radio clubs in communities which lack them. Officers of the organizations will be admitted to the council of the National Wireless Association and arrangements will be made for the clubs to affiliate with military companies as accredited members and officers of signal corps which plan to hold summer military encampments.

The following brief outline of the parliamentary procedure in general practice will serve as a guide to amateurs. The outline for the constitution for a radio club has been made as brief as possible, but nothing essential has been omitted. No reference has been made to by-laws. In reality they are amendments to the constitution and may be adopted from time to time at the business meetings. Amendments generally refer to the duties of committees, enlarging or diminishing their powers.

TEMPORARY ORGANIZATION.—At the first meeting a *temporary organization*—a preliminary step to a permanent organization—should be formed. One of the amateurs present should rise and suggest that a chairman of the meeting be named. It is generally the custom to take a quick vote, and if the majority agree on some one individual he immediately may be considered elected and should take the *presiding officer's chair*. This appointment should be given preferably to one of those who aided in composing the letters to prospective members.

The chairman should be supported by a *recording officer*, whom he may appoint directly. The recording officer, who is known as the *secretary for the temporary organization*, should make a complete record of the proceedings of the meeting.

The chairman should then call the meeting to order. He should deliver a brief address, stating the object of the meeting and inviting discussion on the subject.

An amateur should rise, addressing the chairma (if he is not known he should give his name) and be permitted to have full opportunity to state his views—the possibilities and impossibilities of the enterprise under consideration and the advantage of taking active steps towards forming a club. All of those at the meeting may voice their opinion regarding the subject in this manner

If the consensus of opinion shows that an organization is desired, it is in order for one person present to present a resolution which, for example, may be introduced as follows:

Amateur addresses the Chair: "Mr. Chairman."

The Chairman acknowledges his right to the floor by calling his name: "Mr. Smith."

Mr. Smith, to the Chair: "It seems te be the general wish of those present tonight that a radio club be formed. I therefore propose that active steps be taken at once for the formation of such a club among the amateurs of this city."

The Chairman repeats the motion to the audience, and says: "The motion is now open for discussion."

If no discussion takes place and no objection is offered, the Chairman says:

(1) "All those in favor of the resolution respond by saying aye."
(2) "All those of a contrary mind say nay."

If the ayes and nays seem about equal a vote by count should be taken.

Several committees may now be appointed. It is often customary to appoint a Resolutions Committee first. The members of this committee may be appointed directly by the Chairman, or, if those present so desire, by a general vote.

It is the duty of the *Committee on Resolutions* to draw up a definite statement, placing in the form of a series of resolutions the general desires of the founders of the organization. The committee may withdraw from the meeting in order to decide upon the form in which the resolutions are to be put to the chair and then ask that they be acted upon by those present in the regular manner.

A second committee, to be known as the *Committee on Nominations*, may be formed. The members of this committee may be appointed by the chair if those present at the meeting so desire. It is the duty of the members of the Nominations Committee to suggest or to place before the meeting, for nominations, the names of amateurs for election as *officers of the permanent organization.*

A third committee, to be known as the *Rules and Regulations Committee*, should be appointed. The duties of the members of this committee include drafting a *constitution* and *by-laws* for the permanent organization.

Before the permanent organization is founded, a fourth committee should be formed to determine the *eligibility for membership* of those at the temporary meeting. This committee may have full power to investigate and determine in whatever way it sees fit whether or not those who wish to join are eligible.

It is understood, of course, that the committee will carry out the ideas of the members of the club. It is suggested that no one should be allowed to join the club as a full member *who has not been actively connected with amateur radio telegraphy for at least one year*. It should be further stipulated that in order to be eligible for membership the applicant must be thoroughly familiar with the *United States laws* pertaining to amateur radio telegraphy. (Copies of these rules and regulations can be secured from the Department of Commerce, Washington, D. C., or the district radio inspectors.

All committee reports must be presented to the chair for reading by the recording officer. It is then in order for some one at the meeting to present a motion for adoption or acceptance of the report of the committee. When a motion is offered it must be seconded by another person. After it has been seconded a vote should be taken to determine the general sentiment of those present.

As a rule, it is advisable that the meeting of the temporary organization be held first and the appointment of committees made as suggested.

THE PERMANENT ORGANIZATION.—The affairs of the permanent organization can be handled at a second meeting. This will give the various committees sufficient time to carry on their deliberations properly. It is, however, possible to effect the entire organization at one meeting, although better results will be obtained if the founding of the permanent organization is postponed to a later date.

When the permanent organization is to be effected, the various committees previously mentioned should report to the chairman. Usually the chairman of each committee reads his report before the entire assembly and the chairman of the temporary organization requests that action be taken.

The Membership Committee should offer its report first. It should name those eligible to membership in the club, and a general vote of all present should be taken. If there are any present who are not eligible to membership they should leave before further business is taken up.

The Rules and Regulations Committee should report next, stating clearly the constitution for the club. An outline of a constitution suited to general needs follows:

Article I.—Sec. (1). The name of this association shall be The Radio Club of New York City, or The Cleveland Wireless Club, or The Allied Amateur Radio Clubs of Chicago.

Sec. (2). The object of this club shall be the bringing together of the amateurs of this city who are interested in the advancement of radio telegraphy and desire to become more familiar with the radio art. Progressiveness shall be the keynote of this organization, and a general diffusion of knowledge pertaining to radio telegraphy its endeavor.

Article II, Club Membership.—Sec. (1). The membership of this club shall be divided into two classes—full members and students.

Sec. (2). Full members shall be those who have been actively connected with amateur radio telegraphy for at least one year and are able to receive messages in the Continental telegraph code at a speed of at least five words per minute.

Sec. (3). Students are those who have had no previous connection with amateur radio telegraphy, but are interested in the art and who, in order to familiarize themselves more fully with radio apparatus, desire to join a radio club.

Sec. (4). A full member shall not be less than sixteen years of age, and a student not less than twelve years of age.

Article III, Fees.—Sec. (1). The entrance fee (payable upon admission to the club) shall be $1 for full members and fifty cents for students.

Sec. (2). The annual dues for full members shall be $2, and for students $1.

Article IV, Officers.—Sec. (1). The officers of the club shall be a President, Vice-President and Secretary-Treasurer. The latter office shall be filled by one member.

Sec. (2). The President and Secretary-Treasurer shall be elected for six months and the Vice-President for one year. The President and Secretary-Treasurer shall not be eligible for immediate re-election to the same office.

Sec. (3). The terms of the officers elected at any annual meeting shall begin on the second meeting of the club following the election.

Article V, Election of Officers.—Sec. (1). Election of officers shall take place once every six months.

Article VI, Management of Radio Club.—Sec. (1). The management of the radio club shall be in the hands of the President, Vice-President and Secretary-Treasurer, who, in addition to their regular duties, shall be known as the Board of Directors.

Sec. (2). The Board of Directors shall direct the care and expenditure of the funds of the club, shall receive and pass on all bills before they are paid by the Secretary-Treasurer, and shall decide upon the expenditure of all moneys in various ways.

Sec. (3). The Board of Directors shall from time to time adopt a series of by-laws which will govern the procedure of the various committees which are later to be formed.

Sec. (4). The President shall have general supervision of the affairs of the club under the direction of the Board of Directors. The President shall preside at the meetings of the club and also at the meetings of the Board of Directors.

Sec. (5). The Secretary-Treasurer shall be the executive officer of the radio club, under the direction of the President and Board of Directors. The Secretary-Treasurer must attend all meetings of the radio club and of the Board of Directors, and record the proceedings thereof. He shall collect all membership fees due to the club, and shall give receipts for them. He shall have charge of the books and accounts of the club. He shall present, every three months, to the Board of Directors, a balance sheet showing the financial condition and affairs of the club.

Sec. (6). Three committees shall be formed:

(1). A Library Committee.

(2). A Meetings and Papers Committee.

(3). An Electrical Committee.

It shall be the duty of the Library Committee to keep the members of the club familiar with the latest articles pertaining to wireless telegraphy appearing in various publications, and to see that the literature and books of the club are properly kept on file.

The Meetings and Papers Committee performs the most important function of all. It shall be the duty of the members of this committee to make the meetings of the club of interest to all, particularly as regards intellectual development. It shall be their duty to make the meetings of scientific and electrical interest to the members of the club, and they shall do all in their means to enhance the knowledge of the members of the club in matters pertaining to radio-telegraphy; they shall also see that once each month a paper is read by an amateur member, chronicling interesting experiments which he has performed or suggestions he has to make.

The Electrical Committee shall have direct charge of all the experimental apparatus in use by the club. The members of the committee shall see that the apparatus loaned by various members of the club is well taken care of. The Electrical Committee shall conduct all experiments and shall see that these are performed in a scientific manner.

Article VII, Business Meetings.—Sec. (1). The semi-annual business meeting of this radio club shall be held on the first Tuesday in November and on the first Tuesday in April of each year. At this meeting a report of the transactions of all meetings of the previous year shall be read and the semi-annual election of officers shall take place.

Article VIII, Club Meetings.—Sec. (1). The regular meetings of this club shall be held on Tuesday night every week throughout the year. Every fourth meeting shall be devoted to the reading of a paper on radio telegraphy by one of the members present.

After the constitution and by-laws have been agreed upon and accepted by the members present, it will be in order for the Nominations Committee to present to the chairman a report on the nominees for the various offices to be filled. If the nominees are accepted, a general election by ballot shall take place. These officers should be elected in accordance with the constitution and by-laws adopted.

THE HOME OF THE CLUB.—A radio club should, if possible, maintain quarters of its own. It is possible in the majority of cities to secure a room at a low price in one of the less prominent buildings upon which may be erected an antenna of fair dimensions. If the finances of the organization will not permit this, it is best that the meetings be held at the station of the member having the best facilities for the accommodation of the members and an antenna well suited for their experiments.

THE ANTENNA.—It is particularly important that the club room be located where it will be possible to erect an antenna. There should be two separate and distinct antennae, one of the inverted L, flat-top type, about 80 feet in length by 40 feet in height. This aerial will permit radiation at the wave length of 200 meters to comply with the government law. The second antenna may be swung parallel to it and may be of any length up to 500 feet. The longer antenna should be used for the purpose of receiving the longer waves of the various high power stations located in the United States and abroad. The shorter antenna should be used only for the purpose of sending to and receiving from amateur stations.

INSIDE THE CLUB ROOM.—In the club room there always should be on hand a file containing copies of the latest magazines. The apparatus room should contain a black board to be used in the drawing of circuit diagrams to explain the working of wireless apparatus.

The members of the club should raise a fund to be devoted to the purchase of books dealing with the technical side of wireless telegraphy. These should be added to from time to time until the library is quite complete.

A series of maps showing the location of wireless telegraph stations of the world can be purchased from the United States Department of Commerce. It is suggested, too, that one of the members of the club who has some skill as a draughtsman, draw a map of the section in which the organization is located. The stations of the various members and the distance from the quarters of the club to each station should be indicated on the map.

THE WORKSHOP.—The workshop of the club should adjoin the radio station. The tools and materials for the latter can be contributed by the members of the club or purchased by a fund collected specifically for this purpose. No apparatus should be constructed in the wireless station proper. All work of this nature should be done in the workshop and after the experiments have been completed the apparatus should be taken to the radio room to be tested. This room should contain a full set of electrician's tools, including an electric soldering iron and a first class work bench. Additional material necessary will suggest itself from time to time, and it may be supplied individually or purchased by a fund collected for the purpose.

The radio club, of course, should have a substantial drawing table with a full set of instruments necessary for the drawing of circuit diagrams, the plotting of resonance curves and the laying out of plans for the construction of apparatus.

A COMPLETE EQUIPMENT DESIRABLE.—Interest at the club headquarters will be maintained if the radio station is fitted with a fairly complete equipment. It is somewhat difficult to give advice applicable to each organization as to how much apparatus to install. The question undoubtedly will be governed largely by the amount of funds available. However, every radio club should, if possible, possess the following:

An efficient 200-meter amateur transmitting set.

A transmitting aerial of the proper dimensions for the radiation of energy at this wave length.

A 200-meter receiving set for communication with local enthusiasts.

An accurate wave meter having a range of from 200 to 3,000 meters. If possible a second wave meter having a range of from 3,000 to 10,000 meters should be provided.

An aerial hot wire ammeter.

A supersensitive long distance receiving set capable of giving response to damped and undamped oscillations.

A receiving aerial from 500 to 1,500 feet in length for use with the long distance set.

A buzzer tester.

A complete buzzer practice system.

The following communication f om the Department of Commerce requires the earnest attention of the officers of radio clubs:

"Radio station licenses can only be issued in the name of the club if it is incorpo ated in some state of the United States; otherwise the license must be in the name of some individual of the club which will be held responsible for its operation.

"Radio clubs having a club station should apply to the radio inspector of their district for the assignment of an official call signal which must be used for all radio communication."

CHAPTER III

ELEMENTARY PRINCIPLES OF THE RADIO TRANS-
MITTER—ELEMENTARY THEORY OF DESIGN—
INDUCTANCE COMPUTATIONS—TRANSFOR-
MER DESIGN—THE THEORY OF SPARK
DISCHARGERS

It is difficult to separate the educational requirements of the individual from those of the Radio Club as a whole. In the author's opinion one of the most important functions of the club is to disseminate ideas. Club discussions, the reading of technical papers and the conduction of public experiments, tend to stimulate progress by extending the knowledge of the individual. Often the experimenter is in a quandary on some technical point but after listening to a paper or discussion between members, he goes away with knowledge a point in advance of what he possessed before. Misunderstandings in some technical point are thus frequently brought to light and the benefit to the radio community as a whole cannot be overestimated.

Beginners will soon find out that apparatus for radio transmission and reception involves advanced technical considerations. The working of the apparatus is based upon fundamental electrical principles which, in the design of a set, must be obeyed to the letter if any useful results are expected. A thorough study of the elements of electricity and magnetism is the first essential. Knowledge of the principles of low frequency alternating currents is next in importance. Then comes the study of radio frequency currents. The author may then be pardoned for abruptly dropping the subject of wireless telegraphy from the standpoint of the club as a whole, and treating it technically for the benefit of the individual for he has learned that the experimenter is primarily interested in the design of a good wireless transmitter and receiver. If an individual, through personal study and experimentation, obtains some unusual results the members of the club will soon hear of it and nothing gives the experimenter greater joy than to report the details of some striking achievement to his fellowmen. Thus, what benefits one, benefits all.

A brief treatment of the theory of radio transmission and reception will follow. The design of amateur apparatus will be discussed in an elementary way.

ELECTRIC WAVES AND HIGH FREQUENCY ALTERNATING CURRENTS.—Radio telegraphy is conducted by means of *electric waves* and electric waves are generated by alternating currents. In practice these currents are of very high frequency, particularly the waves of short length used by amateurs.

The *length of an electric wave* is easily determined when the frequency of the alternating current is known. For if we divide the *velocity* of electricity which is computed to be 300,000,000 meters per second, by the *frequency* of the current the result is the *length of the electric wave.* Conversely if we divide the velocity of electricity by the length of the wave we obtain as a result the *frequency* of the alternating current.

Thus amateur stations are limited by U. S. statute to the wave length of 200 meters. The frequency of the current for generating these waves is therefore $\frac{300,000,000}{200}$ or 1,500,000 cycles per second. If we compare this number with the frequency used in ordinary power work—60 cycles per second—it is clear that the apparatus for generating currents of radio frequencies must be of a special type. The laws governing radio frequency currents, however, are not very different from those surrounding low frequency currents except in respect to certain phenomena which have been brought to light since Marconi's basic discovery.

We arbitrarily call current of frequencies above 10,000 *per second radio frequency currents; those below* 10,000 *per second audio frequency currents.* It is a striking fact, of which considerable note will be taken further on, that currents above 20,000 cycles per second are not audible in the telephone receiver, for the ear will generally not respond to sound vibrations above 20,000 per second.

The lowest frequency so far used for wireless transmission is 15,000 cycles and the highest runs into millions. The first named frequency corresponds to the wave length of 20,000 meters, or a wave whose length is approximately 13 miles. On the other hand the wave length of one of Marconi's early types of transmitter has been computed to be a few centimeters and the current frequency runs into billions.

Commercial wave lengths in radio lie between 300 and 20,000 meters; the corresponding frequencies vary from 1,000,000 cycles down to 15,000 cycles.

It is clear that in designing a wireless transmitter for amateur use, the first consideration is the construction of apparatus to generate oscillations at a frequency of 1,500,000 cycles.

THE ELECTRIC WAVE RADIATOR.—The wave maker at the wireless transmitter is called the *antenna* which is an elevated, insulated conductor varying in shape and form. The antenna may consist of a number of vertical wires attached to the top of a mast or tower; or it may have a portion vertical and the remainder horizontal; or it may assume the shape of the ribs of an umbrella.

The antenna is more commonly called the *"aerial"* and thus we have the *"vertical"* aerial, the *"inverted L flat top"* aerial, the *"T flat top"* aerial, the *umbrella* aerial and other modified forms.

The lower end of the aerial is usually connected to earth through the medium of a *good ground plate*, but it is not necessarily so connected. A so-called *ground capacity* or counterpoise may be used. It consists of *several wires spread over the surface of the earth* or even insulated from the earth. A combination of the imbedded earth plate and the counter-poise gives the best results.

It is a fundamental principle of wireless telegraphy that if we generate high frequency currents and cause them to flow in an elevated conductor, more commonly called *an open circuit oscillator or antenna*, electric wave radiation automatically takes place; but for intensive radiation the electrical constants of the oscillator must bear a definite relation to the frequency of the applied current. This relation will be explained in a following paragraph.

The flow of currents in the elevated conductor or capacity is accompanied by electromagnetic and electrostatic fields, a portion of which is detached from the radiator. These two fields constitute the electric wave. Fig. 11b shows the electrostatic field for a few cycles of aerial current and Fig. 11c a complete cycle of the electromagnetic field.

In transit these two fields are at right angles to each other and to the direction the wave is travelling. They are radiated outward at a velocity of 186,000 miles per second, or 300,000,000 meters. Actually the phenomena surrounding the detachment of the two fields are more complicated than they appear here but the two drawings serve to indicate the process of wave radiation in an elementary way.

If another elevated capacity or conductor (known as the receiving aerial) be erected miles distant, the magnetic and static components of the advancing wave motion act to induce in the aerial currents of the same frequency as flow in the transmitter aerial. The receiving aerial must have the same natural time period of electrical oscillation as the transmitter aerial, to receive any considerable distance. In other words the receiver must be *tuned* to the transmitter as will be explained in detail in a following paragraph.

GENERATORS OF RADIO FREQUENCY CURRENTS.—We may now center our attention on apparatus for generating high frequency currents. The lowest frequency so far employed for practical electric wave transmission, to the author's knowledge, is 15,000 cycles, the highest a little less than 3,000,000 cycles. Theoretically, electric wave radiation occurs at all frequencies from the lowest to the highest, but aerials of enormous length would be required to radiate at frequencies below 10,000 per second. On the other hand the wave radiator for very short wave lengths must be exceedingly small. During the European war very successful communications were carried on over short distances at the wave length of 3 meters. The frequency of the antenna current was 100,000,000 cycles!

Frequencies up to 200,000 cycles may be generated by dynamos. An example of such machines is the 2 kw. *Alexanderson radio frequency alternator*, the armature of which rotates 20,000 r.p.m.! The design of such a dynamo introduces many difficult mechanical problems and the construction is very expensive. Besides this, a current of 200,000 cycles would radiate at the wave length of 1500 meters, which is more than seven times the wave length alloted to amateurs, i.e.,—200 meters.

The *Poulsen arc generator* works well at frequencies up to 200,000 but is unsuitable at 1,500,000 cycles required for amateur transmission.

The *vacuum tube generator* works well at all frequencies from ½ cycle to 20,000,000 per second. There is every reason to believe that it will be used by amateurs in increasing numbers.

Spark systems of transmission are mostly used among amateurs and will first be considered. The apparatus in this method is not so complicated as that of other systems and the material for construction is more readily obtained by the amateur. The spark transmitter requires neither the elaborate machinery nor the technical skill demanded by other systems.

PRODUCTION OF DAMPED ELECTRICAL OSCILLATIONS.—
An electrical oscillation circuit possesses the qualities of *inductance, capacitance* and *resistance.* The last named quality is undesirable, but an invariable accompaniment of an electrical circuit.

A circuit made up of an *inductance and a capacitance* in series will, if impulsed periodically by an externally applied electromotive force, *oscillate at a radio or an audio frequency,* depending upon the magnitude of the product of the inductance and the capacity.

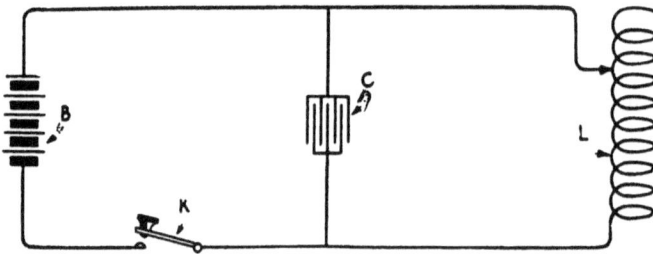

Fig. 8. Simple oscillation circuit illustrating the production of radio frequency currents.

Fig. 8 represents such a circuit. *L* is a simple *coil of wire* and *C* a *condenser* consisting of interleaved parallel conducting plates such as copper or tin foil insulated from one another by air spaces, by sheets of glass or mica, or by any of the well known insulators. An *impulsing circuit* comprising a battery *B* and a key *K,* is shunted across *L.* Upon closing *K,* current flows through *L* and a magnetic field encircles the turns of the coil. The condenser *C* receives a charge varying as the e.m.f. of the battery. Neither the current in the coil nor the charge in the condenser reaches its maximum value instantaneously. A certain interval of time is required to fulfill these conditions.

The condenser exerts a back pressure or e.m.f. on the battery *B,* but since their e.m.f.'s are equal the condenser cannot discharge until the battery *B* is disconnected. If key *K* is then opened, the magnetic field about *L* collapses, induces an e.m.f. in *L* which gives a further charge to *C* after which *C* begins to discharge back through *L.* When *C* is completely discharged the field created around *L* collapses setting up an e.m.f. which charges *C* in the opposite direction from that in the first instance but with less intensity, for the energy supplied originally to the circuit is gradually dissipated.

Several cycles of current traverse circuit *L C* before the energy of the original charge is completely used up. Oscillations of decaying amplitude are generated as shown in Fig. 9. If L and C are properly chosen the current may oscillate at frequencies ranging from audibility to a million or more.

Such groups of oscillations are called *damped oscillations* and the rate at which the amplitudes decrease is expressed by the "logarithmic decrement." If K is opened and closed rapidly, say 500 times per second, 500 groups of damped oscillations will be generated in the circuit.

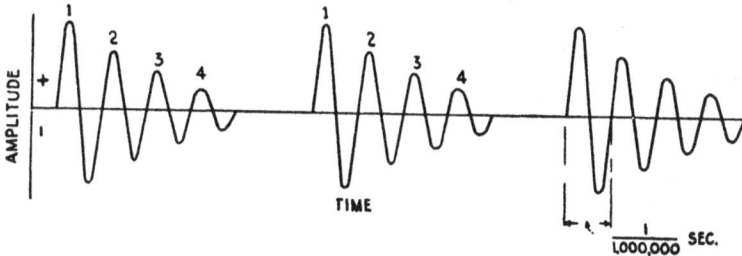

FIG. 9. Graphs showing the decaying groups of oscillations generated by oscillation circuits when impulsed by audio frequency voltages.

The total energy in watts stored in $L\,C$, Fig. 8, during the charging period is the sum of the energies stored in the condenser and coil, that is

$$W = \frac{C\,E^2}{2} \text{ for the condenser,} \tag{1}$$

and

$$W = \frac{L\,I^2}{2} \text{ for the coil.} \tag{2}$$

C is the capacitance of the condenser in farads, L the inductance of the coil in henries, I the final current in amperes flowing through L, and E the electromotive force in volts. It is evident that the energy in the condenser increases as the square of the voltage, and the energy in the coil as the square of the current. Powerful oscillations are produced in radio telegraphy by high voltages—15,000 to 30,000 volts.

We cannot produce powerful oscillations in the circuit of Fig. 8, for the voltage drop across the inductance of a radio frequency coil is con-

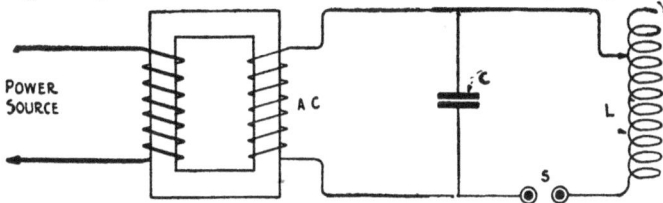

FIG. 10. Circuit for the production of powerful radio frequency oscillations. The condenser C is energized by a high voltage transformer. The circuit $L\,C\,S$ is known in radio telegraphy as the closed oscillation circuit.

siderable and the available e.m.f. to charge the condenser therefore relatively small. For effectiveness the inductance should be discon- nected from the condenser during the charging period and re-connected for discharge.

If a high voltage charging e.m.f. is employed, a special switching arrangement is not necessary. We may include a spark gap S in series with the inductance and condenser as in Fig. 10 and connect the condenser to the secondary of a high voltage transformer. When the potential difference across the condenser terminals reaches a maximum, the gap S, if of the correct length, will be ruptured, and will become temporarily conductive permitting the passage of a few cycles of radio frequency currents around the circuit $L\,C\,S$.

If condenser C is connected to a 500 cycle high voltage transformer and the gap discharges the condenser once for each half-cycle of the charging current, 1000 groups of radio frequency oscillations will flow in circuit $L\,C\,S$. The frequency of the oscillations in $L\,C\,S$, may be varied either by a change of L or C.

A reduction of the capacitance of C increases the oscillation frequency. Cutting out turns at L effects the frequency in the same way. Ignoring resistance, the oscillation frequency of the circuit $L\,C$ is determined from

$$N = \frac{1}{2\,\pi\,\sqrt{L\,C}} \tag{3}$$

L is the inductance of the coil in henries and C the capacitance of the condenser in *farads*. Hence if $L = 0.0001$ henry and $C = 0.0,000,001$ farad,

$$N = \frac{1}{6.28\;\sqrt{0.0001 \times 0.0,000,001}} = 5033 \text{ cycles.}$$

A more practical unit for capacity is the *microfarad*, $= \dfrac{1}{1,000,000}$ farad.

The *microhenry* and the *centimeter* also are more practical units to express inductance. 1 microhenry $= \dfrac{1}{1,000,000}$ henry and 1 centimeter

$= \dfrac{1}{1,000,000,000}$ henry.

If L be expressed in centimeters and C in microfarads, L in formula (3) must be divided by one billion and C by one million. It then becomes

$$N = \frac{5,033,000}{\sqrt{L\,C}} \tag{4}$$

Again if L be expressed in microhenries and C in microfarads, then

$$N = \frac{300,000,000}{1884\;\sqrt{L\,C}} \tag{5}$$

It is important to note that the frequency of circuit $L\,C\,S$ has nothing to do with the frequency of the charging current (which impulses the condenser). No matter whether C is charged 100 or 1000 times per second, the frequency of $L\,C\,S$ varies inversely as the product of the inductance and capacitance of the circuit.

Suppose we desire to design circuit $L\,C\,S$ to oscillate at approximately 1,500,000 cycles. Letting $L=1840$ centimeters and $C=0.006$ mfd. (an average value for amateur transmitters) we have,

$$N = \frac{5,033,000}{\sqrt{1840 \times 0.006}} = \frac{5,033,000}{3.323} = 1,514,500 \text{ cycles.}$$

By increasing the inductance of L slightly, the frequency will be lowered to 1,500,000 cycles. The circuit $L\,C\,S$ in wireless telegraphy is called the *closed oscillation circuit*.

THE OPEN CIRCUIT OSCILLATOR.—Just as we may generate oscillations of a predetermined frequency in the closed oscillation circuit, so we may generate them directly in what is termed the *open oscillation circuit*. This circuit comprises, the *aerial*, the *earth plate*, and the local *tuning appliances* for varying the wave length. In contrast to the closed oscillation circuit, the open circuit, when set into oscillation, *radiates powerful electromagnetic waves*.

The closed oscillation circuit is a feeble radiator unless an enlarged closed circuit loop aerial is employed, such as will be described in Chapter XII.

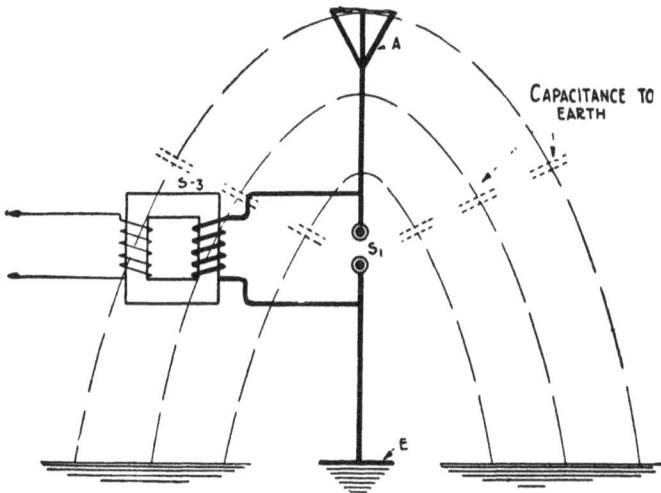

Fig. 11a. Illustrating the electrostatic capacitance of a wireless telegraph aerial

The inductance and the capacitance of the closed oscillation circuit, Fig. 10, are concentrated in the coil L and the condenser C, respectively. The open oscillation circuit external to the tuning apparatus has *distributed inductance* and *distributed capacitance*. Such a circuit is shown in Fig. 11a. The inductance of the wire $A\,E$ lies in its ability to store up energy in the form of a magnetic field; and its capacitance in its ability to store up energy in the form of an *electrostatic* field. But these energies are distributed throughout the length of the vertical wire, and not concentrated or lumped in a small space as in Fig. 10.

We may, in fact, look upon the *aerial* of Fig. 11a as one plate of an *enlarged condenser*, the earth being the *opposite plate*, and the *intervening air* the *dielectric*. Due to the large separation of the two sides of the condenser, we may expect the capacity of the average antenna to be relatively small. Normal values for amateur transmitting aerials lie between 0.00025 and 0.0006 microfarad.

In Fig. 11a the spark discharge gap S_1 is connected to the secondary coils S-3 of a high voltage transformer. The primary coil is connected to a 60 to 500 cycle source, at potentials between 110 and 500 volts. The secondary potential may be 20,000 volts.

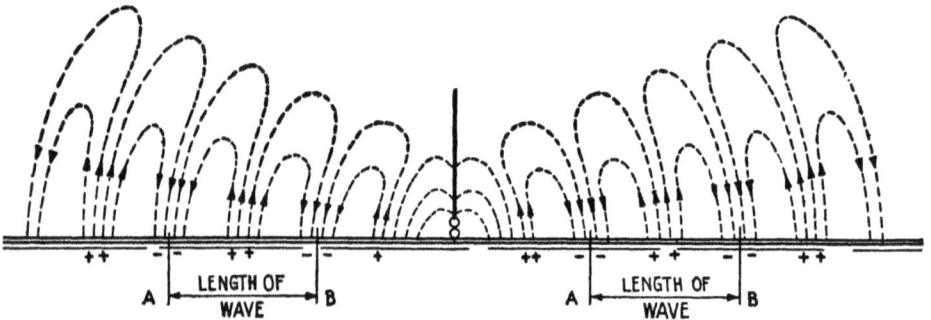

FIG. 11b   Detached loops of the electrostatic component of an electric wave motion

If we separate the spark electrodes of S_1 to the correct sparking distance, and close the circuit to the primary coil, a series of sparks will discharge across S_1; and, as in the case of the closed circuit, the open circuit oscillates at a radio frequency determined by the product of the inductance and capacitance of the circuit.

Just before the spark discharges the space between the aerial and earth is filled with electrostatic lines of force. The potential difference at the spark gap S_1 ionizes the air, making it conductive and allowing the stored-up energy to discharge across the gap. Only part of the stored-up electrostatic field contributes to the energy of the spark. The remainder is radiated in the form of electrostatic loops as in Fig. 11b. Magnetic fields also accompany the current oscillation as in Fig. 11c, and part of this field also is detached from the aerial. Both fields, as already mentioned, are radiated outward at a velocity of 186,000 miles per second (or 300,000,000 meters).

The aerial current oscillates through a very few cycles, the energy being dissipated in wave radiation, heat losses due to resistance, and heat due to absorption by obstacles in the dielectric medium.

WAVE LENGTH.—The wave length of a simple vertical aerial wire is four times its natural length. Thus a rod 150′ high would radiate a wave 600′ in length or 184 meters. We would then say that the natural or fundamental wave length of the aerial is 184 meters.

For an aerial with 3 or 4 wires spaced a few feet, the factor of 4 no longer applies, for we have thus slightly increased the electrostatic capacitance and decreased the inductance of the system. We therefore must determine the wave length of the aerial from knowledge of its inductance and capacitance, or by means of a wave meter.

The wave length of an open or closed oscillation circuit may be increased by adding a coil of wire in series, or decreased by a *series condenser*. All transmitting and receiving aerials have a coil at the base and many stations are equipped with a series condenser. Simple as well as more scientific means of determining the wave length will be described on pages 293 to 297.

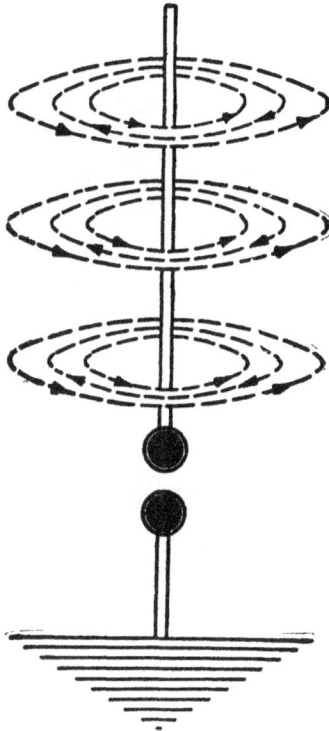

Fɪɢ. 11c.   The magnetic component of an electric wave motion for one cycle of antenna current.

PRACTICAL DAMPED WAVE TRANSMITTERS.—Marconi's early type of transmitter is substantially that shown in Fig. 12. The diagram perhaps represents the simplest possible type of electric wave generator that may be constructed by the amateur. The aerial A to E, includes a loading coil L-1, a spark gap S-1 and is connected to earth at E.   P S is an induction coil with a secondary voltage of, say, 30,000 volts. The primary is fed by a 6 to 30 volt storage battery. Some induction coils are constructed to operate off 110 volts d.c.

An automatic interrupter I is mounted on the end of the core. When key K is closed the interrupter breaks the primary current and high voltage currents are induced in the secondary S. These currents charge the antenna, the stored-up energy of which discharges across the spark gap S-1, producing a few cycles of radio frequency oscillations for each discharge.

The wave length of the system may be increased by cutting in turns at the coil L-1. The coil consists of a few turns of a *copper conductor of low resistance*, such as *stranded wire, copper tubing* or *copper strip* wound in the form of a *helix* or a *"pancake."*

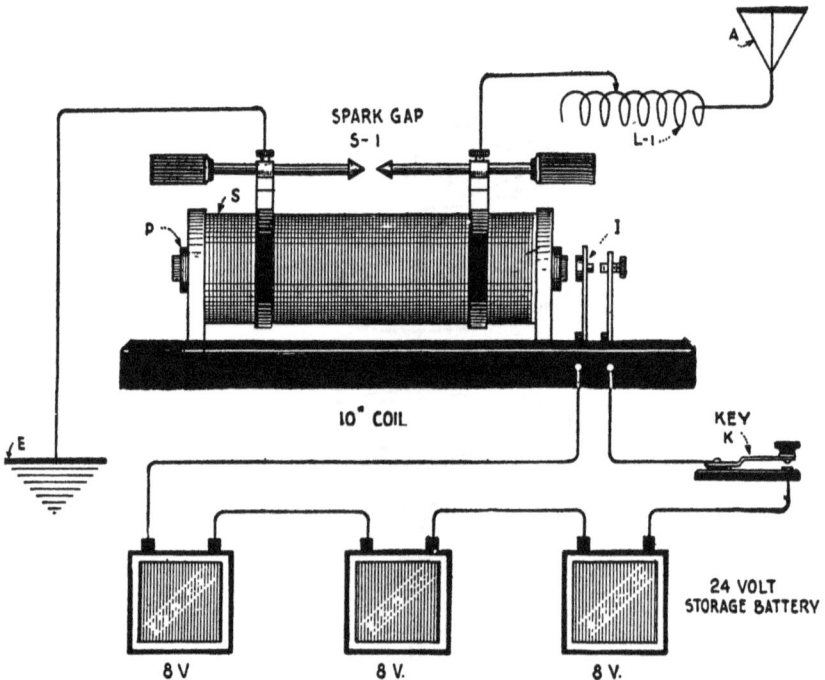

Fig. 12. The circuits of Marconi's early type of wireless telegraph transmitter   The antenna circuit is set into oscillation by an induction coil which may be fed by a storage battery or a d.c dynamo

The spark electrodes are preferably of *zinc*, although *brass* or *copper* may be used.   Electrodes $3/16$ of an inch in diameter are sufficient for small coils.   For large coils, cooling flanges are attached to the spark electrodes.

If L-1 is cut out of the circuit, the aerial being worked at its natural wave length, this transmitter radiates a so-called *"broad"* wave; meaning that the wave emission will interfere with receiving stations, somewhat off tune with the natural oscillation frequency of the transmitter aerial. With a few turns of L-1 cut in, the statement above does not necessarily apply.   The radiated wave may possess a degree of "sharpness" that compares very favorably with more modern apparatus using coupled circuits.

The government authorities are not inclined to grant a license to operate the transmitter in Fig. 12, unless it be of a low power and located in a region where it is not apt to interfere with the working of government or commercial stations.

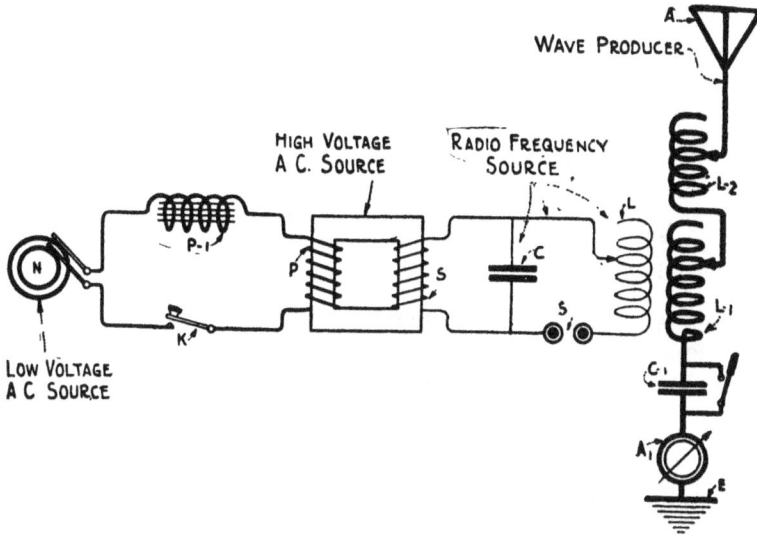

FIG. 13. Fundamental circuits of Marconi's inductively coupled transmitter which is almost universally employed in commercial and amateur practice. Powerful oscillations are first generated in the closed oscillation circuit $L\,C\,S$ and then fed to the antenna circuit through the oscillation transformer L, L-1.

MARCONI'S INDUCTIVELY COUPLED TRANSMITTER.—This type of transmitter is employed almost universally. The fundamental circuit is shown in Fig. 13. Powerful radio frequency currents are first generated in the circuit $L\,C\,S$ and then transferred by electromagnetic induction to the aerial or *open circuit*, A, L-1, L-2, C-1, A-1, E. N is a *low frequency generator*—60 to 1000 cycles, P-1 a *primary reactance coil*, and P the *primary of a closed core transformer*. The e.m.f. of the alternator may vary from 110 to 500 volts.

The secondary voltage of the transformer may vary from 2500 to 30,000 volts according to design. Fifteen thousand volts is more generally used.

C is a *high voltage condenser* which in the amateur station rarely exceeds 0.008 mfd. L is the *primary coil* of the *oscillation transformer* L, L-1. Its value for the amateur set rarely exceeds 6 microhenries. The inductance of L-1, the secondary, may lie between 15 and 25 microhenries. L-2, the antenna *loading coil*, is not required for the general amateur station except for very short aerials. Twenty to 30 microhenries are generally sufficient.

C-1, the *short wave condenser*, is only employed when the natural wave length of the aerial exceeds 200 meters. A condenser of 0.0002 to 0.0005 microfarad is about correct for the amateur aerial where the fundamental wave length exceeds 200 meters.

A-1, the aerial ammeter is used to measure the aerial current and to determine resonance adjustments. A maximum scale reading of 5 amperes suffices for the set of the average experimenter.

THEORY OF OPERATION.—In brief, the circuit in Fig. 13 operates as follows: In general, when the key *K* is closed, current from the generator flows through the primary *P*. The resulting magnetic lines of force thread through *S*, the secondary, inducing therein high voltage currents of the same frequency as the primary current.

With a 60 cycle transformer *C* is charged 120 times per second, and with a synchronous spark gap *S*, 120 sparks occur. One hundred and twenty groups of radio frequency currents are generated in *L C S* per second and since *L* is in inductive relation to *L*-1, 120 groups of radio frequency currents are induced in the aerial system.

Some of the energy of the antenna currents, as already explained, is converted into an electric wave motion.

THE PHENOMENA OF REACTANCE.—A striking phenomenon of radio frequency currents is that of *electrical resonance*. The process of bringing two circuits into resonance is called *tuning*.

It is hoped that the explanation of the reactance effects of a coil and a condenser here given and in the paragraphs following, will aid the amateur in comprehending what adjustments are necessary to effect resonance in alternating current circuits. If a coil of wire is connected first to a source of 110 volts a.c. and then to a source of 110 volts d.c., it will draw a great deal more current in the latter case than in the former. The *self-induction* of the coil to rapidly changing currents causes it to "choke" their flow. The constantly changing magnetic field around a coil carrying alternating currents generates a counter e.m.f. within the coil which acts to *impede* the rise of current and which is sometimes called *reactance voltage*. We say then that a circuit possesses so much *reactance* and we express the opposing effects of reactance in *ohms*.

The reactance of a coil increases directly with increase of frequency. A coil which exhibits negligible reactance to a current of 60 cycles will offer very appreciable reactance to radio frequency currents. If it is desired to build up the currents in radio frequency circuits to an appreciable value the reactance of a coil must be neutralized. As will presently be seen, *the reactance of a coil, in radio frequency circuits, is neutralized by the opposite reactance of a condenser connected in series with the coil*.

In d.c. circuits we are concerned mostly with their resistance, but in a.c. circuits reactance plays such an important part (as well as the resistance) that it must be given very serious consideration. Resistance, it must be remembered, entails a *loss of energy in the form of heat*, but reactance occasions no loss of energy in that way; it compels the application of a higher e.m.f. to a circuit to pass a given amount of current through it. It is well for the student to gain a clear understanding of the relative importance of reactance, resistance and impedance. This will be discussed more in detail in paragraphs following.

FORMULAE FOR REACTANCE.—The reactance of a circuit is expressed in ohms. Letting N = the frequency of the applied e.m.f.,

L = the inductance of the circuit in henries, and X_1 = reactance in ohms, the reactance of an inductance is expressed

$$X_1 = 2\pi N L \qquad (6)$$

It can be shown also that the reactance of a capacitance is inversely proportional to the frequency of the applied e.m.f., or

$$X_c = \frac{1}{2\pi N C} \qquad (7)$$

where C = capacity in farads.

Since in a series circuit the reactance of a condenser is opposite to that of a coil, as a means of distinction inductance reactance is called *positive reactance*, and capacitance reactance is termed *negative reactance.*

If in an alternating current circuit inductive reactance predominates, we obtain the resulting reactance by subtracting the capacitive reactance from the inductive reactance. The resulting figure denotes *positive* reactance. Conversely, if capacitive reactance predominates, we subtract the inductive reactance from the capacitive reactance and the resulting figure denotes *negative* reactance.

If in any circuit to which is applied an e.m.f. of a given frequency,

$$2\pi N L = \frac{1}{2\pi N C} \qquad (8)$$

the reactance is zero and the strength of the current in such a circuit is governed solely by its *resistance.* This circuit under closer analysis will be found to be *resonant* to the impressed frequency and in radio telegraphy it is called a *"tuned"* circuit.

Any oscillation circuit may be placed in resonance with a particular impressed frequency by selecting a capacitance and an inductance of such magnitude that their reactances are equal.

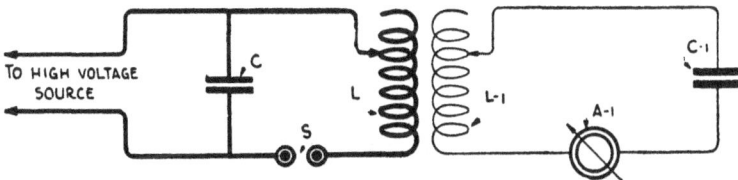

FIG. 14. Illustrating how radio frequency circuits may be placed in electrical resonance by the aid of a hot wire ammeter.

DETERMINATION OF RESONANCE BY CALCULATION.—Using formulae (6) and (7), the reactance of the circuit L-1, C-1 in Fig. 14 may be calculated. Circuit $L C S$ generates radio frequency currents at a frequency, let us assume, of 1,500,000 cycles. Circuit L-1, C-1, A-1 in inductive relation to $L C S$ contains the hot wire ammeter A-1 (scale 0 to 5 or 0 to 10 amperes).

If condenser C-1 is taken out of the circuit and the leads thereto connected together, and circuit $L\,C\,S$ is set into oscillation, the ammeter A-1 will give but a small deflection.   The reason for this is that the *inductive reactance* of L-1 is very high for such frequencies but, on the other hand, if the reactance of C-1 equals that of L-1, the ammeter will give a higher reading, for the amplitude of the current in the circuit is then governed by the *resistance* of the circuit.

Using formula (6) assume that L-$1 = 0.00001$ henry and N, the oscillation frequency $= 1,500,000$ cycles, the reactance of L-1,

$$X_1 = 6.28 \times 1,500.000 \times 0.00001 = 94.2 \text{ ohms.}$$

Giving C-1 a capacity of 0.000,000,001,126 farad, its reactance

$$X_c = \frac{1}{6.28 \times 1,500,000 \times 0.000,000,001,126} = 94.2 \text{ ohms.}$$

Hence, the capacitive and inductive reactance are equal and opposite at the frequency of 1,500,000 cycles.   The two circuits in Fig. 14 are then *in resonance*, that is, they have the same *natural frequency of oscillation*.

We may prove this by formula (4).   Remembering that 0.000,000,-001,126 farad $= 0.001,126$ mfd. and 0.00001 henry $= 10,000$ centimeters, we may substitute these values in formula (4).

$$N = \frac{5,033,000}{\sqrt{L\,C}}$$

or

$$N = \frac{5,033,000}{\sqrt{10,000 \times 0.001126}} = 1,500,000 \text{ cycles.}$$

Knowing the reactance of the coil in a series circuit for a   given frequency, the capacity of a condenser for the same reactance is found by transposing (7), viz:

$$C = \frac{1}{2\,\pi\,X\,N} \tag{9}$$

To find L, when C for a given frequency is known,

$$L = \frac{X}{2\,\pi\,N} \tag{10}$$

TUNING IN PRACTICE.—Fortunately for the amateur experimenter, he does not have to go through the preceding calculations to obtain resonance in established radio transmitting circuits, for two circuits of radio frequency may be placed in resonance by means of a *hot wire ammeter* alone.   Coupled circuits may be tuned in this way, provided the inductance and the capacitance in both circuits are of such magnitude that resonance is possible. The chief disadvantage of tuning by hot wire ammeter is that one does not know the frequency of the circuits and consequently cannot tell the wave length—a matter of prime importance in view of the restricted amateur wave length.

The physical operation in tuning circuit L-1, C-1 to circuit L C S, Fig. 14, is to *set the tap on inductance L-1 at some point and vary C-1 until the meter A-1 reads a maximum*. Similarly C-1 may be set at some capacity, and L-1 varied until meter A-1 reads a maximum. Likewise within limits L-1 and C-1 may be set at some definite value, and either L or C varied until A-1 shows a maximum.

Precisely the same method is employed in tuning to resonance the closed and open circuits of Fig. 13. The variable elements in the antenna circuit are L-1 and L-2. In the closed circuit C is usually fixed and L is varied by the tap. Circuit L C S may be set to some wave length by a *wavemeter*, and then L-1 or L-2 varied until A-1 reads a maximum. The antenna circuit is then in resonance with the closed or spark gap circuit. Tuning in practice is preferably carried on by means of a wavemeter, the use of which will be explained in Chapter XI.

The amateur must keep the following clearly in mind. If two oscillation circuits (L C and L-1, C-1) are in inductive relation, one transferring energy to the other, no matter how large or small may be the values of L or C, L-1 or C-1, if

$$L \times C = L\text{-}1 \times C\text{-}1$$

then the circuits are in *electrical resonance* and the most effective transfer of energy will take place.

The method of determining the values of L and C for any particular wave length (or frequency) will be explained below.

Referring again to the circuits of Fig. 14. When the generating circuit impresses high frequency e.m.f.'s on the next circuit, the reactance of L-1 at resonance is equal to that of C-1. Say then, that C-1 is set at some capacitance below that necessary for resonance, and gradually increased to resonance and beyond. At the lower values of capacity, capacitive reactance predominates, but as it is increased to resonance, the capacitive reactance gradually decreases until at resonance it equals the inductive reactance of L-1. If now, the capacitance of C-1 is increased further, inductive reactance predominates, the capacitive reactance becoming less and less.

When the circuits L-1, C-1 and L C S are in resonance with each other, either will separately oscillate at substantially the same frequency, if impulsed periodically by an external e.m.f.

PHASE ANGLE.—When two alternating currents in a given circuit or in separate circuits reach their positive and negative amplitudes simultaneously, we say that the currents *are in phase*. If one current reaches, let us say, a positive maximum a little later than the second current reaches the same maximum, we say the currents are *out of phase*. Now a difference of phase of one complete cycle is regarded as equivalent to the angle of the whole circumference of a circle or 360°; therefore a difference of phase of a quarter cycle is equivalent to 90°, and of a half cycle to 180°.

The term phase angle is not only convenient to express the difference in phase between two different currents, but also to express the *angle of lead or lag* in circuits in which the applied e.m.f. and the current do not reach their maximum amplitudes at the same instant.

LAG AND LEAD IN ALTERNATING CURRENT CIRCUITS.—

When a coil of wire is traversed by alternating currents, its self induction tends to prevent the rise of the current, whereas a condenser under the same conditions has the opposite effect, i.e., it assists the current to reach its maximum amplitude sooner than it would were the capacity not present.

This may be summed up by saying that the reactance of a coil tends to make the current lag behind the impressed e.m.f., and that the reactance of a condenser causes the current to lead the applied e.m.f. In the first case, we have what is called a *lagging current* and in the last case a *leading current*. We say in such circuits that there is a *phase displacement* and we express such displacements in *degrees*.

If a circuit possessed inductance only (if such were possible), the current would lag 90° behind the impressed e.m.f. If a circuit contained capacitance only, the current would lead the impressed e.m.f. by 90°. In practice, a displacement of 90° cannot be obtained but the condition can be approached.

It is now clear that in any circuit containing a condenser and a coil in series, their reactances are opposite and if they were made just equal they would neutralize; that is, there would be neither "lag" nor "lead" and the current and the applied e.m.f. would build up in phase.

HOW TO CALCULATE THE ANGLE OF LAG OR LEAD.—The

angle of lag and lead may be calculated if we know the capacitance, inductance, resistance and the frequency of the e.m.f. applied to the circuit. This computation is of little interest to the average amateur experimenter but for those who may care to carry it out the following brief analysis is given.

The tangent of the angle of lag is the ratio of the inductance reactance to the resistance of the circuit, or

$$\tan \theta = \frac{\text{reactance}}{\text{resistance}} = \frac{2 \pi N L}{R} \tag{11}$$

The tangent of the angle of lead is the ratio of the capacitance reactance to the resistance of the circuit, or

$$\tan \theta = \frac{\text{reactance}}{\text{resistance}} = \frac{1}{2 \pi N C} \div R \tag{12}$$

In both formulae $\theta =$ the phase angle.

The angle corresponding to any value of tan θ may be found in a table of sines, cosines and tangents. Suppose, for example, tan θ as found by either formula (11) or (12) is 1.16. By referring to the above mentioned table of sines and cosines it will be found that 1.16 is the tangent of the angle of 49°. We would then say that the angle of lag or lead (depending upon whether we are considering inductive or capacitive reactance) is 49°, that is, the current leads or lags behind the impressed e.m.f. by 49°.

If a circuit has both capacitive and inductive reactance, the smaller value should be subtracted from the larger value and the result inserted in the numerator of formula (11), that is

$$\tan \theta = \frac{X}{R} \tag{13}$$

where X is the resulting reactance.

PHASE ANGLE AND ITS RELATION TO POWER.—Although the amateur experimenter may not be inclined to give the matter of phase angle much consideration, he is compelled to do so when determining the power of an alternating current as in transformer circuits.

In d.c. circuits, the power consumption is determined as in Fig. 15, where G is a d.c. dynamo, R, a load, V, a voltmeter and A, an ammeter. The power in watts is expressed:

$$W = I \times E \tag{14}$$

Where I = current in amperes

         E = e.m.f. in volts.

Then if E = 100 volts, I = 10 amperes,

         W = 100 × 10 = 1000 watts = 1 kilowatt.

Now in a.c. circuits when the current either leads or lags behind the voltage, *the true watts cannot be measured by a voltmeter or an ammeter*, for if curves of e.m.f. and current out of phase are plotted, it will be found that there are instants during a complete cycle when the volts are directed in one way and the current the other. If power curves (which are the product of the effective volts and amperes at various instants during the complete cycle) are plotted it will be found that the resulting power is less than would be obtained if the e.m.f. and current were in phase.

In order to measure the a.c. power of a circuit such as that in Fig. 16, it is necessary to multiply the product of the readings of the voltmeter and ammeter by some factor which takes into account the *phase relation* of the e.m.f. and current in the circuit. The quantity by which formula (14) is multiplied to get the true power in watts is called the *power factor* which is equal to cos θ, where θ as before is the phase angle; that is, the cosine of this angle gives us the multiplier for the a.c. power formula. For a.c. circuits then,

$$W = I \times E \times \text{Cos } \theta \tag{15}$$

where Cos θ = power factor which also = $\dfrac{R}{Z}$

where Z = the impedance of the circuit in ohms.

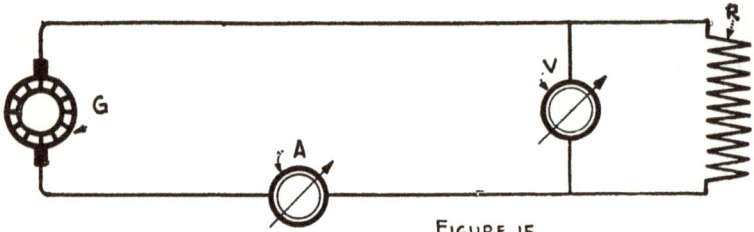

FIGURE 15

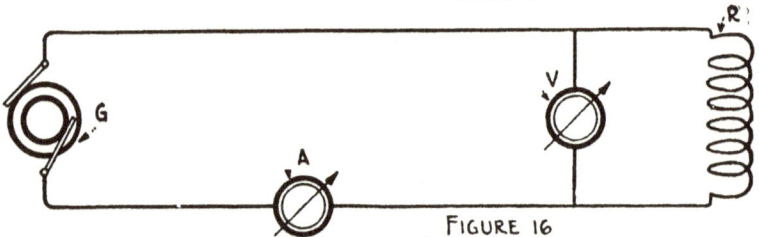

FIGURE 16

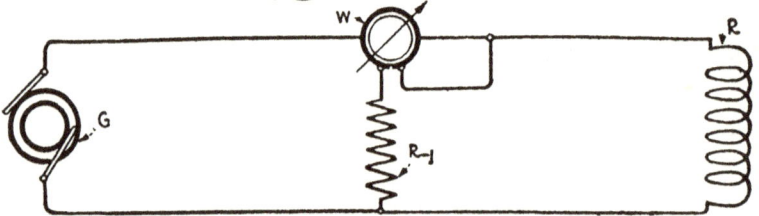

FIGURE 17

Fig. 15. Showing the use of an ammeter and a voltmeter for measuring the power of d c circuits.
Fig. 16. Showing how the k v.a reading of an alternating current power circuit may be determined.
Fig. 17. Showing the position of a wattmeter in an alternating current circuit for measuring the power consumption in watts.

The product of the readings of the voltmeter V and the ammeter A in an inductive circuit gives a resultant called the *apparent watts*.

On the other hand, the wattmeter connected as in Fig. 17 gives the *true watts* because the wattmeter is designed to be independent of phase angle.

If I and E are read as in Fig. 16 and W is determined as in Fig. 17, we may obtain the power factor as follows:

$$\text{Power factor} = \frac{\text{true watts}}{\text{apparent watts}} \qquad (16)$$

For example, if, in Fig. 16, $I = 20$ amperes and $E = 110$ volts, the apparent watts $= 20 \times 110 = 2200$ watts. If, by means of the wattmeter connected in the same circuit it is found that $W = 1200$ watts, then

$$\text{power factor} = \frac{1200}{2200} = 0.54 = 54\%.$$

The apparent power in any circuit is often referred to as the *k.v.a.*

reading, that is the kilowatts obtained by multiplying the pressure by the current as read from instruments in the circuit.

While the apparent watts and the true watts in a.c. circuits may differ considerably, the amateur must take care to design his apparatus to handle the full current as dete mined from the k.v.a. reading, for the heating effect of the current is there, although the full power of the current is not available for a useful purpose.

OHM'S LAW FOR ALTERNATING CURRENTS.—In d.c. circuits, Ohm's law is expressed $I = \dfrac{E}{R}$, meaning that the current in any circuit is directly proportional to the applied electromotive force and inversely proportional to the resistance. In alternating current circuits (to express the true relation between the voltage and the current) the effects of reactance must be taken into account as well as resistance. For alternating currents,

$$I = \frac{E}{Z} \tag{17}$$

where Z = the impedance of the circuit.

The combined opposition of *reactance* and *resistance* to the flow of an alternating current is called *impedance*, which, like reactance and resistance, is expressed in ohms.

Letting Z = impedance, X = reactance, and R = resistance, it can be shown that

$$Z = \sqrt{R^2 + X^2} \tag{18}$$

Formula (17) may then be written:

$$I = \frac{E}{\sqrt{R^2 + X^2}} \tag{19}$$

Now if a circuit possesses inductive and capacitive reactance formula (18) becomes

$$Z = \sqrt{R^2 + \left(2\pi N L - \frac{1}{2\pi N C}\right)^2} \tag{20}$$

or

$$Z = \sqrt{R^2 + (X_1 - X_c)^2}$$

It is now clear that if $2\pi N L$—the inductive reactance, is equal to $\dfrac{1}{2\pi N C}$ —the capacitive reactance, the reactance expression in the denominator of equation (20) may be eliminated. Then $Z = \sqrt{R^2}$ or $Z = R;$ that is, the impedance of the circuit is equal to its resistance.

This is what may be expected in a *resonant* circuit. Formula (19) for resonance circuits then becomes $I = \dfrac{E}{\sqrt{R^2}}$ or $I = \dfrac{E}{R}$ which is the same as the expression for direct current.

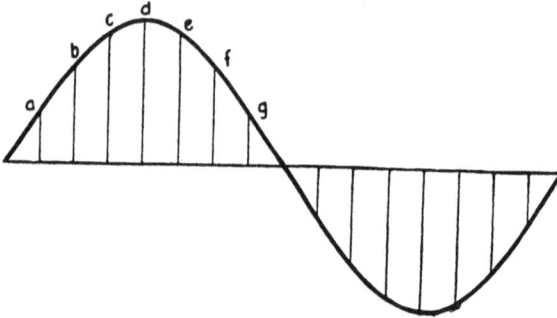

Fig. 17a.   Graph of a cycle of an alternating current following the curve of sines. If the maximum amplitude represents an e.m.f. of 100 volts, the effective value=100 x 0.707=70 volts.

EFFECTIVE AND MAXIMUM VALUES OF ALTERNATING CURRENTS.—It will be evident from the current curve of Fig. 17a, that an alternating current undergoes periodic changes in amplitude and in direction. Say, as an illustration, that the current during each cycle rises from zero to a positive maximum of 15 amperes and then falls to zero; and for the next half cycle, reverses its direction and goes through the same set of values. The question arises, how are we to determine the effect of a current undergoing such a continual change in amplitude? It is clear that we must take some sort of an average value to determine the effectiveness of an alternating current in order that we may compare it with a direct or steady current.

The effectiveness of the alternating current is defined in terms of its relative heating effect compared to the heating effect of a direct current. The heat developed by a direct current is at every moment proportional to the square of the current at that moment, and is therefore constant. In the case of an alternating current, the average heating effect is proportional to the average of the squares of all the values of the current during a cycle (such as may be equal to the points A, B, C, D, E, F, etc., in Fig. 17a). The average in the case of a sine current is the same value that would be obtained by taking one half of the square of the maximum current during the cycle.

It is now clear that a direct current, the square of which is equal to the average of the squares of an alternating current over a complete cycle, produces the same heat as the alternating current. This is the *effective value* of the alternating current and since its square is equal to ½ the square of the maximum value, the effective value

$I_e = \dfrac{I^2_m}{2}$ where I_m = the maximum value. This may be written $I_e = \dfrac{I_m}{\sqrt{2}}$

which is the same as the maximum value multiplied by 0.707.

This means, then, that if the maximum amplitude of a cycle in a sine wave alternating current is 15 amperes, its effective value is 15×0.707 $= 10.6$ amperes. That is, an alternating current that rises and falls uniformly between a value of $+15$ and -15 amperes, produces the same heating effect as a direct current of 10.6 amperes.

Alternating current voltmeters and ammeters give the *effective values* of alternating currents not their maximum values.

In the primary circuits of his high voltage transformer the amateur experimenter is concerned with the *effective values* of the current and voltage. But in the secondary circuit, knowledge of the *maximum* *e.m.f.* is essential because the discharge at the spark gap is related to the maximum e.m.f.

The effective value of an alternating current is sometimes called the *"root mean square"* (r.m.s.) value because the effective value of a sine wave is equal to the square root of the average of the squares of current taken at all points throughout the cycle.

It is well to remember then that when connection is made to a 110 volt a.c. source, the maximum value of the e.m.f. per cycle is $\dfrac{110}{0.707}$

$= 155$ volts.

When the amateur purchases a 15,000 volt transformer, for example, he should ascertain from the maker whether this represents the r.m.s. value or the maximum value per cycle. For if it is the maximum value, the effective or r.m.s. value, in a sine wave, is $15,000 \times 0.707 = 10,605$ volts.

DETERMINATION OF THE WAVE LENGTH OF AN OSCILLA-TION CIRCUIT.—Knowing the values of inductance and capacitance in a closed oscillation circuit, i.e., a circuit with *lumped* or *concentrated* *inductance and capacitance*, the amateur may calculate the equivalent wave length by a simple formula. The computation is a little more complex when applied to the *open* or *antenna circuit*, due to part of the inductance and capacitance being *distributed* throughout its length and the remainder *lumped*. The experimenter who does not lean toward preciseness may apply the *wave length formula for lumped circuits* to the open circuit. The per cent error for the open circuit is roughly negligible.

We have first the *fundamental wave length* formula:

$$\lambda = \frac{V}{N} \tag{21}$$

where λ stands for wave length, $V =$ the velocity of electromagnetic waves (300,000,000 meters per second) and $N =$ the oscillation frequency.

Hence if the antenna circuit of the amateur's transmitter oscillates at 1,500,000 cycles,

$$\text{wave length} = \frac{300,000,000}{1.500.000} = 200 \text{ meters.}$$

If

$$\lambda = \frac{V}{N}$$

then

$$N = \frac{V}{\lambda}$$

But, according to formula (3), $N = \dfrac{1}{2\pi \sqrt{LC}}$ where $L =$ inductance in henries and $C =$ capacity in farads.
Hence

$$\frac{V}{\lambda} = \frac{1}{2\pi \sqrt{LC}}$$

or

$$\lambda = 2\pi V \sqrt{LC} \qquad\qquad (22)$$

This is the *fundamental formula* for determining the wave length when the *lumped* inductance and capacitance of a circuit are known.

The units, the *farad* and the *henry*, are too large for radio circuits in every day practice and if used would involve long decimal expressions. For example, letting C in Fig. 13 = one ten-billionth of a farad and $L = 16$ millionths of a henry, then
$\lambda = 6.28 \times 300,000,000 \times \sqrt{0.000016 \times 0.0,000,000,004} = 151$ meters.

The more serviceable units for radio calculations are the *microhenry*, the *millihenry* or the *centimeter* for inductance, and the *microfarad* or the *micro-microfarad* for capacitance.

1,000,000 microhenries	= 1 henry
1,000 millihenries	= 1 henry
1,000,000,000 centimeters	= 1 henry
1,000,000 microfarads	= 1 farad
1,000,000,000,000 micro-microfarads	= 1 farad.

If L is expressed in *centimeters* and C in *microfarads*, L in formula (22), must be divided by one billion and C by one million. The formula then becomes

$$\lambda = 59.6 \sqrt{LC} \qquad\qquad (23)$$

Similarly if L is expressed in microhenries and C in microfarads, L in formula (22) must be divided by one million, and similarly C. The formula then becomes

$$\lambda = 1884 \sqrt{LC} \qquad\qquad (24)$$

Formulae (22), (23) and (24) are, as mentioned above, strictly applicable to lumped circuits, but many experimenters for a first approximation are content to employ them for the open circuit as well.

Some confusion has existed regarding the computation of the wave length of the antenna circuit, with or without a condenser or a loading coil at the base. The matter has been cleared in Bureau of Standards Bulletin No. 74 and should be consulted by the reader.

In the first place, we may assume a uniform current and voltage distribution throughout an aerial, a condition which does not exist in practice, because of the high frequency of the currents employed. Generally the current is a maximum at the base or grounded portion, and zero at the top or free end. Conversely, the voltage reaches a maximum at the top or free end, and is zero at the earth connection.

However if L_0 is the inductance of the flat-top portion for uniform current distribution, and C_0 the capacitance for uniform voltage distribution, and an inductance L_1 is inserted in the lead-in, formula (22) expressed precisely reads as follows: (The lead-in is assumed to be free from inductance and capacitance except the inductance of L_1).

$$\lambda = \frac{2\,\pi}{K}\ V\ \sqrt{L_0\,C_0}$$

or

$$\lambda = \frac{59\ 6}{K}\ \sqrt{L_0\,C_0} \qquad\qquad (25)$$

Here K is a correction term (first given by Dr. Cohen), the ratio of the inductance of the loading coil at the base to the distributed inductance of the antenna. Values of K for various ratios of $\dfrac{L_1}{L_0}$ are given in Fig. 18.

Now, if there is no lumped inductance at the base, i.e., the antenna has purely distributed *inductance* and *capacitance*,

$$\frac{L_1}{L_0}=0;\ \text{and from Fig. 18, } K=1.57$$

hence

$$\lambda = \frac{6\ 28}{1.57}\ V \sqrt{L\,C}$$

or

$$\lambda = 4\ V \sqrt{L\,C} \qquad\qquad (26)$$

TABLE I

$\dfrac{L_1}{L_0}$	K	$\dfrac{L_1}{L_0}$	K
0.0	1.571	3.1	0.539
.1	1.429	3.2	.532
.2	1.314	3.3	.524
.3	1.220	3.4	.517
.4	1.142	3.5	.510
.5	1.077	3.6	.504
.6	1.021	3.7	.4977
.7	0.973	3.8	.4916
.8	.931	3.9	.4859
.9	.894	4.0	.4801
1.0	.860	4.5	.4548
1.1	.831	5.0	.4330
1.2	.804	5.5	.4141
1.3	.779	6.0	.3974
1.4	.757	6.5	.3826
1.5	.736	7.0	.3693
1.6	.717	7.5	.3574
1.7	.699	8.0	.3465
1.8	.683	8.5	.3366
1.9	.668	9.0	.3275
2.0	.653	9.5	.3189
2.1	.640	10.0	.3111
2.2	.627	11.0	.2972
2.3	.615	12.0	.2850
2.4	.604	13.0	.2741
2.5	.593	14.0	.2644
2.6	.583	15.0	.2556
2.7	.574	16.0	.2476
2.8	.564	17.0	.2402
2.9	.556	18.0	.2338
3.0	.547	19.0	.2277
		20.0	.2219

Fig. 18. A table of constants showing the values of K for various ratios of L_1/L_0 where L_1=the inductance of the antenna system and L_0=the inductance of a loading coil at the base.

The above formula to correspond with (23) should read

$$\lambda = 38 \sqrt{LC} \qquad (27)$$

and to correspond with (24)

$$\lambda = 1200 \sqrt{LC} \qquad (28)$$

Formulae (26), (27) and (28) are the expressions for a *plain aerial circuit*.

It has been shown that wave length computations of loaded antennae, using the formula for lumped circuits, are sufficiently accurate when C_0, the capacitance of an aerial for uniform voltage distribution, is employed as the capacitance for alternating currents of any frequency. But L_0, the inductance for uniform current distribution, is not the inductance for any frequency.* However, $\dfrac{L_0}{3}$ may be taken as the inductance for alternating currents, and the two values $\left(C_0 \text{ and } \dfrac{L_0}{3}\right)$ may then be inserted in the formulae (23) or (24) with practically as accurate results as are obtained from the use of formula (25).

---

* See "Radio Instruments and Measurements," pp 69 to 81.

Hence, if L_1 is the low frequency inductance of the loading coil at the base of the aerial, $\dfrac{L_0}{3}$ the low frequency inductance of the antenna itself, and C_0 its capacitance, we may use the relation

$$\lambda = 59.6 \sqrt{\left(L_1 + \frac{L_0}{3}\right) C_0} \qquad (29)$$

If L_1 is $3\frac{1}{2}$ times $\dfrac{L_0}{3}$, the above formula is within a negligible percentage as accurate as formula (25).

If then we call L_a the low frequency inductance, C_a the low frequency capacity, and L_1 the low frequency inductance of the loading coil, formula (29) becomes

$$\lambda = 59.6 \sqrt{(L_1 + L_a) C_a} \qquad (30)$$

In general the low frequency value for the inductance of a coil may be used for all frequencies.

As will be shown on page 297 formula (30) may be employed to determine the effective inductance and capacity of an aerial by inserting two loading coils at the base and noting the corresponding wave lengths on a wave meter. This data is then substituted in formula (67) page 297.

DATA FOR THE AMATEUR TRANSMITTER.—The builder of amateur apparatus desires to determine the electrical dimensions of the inductances and capacities in the closed and open circuits, and the winding data and core dimensions of the high voltage transmitter. It is the intention to show first, in an elementary way, how these values are related and then to describe in detail the construction of the actual apparatus.

The magnitude of the electrical constants of the amateur's transmitter are limited to certain values by reason of the enforced use of the 200 meter wave. Haphazard design and construction is not permissible, if maximum efficiency is the goal and the law is to be obeyed.

Take first the matter of *power consumption* in the transformer circuits. As a first consideration we want the condenser C of Fig. 13 for amateur wave lengths, to be just as large as possible in order that the set will absorb a fair amount of power at comparatively low voltages. The energy taken by the condenser circuit may be computed by the following formula: (This equation does not take into consideration certain transient phenomena in spark gap circuits and requires considerable modification in practice).

Letting W = power in watts, E = spark voltage in kilovolts, N = spark frequency (twice the frequency of the charging current), and C = capacitance of the condenser in microfarads, then

$$W = \frac{C E^2 N}{2} \qquad (31)$$

Say then that in Fig. 13, the transformer potential with the condenser connected is 20,000 volts, the spark frequency 120 cycles (power frequency = 60 cycles) and the capacitance of the condenser, 0.01 microfarad, then

$$W = \frac{0.01 \times 20^2 \times 120}{2} = 240 \text{ watts}$$

and if N be increased to 500 cycles;

$$W = \frac{0.01 \times 20^2 \times 1000}{2} = 2000 \text{ watts or 2 kw.}$$

This shows that, all other conditions remaining equal, the power increases with the frequency of the charging source*. Actually the power may exceed or be less than the value given by the above formula; for one thing when the condenser discharges across the gap, the transformer is on short circuit, and whether this will cause it to draw more power depends upon the amount of magnetic leakage in the transformer circuits and whether or no the transformer operates at resonance.

It will now be clear that at the wave length of 200 meters, and with a current source of 60 cycles, the amateur cannot utilize the power input of 1 kw. permitted by law, unless *very high voltages* are employed. For the maximum condenser capacitance that may be employed for 200 meter working is 0.01 mfd. as will be explained in the following paragraph. Such high voltages are disastrous to insulation and require an abnormally long spark gap which tends to puncture the dielectric of the condensers and leads to all-around inefficiency.

To find the capacitance of the secondary condenser for a given power, voltage, and frequency, we may transpose formula (22) viz:

$$C = \frac{2W}{E^2 N} \tag{32}$$

As an example let $E = 15,000$ volts, $N = 1000$ sparks (500 cycles) and $W = 500$ watts, find the required capacity.

$$C = \frac{2 \times 500}{15^2 \times 1000} = 0.0044 \text{ mfd.}$$

DETERMINING THE INDUCTANCE OF THE PRIMARY OF THE OSCILLATION TRANSFORMER.—The primary of the oscillation transformer requires an inductance of at least 1000 centimeters to transfer an effective amount of energy to the antenna circuit. The following computation will show that 0.01 mfd. is the maximum permissible capacitance in the closed circuit for the 200 meter wave.

---

*Assuming synchronous discharges.

Letting 0.01 mfd. = the capacitance of the condenser, let us determine the value of L in Fig. 13. L in practice should not be less than a single turn of wire of not too small diameter. Transposing formula (23)

$$L = \frac{\lambda^2}{3552 \times C} \tag{33}$$

or

$$L = \frac{200^2}{3552 \times 0.01} = 1126 \text{ centimeters}$$

$$= 1.12 \text{ microhenries} = 0.00112 \text{ millihenry.}$$

This in practice, as will be found by inductance computations, means that the connections between the condenser, spark gap, and primary inductance of the oscillation transformer must be extremely short, and that the average amateur's oscillation transformer could not have more than one small turn of wire in the primary. The capacity, 0.01 microfarad, obviously is the highest value for the closed circuit condenser that the amateur can use.

A condenser of 0.01 microfarad requires such compact mounting of the apparatus to keep the inductance of the connecting leads in the closed oscillation circuit at a minimum, that it is more practical to reduce the capacitance to, say, 0.008 microfarad. Then for the 200 meter wave the primary inductance,

$$L = \frac{200^2}{3552 \times 0.008} = 1407 \text{ centimeters,}$$

a slight increase over the former case permitting the use of longer leads for connecting up the apparatus in the closed circuit.

CALCULATION OF CAPACITY.—To determine the capacitance of a condenser such as used in spark gap circuits, we may use the formula,

$$C = 0.0885 \, K \, \frac{S}{r} \tag{34}$$

where C is measured in the unit micro-microfarad which is $\dfrac{1}{1,000,000}$

of a microfarad (mfd.) $= \dfrac{1}{1,000,000,000,000}$ farad that is, a micro-micro-

farad $= \dfrac{1}{1,000,000,000,000}$ farad

S = surface area of one plate in square centimeters
r = thickness of dielectric in centimeters
K = a constant—the inductivity of the dielectric which varies with the insulating material as in the following table.

Table II

Dielectric	Value of K
Flint glass, double extra dense	10.10
" " very dense..	7.40
" " light..............................	6.85
" " very light...	6.57
Mica sheet, pure.	4.00 to 8.00
Glass, common (radio frequency)	3.25 to 4.00
" " (audio frequency)	3.02 to 3.09
Paraffined paper	3.65
Air at ordinary pressure (standard)	1.00

FIG. 19. Table showing the inductivity constants of various dielectric substances. This table is useful in the calculation of condenser capacities.

An average value of K for ordinary glass is 6.

Let us take a pane of common glass 14″ x 14″, cover it on both sides with tin or lead foil 12″ x 12″ and calculate its capacity by the foregoing formula. Remembering that 1 inch = 2.54 centimeters:

$$S = (12 \times 2.54)^2 = 929$$
$$r = (\tfrac{1}{8} \times 2.54) = 0.317$$
$$K = 6$$

$$C = 0.0885 \times 6 \times \frac{929}{0.317} = 1554 \text{ micromicrofarads} = 0.0016 \text{ mfd. (approx.) per plate.}$$

Five plates in parallel = $5 \times 0.0016 = 0.008$ mfd., the capacitance desired as in the preceding paragraph. If the potential of the transformer exceeds 15,000 volts, a series—parallel connection must be employed.

For two condensers in series,

$$C = \frac{1}{\frac{1}{C_1} + \frac{1}{C_2}} \tag{35}$$

where C_1 and C_2 are the total capacity of each bank. This formula indicates that the capacity of two equal condenser banks in series is one-half that of one bank. Therefore with the series parallel connection, the number of plates used in a simple parallel connection must be multiplied by 4 to obtain the same capacity as that of a single bank. That is, we must make 20 plates of the above dimensions, place 10 plates in parallel in each bank and connect the two banks in series. Following formula (35),

$$C = \frac{1}{\frac{1}{0.016} + \frac{1}{0.016}} = \frac{1}{\frac{2}{0.016}} = 1 \times \frac{0.016}{2} = 0.008 \text{ mfd.}$$

If the conducting surface of the condenser A and the thickness of the dielectric t are expressed in inches, formula (34) becomes

$$C = \frac{K A 2248}{t \times 10^{10}} \tag{36}$$

$10^{10} = 10,000,000,000$
C = capacity in microfarads
A = area of the dielectric (between conducting surfaces) in square inches.

Owing to the different values of K for various grades of glass, the value of C obtained by (**34**) or (**36**) will only be correct when K is accurately known. For the closed oscillation circuit the inductance of the primary L may be varied slightly until the desired frequency is obtained (note formula (**4**) page 24) and any inaccuracies in the capacitance computations may thus be compensated for. The amateur may employ (**34**) or (**36**) for calculating the capacitance of condenser plates of other dimensions.

For a condenser of any number n of similar plates, alternate plates connected in parallel, formula (**34**) becomes

$$C = 0.0885 \ K \ \frac{(n-1)S}{r} \tag{37}$$

For $\frac{1}{2}$ kw. 500 cycle transmitters, C is usually 0.006 mfd. For $\frac{1}{4}$ kw. 500 cycle transmitters, C is generally 0.004 mfd.; 0.008 mfd. is a good average for the amateur set operated from 60 cycles at powers between $\frac{1}{2}$ and 1 kw.

In summary, it is now clear that if the capacity C of the closed oscillation circuit for any given wave length is first determined upon, L may be found by formula (**33**) here repeated:

$$L = \frac{\lambda^2}{3552 \ C}$$

If L is decided upon for any given wave length, C is found by the following:

$$C = \frac{\lambda^2}{3552 \times L} \tag{38}$$

Assume $\lambda = 200$ meters and $L = 3000$ cms. Find the value of C.

$$C = \frac{200^2}{3552 \times 3000} = 0.0037 \text{ mfd.}$$

DESIGN OF THE OPEN OR ANTENNA CIRCUIT.—The inductance and the capacitance of the plain aerial circuit without a coil or condenser at the base, are distributed throughout its length. Assume an aerial, which as determined by measurement, has inductance of 50,000 cms. and capacitance of 0.0005 mfd. According to the Cohen formula (**27**),

$$\lambda = 38 \ \sqrt{50,000 \times 0.0005} = 190 \text{ meters.}$$

In case a loading coil is inserted at the base to raise the wave length, we may use the formula (30) for lumped circuits, being sure to draw the distinction between the inductance for uniform current distribution and the inductance for low frequency alternating currents: that is, in (30)

$$\frac{L_0}{3} = L_a.$$

Assume for an amateur's aerial, L_1 the secondary of the oscillation transformer $= 10,000$ cms.; L_a, the low frequency inductance of the antenna $= 17,500$ cms.; and C_a the capacitance of the antenna $= 0.00041$ mfd. Then from formula (30)

$$\lambda = 59.6 \sqrt{(10,000+17,500) \times 0.00041}$$
$$= 59.6 \sqrt{11.27}$$
$$= 200 \text{ meters approx.}$$

We may, as a matter of illustration, compare the Cohen wave length formula for the open or antenna circuit with the formula for lumped constants (30). Take, for example, a flat top aerial 40′ in height, 60′ in length, composed of four wires spaced 2′. According to the table in Fig. 20,

$$L_0 = 35,000 \text{ cms.}$$
$$C_0 = 0.000258 \text{ mfd.}$$

For the fundamental wave length according to (25)

$$\lambda = \frac{59.6}{K} \sqrt{L_0 C_0}$$

Since there is no lumped inductance at the base $\frac{L_1}{L_0} = 0$, and from the table Fig. 18, $K = 1.57$. Hence,

$$\lambda = \frac{59.6}{1.57} = \sqrt{0.000258 \times 35,000} = 38 \sqrt{9.03} = 115 \text{ meters (approx.).}$$

Let us now insert a coil of 10,000 centimeters at the base. Then $\frac{L_1}{L_0} = \frac{10,000}{35,000} = 0.286$. From the table Fig. 18 $K = 1.21$ (approx.). Then, according to the Cohen formula,

$$\lambda = \frac{59.6}{1.21} \sqrt{0.000258 \times 35,000} = 139.5 \text{ meters (approx.).}$$

According to formula (30)

$$\lambda = 59.6 \sqrt{(L_1+L_a)C_a}$$

$$L_1 = 10,000 \text{ cms. } L_a = \frac{L_0}{3} = \frac{35,000}{3} = 11,666 \text{ cms.}$$

$C_a = 0.000258$ mfd. Hence,

$$\lambda = 59.6 \sqrt{(10,000+11,666) \times 0.000258} = 141 \text{ meters (approx.).}$$

The result disagrees with the Cohen formula by less than 2 meters. For larger loading coils at the base the error becomes proportionately less as will be found from further use of the table Fig. 18.

DETERMINING THE SECONDARY OR ANTENNA LOADING INDUCTANCE FOR A GIVEN WAVE LENGTH.—Taking the antenna of the dimensions cited in the previous paragraph, we may determine the amount of lumped inductance to be inserted at the base, in order that the antenna will radiate at 200 meters.

Remembering $L_a = 11,666$ cms.; $C_a = 0.000258$ mfd. and by transposing (21)

$$L_1 = \frac{\lambda^2}{3552\ C_a} - L_a \tag{39}$$

Hence

$$L_1 = \frac{200^2}{3552 \times 0.000258} - 11,666 = 32,483 \text{ cms.}$$

10,000 cms. could be allotted to the secondary coil and the remaining 22,000 cms. to the aerial tuning inductance or loading coil.

DETERMINING THE WAVE LENGTH OF AN AERIAL FROM ITS DIMENSIONS.—It is scarcely worth while for the amateur to calculate the wave length of an antenna from its dimensions. Elaborate and intricate equations have been developed for determining the capacitance and inductance of aerials per centimeter length, but they are laborious and do not take into account the effect of local obstacles, such as trees, buildings, roofs, etc., all of which make such computations somewhat inaccurate.

If the reader doubts this, let him refer to the tuning records of the Marconi Company and note the widely different wave lengths obtained from antennae of nearly identical dimensions. The equations are based upon ideal conditions which are not found in practice. Moreover the disposition of the lead-in introduces an uncertainty into the wave length equation.

There is a simple rule for calculating the wave length that applies roughly to flat top aerials. If the antenna wires are spaced no more than 3', multiply the total length of the antenna by 4.7. Thus, if the total length of the aerial from the earth plate to the free end is 100', the wave length $= 100 \times 4.7 = 470'$. And since 1 meter $= 3.25'$, $\dfrac{470}{3.25}$

$= 145$ meters approximately. In other words the total length of the aerial in feet multiplied by the factor 1.44 will give the wave length in meters. This rule of course does not take into account a loading coil at the base. It is simply a crude determination of the fundamental wave length.

Table III *

HORIZONTAL LENGTHS

H ft.	40 ft.		60 ft.		80 ft.		100 ft.		120 ft.	
	C mf.	L cm.	C mf.	L cm.	C mf.	L cm.	C mf.	L cm.	C mf.	L cm.
30	.000186	22430	.000252	28230	.000331	34010	.000395	39770	.000456	45610
40	.000190	28900	.000258	35000	.000324	41100	.000392	47200	.000459	53310
60	.000213	42180	.000276	48800	.000337	55460	.000400	62090	.000463	68700
80	.000241	55410	.000300	62400	.000360	69320	.000418	76300	.000478	83300
100	.000268	69000	.000325	76260	.000382	83500	.000439	90750	.000496	98020

FIG 20.   Table giving the total inductance and capacitance of 4-wire aerials of the inverted "L" type.

Table IV *

HORIZONTAL LENGTHS

H ft.	60 ft.		80 ft.		100 ft.		120 ft.		140 ft.	
	C mf.	L cm.	C mf.	L cm.	C mf.	L cm.	C mf.	L cm.	C mf.	L cm.
30	.000252	15050	.000334	16530	.000395	18000	.000456	19480	.000555	20950
40	.000258	21000	.000324	22580	.000392	24150	.000459	25740	.000528	27320
60	.000276	33790	.000337	35460	.000400	37150	.000463	38820	.000522	40500
80	.000300	46530	.000360	48330	.000418	49850	.000478	51870	.000538	53630
100	.000325	59870	.000382	61690	.000439	63430	.000496	65340	.000553	67180

H ft.	160 ft.		180 ft.		200 ft.		240 ft.	
	C mf.	L cm.	C mf.	L cm.	C mf.	L cm.	C mf.	L cm.
30	.000629	22430	.000702	23900	.000775	25380	.000923	28330
40	.000599	28900	.000664	30500	.000731	32050	.000867	34720
60	.000582	42180	.000645	43860	.000706	45550	.000830	48890
80	.000597	55380	.000654	57190	.000713	58950	.000831	62490
100	.000610	69000	.000667	70840	.000724	72680	.000838	76310

FIG. 21.   Table giving the total inductance and capacitance of 4-wire "T" aerials.

It is preferable in any case for the amateur to purchase a calibrated wave meter. By means of this instrument he can measure the fundamental wave length in a few minutes with greater accuracy than it can be calculated; and moreover, he can measure the inductance and capacitance of his aerial in a much shorter time than he could calculate it. This data enables him to determine the required antenna inductance for any desired wave length.

---

*Calculated by formulae given by G. W Howe, London Electrician, August and September 1914. A simple explanation of Howe's formulae appears on pages 46 to 61 in the book "Calculation and Measurement of Inductance and Capacity" by W H. Nottage.

The experimenter interested in making the calculation of the inductance and capacitance of the flat top portion of an aerial should consult pages 247, 248, 249, and 250 of Circular No. 74,* Bureau of Standards, giving formulae for calculating the inductance per centimeter length of the flat top portion of an antenna; and on pages 239 and 240 he will find formulae for determining the capacity of the flat top per centimeter length. The inductance calculations involve the determination of the inductance of a grounded horizontal wire, then the mutual inductance of two ground parallel wires, and finally the inductance of N grounded wires in parallel. The value so obtained is the inductance

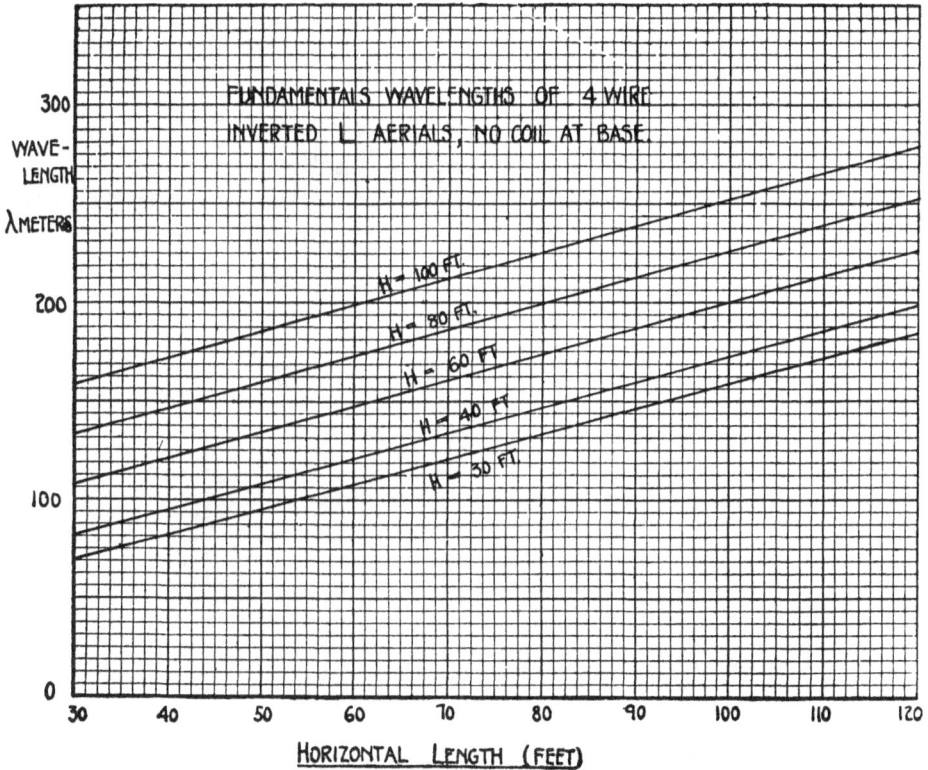

FIG. 22.   The wave lengths of 4-wire "L" aerials with no loading at the base.

for uniform current distribution. When a loading coil is inserted at the base of the antenna and the formulae for the wave lengths of lumped circuits is employed, the value of L determined as above, divided by 3, that is $\frac{L_0}{3}$, eliminates the error which otherwise would exist when using the formula for lumped circuits, viz., $\lambda = 59.6 \sqrt{L\,C}$.

* Entitled "Radio Instruments and Measurements."   Wireless Press, Inc., 233 Broadway, N. Y.

WAVE LENGTH TABLES.—Employing the formulae developed by G. W. Howe for determining the capacitance and inductance of horizontal aerials, A. S. Blatterman has prepared tables giving these values for four-wire inverted "*L*" and "*T*" aerials, composed of 4, No. 14 wires spaced 2' apart. For the *L* aerials, the table in Fig. 20 covers heights from 30' to 100' and lengths from 40' to 120'. For the "*T*" aerials, the table in Fig. 21 covers heights from 30' to 100' and lengths from 60' to 240'.

The curves of Fig. 22 show the fundamental wave lengths of "*L*" aerials without a loading coil at the base, and those of Fig. 23 the wave lengths of the same aerials with 10,000 centimeters at the base. The

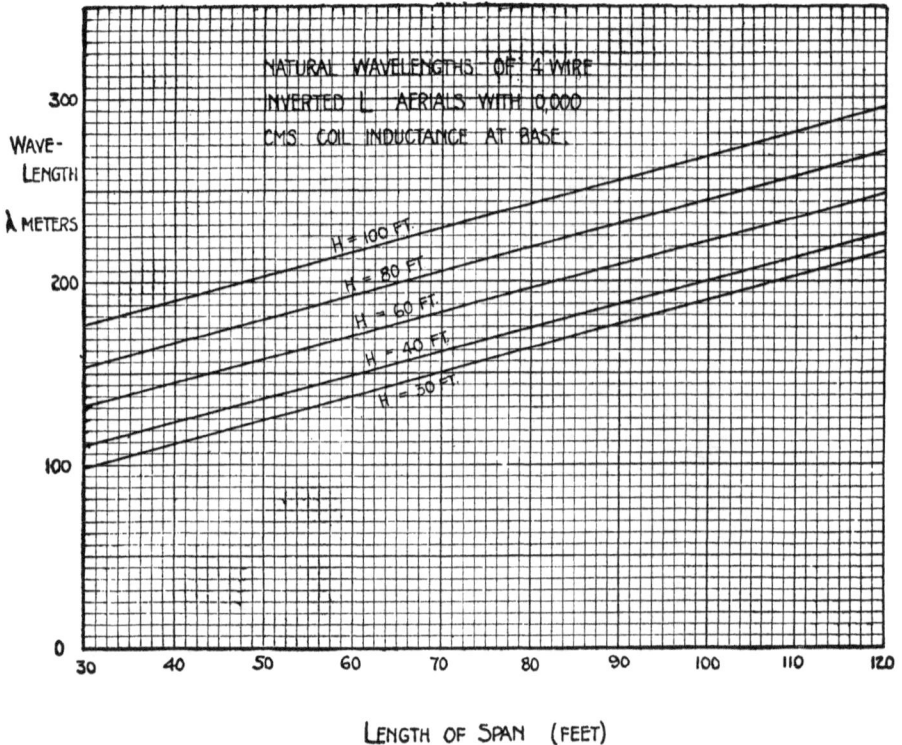

FIG. 23. The wave lengths of 4-wire "L" aerials with a loading of 10,000 centimeters at the base.

value of 10,000 centimeters may conveniently represent the secondary inductance of the oscillation transformer. Fig. 24 gives the fundamental wave lengths of "*T*" aerials of various dimensions, and Fig. 25 their wave lengths with 10,000 centimeters at the base.

These data do not check up identically with the practical antennae of the given dimensions, but the figures are sufficiently close for approximate computations.

The curves of Fig. 23 are perhaps of the most value to the amateur who has not yet erected a transmitting aerial, for they give him the wave lengths of "L" aerials of different heights and lengths with a loading coil of 10,000 centimeters at the base.

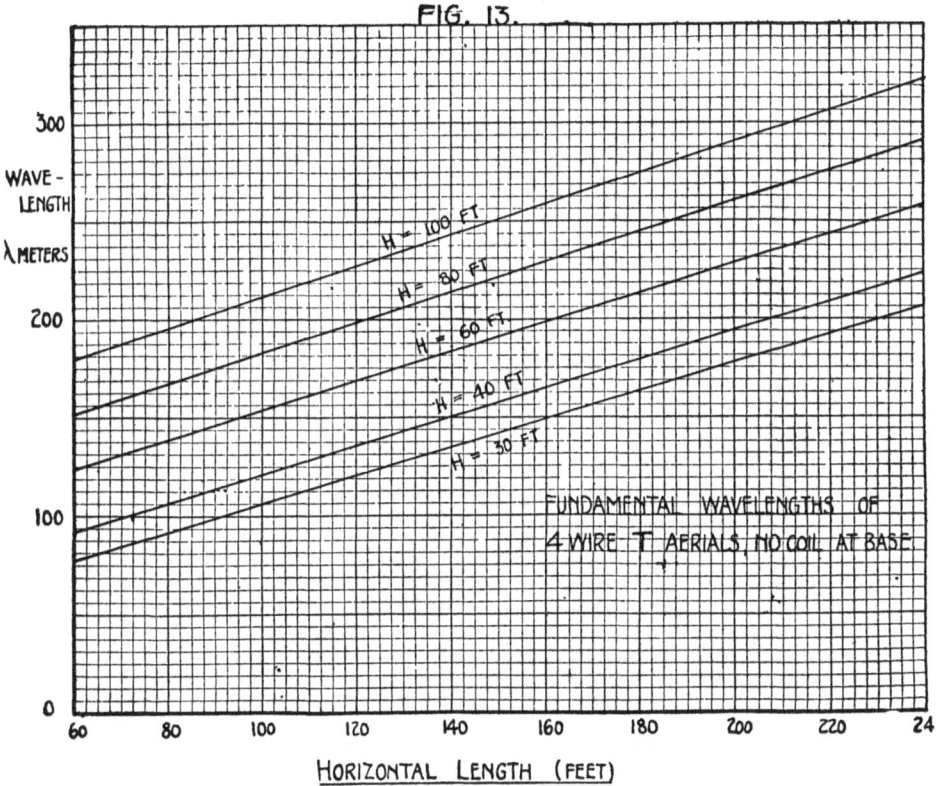

FIG. 13.

FIG. 24. The wave lengths of 4-wire "T" aerials with no loading at the base.

DIMENSIONS OF AERIALS FOR 200 METERS.—It will be found from the curve Fig. 23 that four wire "L" aerials, with 2 foot spacing, of the following dimensions will radiate at 200 meters, with 10,000 centimeters at the base.

TABLE V

Length of flat top	Height
107.5 feet	30 feet
1C0 "	40 "
82.5 "	60 "
66 "	80 "
47.5 "	100 "

FIG. 24a. Dimensions of aerials suitable for 200 meters.

HOW TO CALCULATE THE INDUCTANCE OF AN OSCILLA-TION TRANSFORMER.—Referring now to Fig. 26, there is shown diagrammatically the fundamental circuits of the amateur's wireless transmitter, marked with the values of inductance and capacity in the closed and open circuits which will cause these circuits to resonate at a frequency equivalent to 200 meters. Let the beginner understand that other values of inductance and capacitance will also afford the wave length of 200 meters as should be clear from the curves of Figs. 22 to 25. These particular values were selected to show the method of computation.

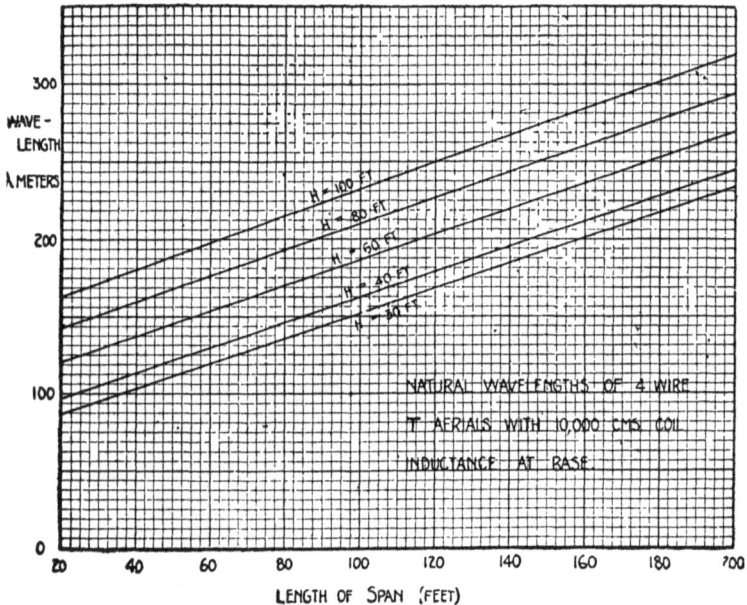

FIG. 25.   The wave lengths of 4-wire "T" aerials with a loading of 10,000 centimeters at the base.

As an illustrative example let us determine the dimensions of inductances L and L_1 in Fig. 26 to equal 1400 and 32,480 centimeters respectively. If the *mean diameter and length* of a coil are first decided upon and the inductance required is known, we may determine the required number of turns by transposing the *inductance formula* to follow. But rather than do this a primary and secondary coil will be selected which are known to be near to the inductance desired and their inductances computed in accordance with the formula. This will show the experimenter how to use the formula so that he may calculate the inductance of any primary and secondary coils he may have at hand. One or two trial computations will reveal the correct dimensions for any given inductance in practice.

Nagaoka's formula for computing the inductance of coils may be expressed as follows:

$$L = 4\pi^2 \frac{a^2 n^2}{?} K \qquad (40)$$

Where

L = inductance in centimeters
a = mean radius of the coil in centimeters
n = total number of turns
b = equivalent length of coil in centimeters

K = a factor varying as $\dfrac{2a}{b}$

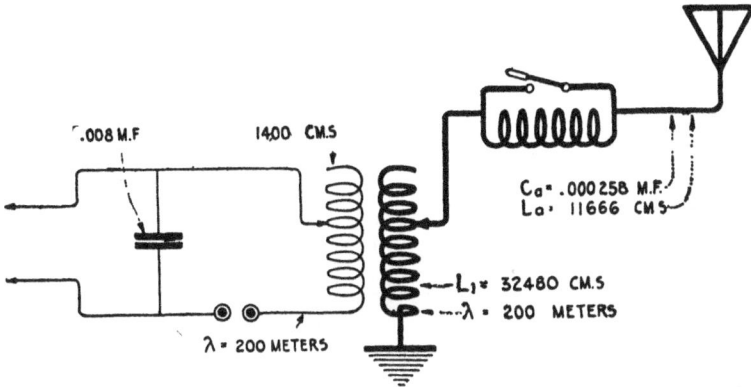

FIG. 26. Showing the electrical constants of the closed and open circuits of the amateur transmitter for the wave length of 200 meters.

The values of K for the ratio $\dfrac{2a}{b}$ appear in Fig. 27 (taken from page 224, Bureau of Standards Bulletin No. 169, Vol. 8, No. 1).

Adopting a coil of dimensions which is known to be suitable for the primary of Fig. 26, we will make it of three turns of copper tubing $\frac{1}{4}''$ in diameter, as a conductor of low resistance is essential to this circuit. The mean diameter will be $7''$ and the distance from center to center of each turn will be $1''$, that is, the *pitch* of the winding is $1''$.

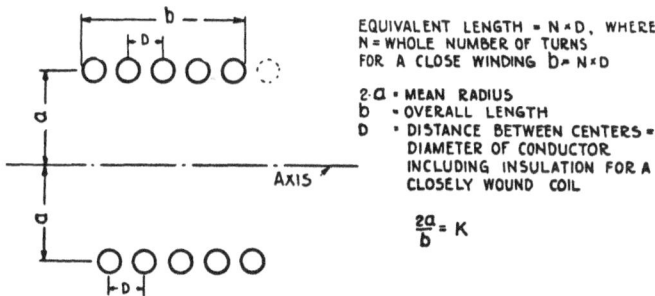

FIG. 28. Showing how the dimensions of a coil are related to Nagaoka's and Lorenz's inductance formulae.

Fig. 28 shows how the dimensions of the coil are connected with the formula; a, the mean radius is the distance from the center of the coil to the center of the wire; $2a$ is the mean diameter; b is the overall length which is $n \times D$, where n is the whole number of turns and D the pitch of the winding.

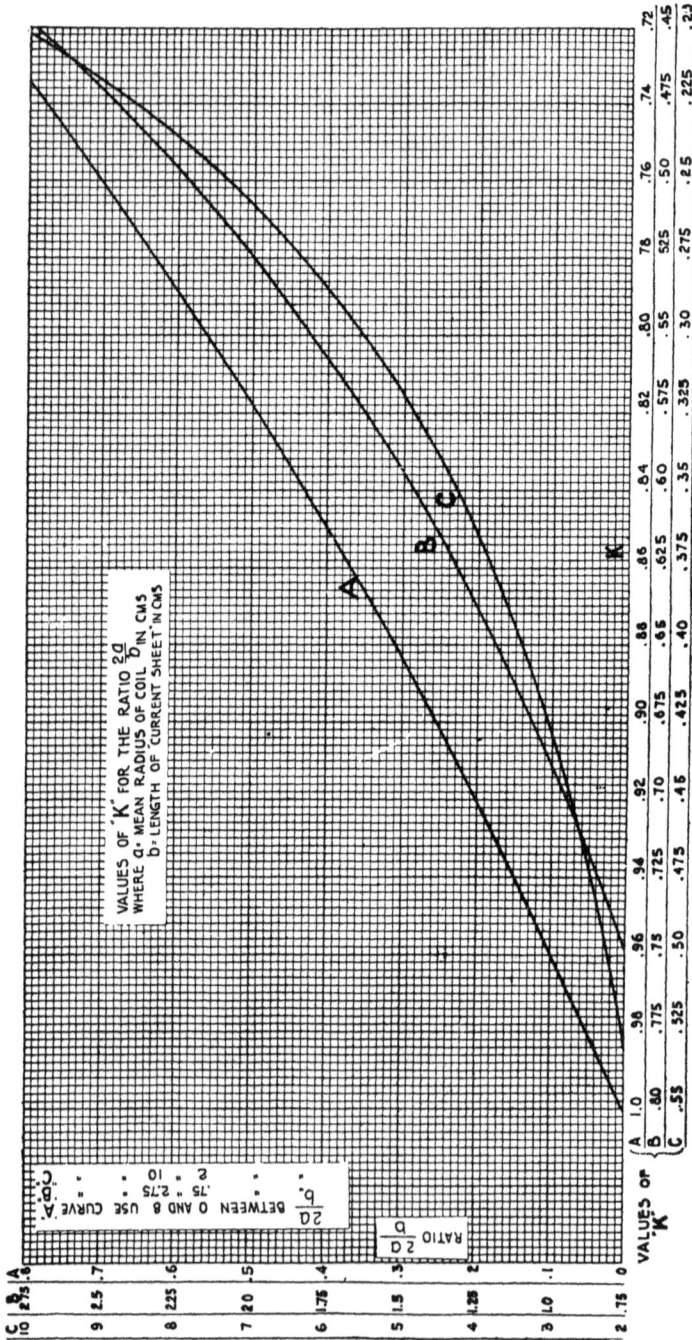

FIG 27   Values of K for the ratio 2a/b, where a is the mean radius, and b the length of the coil

The pitch of the coil above is $1''$, that is, it is $1''$ from the center of wire No. 1 (Fig. 28) to wire No. 2. The "equivalent length" is therefore 3×1 or $3''$.

Using the notations in formula (40) and remembering that $1'' = 2.54$ cms.

$$n = 3$$
$$b = (3 \times 2.54) = 7.62$$
$$a = (3\frac{1}{2} \times 2.54) = 8.89$$
$$4\pi^2 = 39.47$$
$$\frac{2a}{b} = \frac{7}{3} = 2.33.$$

Entering Fig. 27, following the line leading from the notation 2.33 in the B column to curve B, and thence downward to the horizontal axis (using the B notations), we find $2.33 = 0.49$ (approx.) the value of K. Hence,

$$L = 39.47 \times \frac{(8.89)^2 \times (3)^2}{7\ 62} \times 0.49 = 1805 \text{ cms.}$$

a value in excess of the desired primary inductance but preferably so, as will be explained.

USE OF THE CORRECTION TERM.—The inductance computed by formula (40) is termed the "current sheet" inductance, based upon the assumption of a coil wound with infinitely thin tape, the turns of which touch but are not in electrical contact. The "current sheet" length of a coil is equal to the whole number of turns multiplied by the pitch of the winding. Dr. Rosa has given a *correction factor* for the practical coil, which is negligible with some coils and important with others.

The result obtained by formula (40) is sufficiently accurate for the amateur's needs, but if he desires greater precision he should apply the correction formula, the use of which follows:

Letting

L_s = current sheet inductance as in (40)
L = true inductance
L_c = the correction value, then
$L = L_s - L_c.$

In other words, L_c must be subtracted from L_s as obtained from (40). Now

$$L_c = 4\pi\, a\, n\, (K_2 + K_3) \tag{41}$$

Here, K_2 is a constant $\frac{d}{D}$, where D is the pitch of the winding (the distance from center to center of successive turns) and d the diameter of the wire. K_2 takes into account the difference between the true induct-

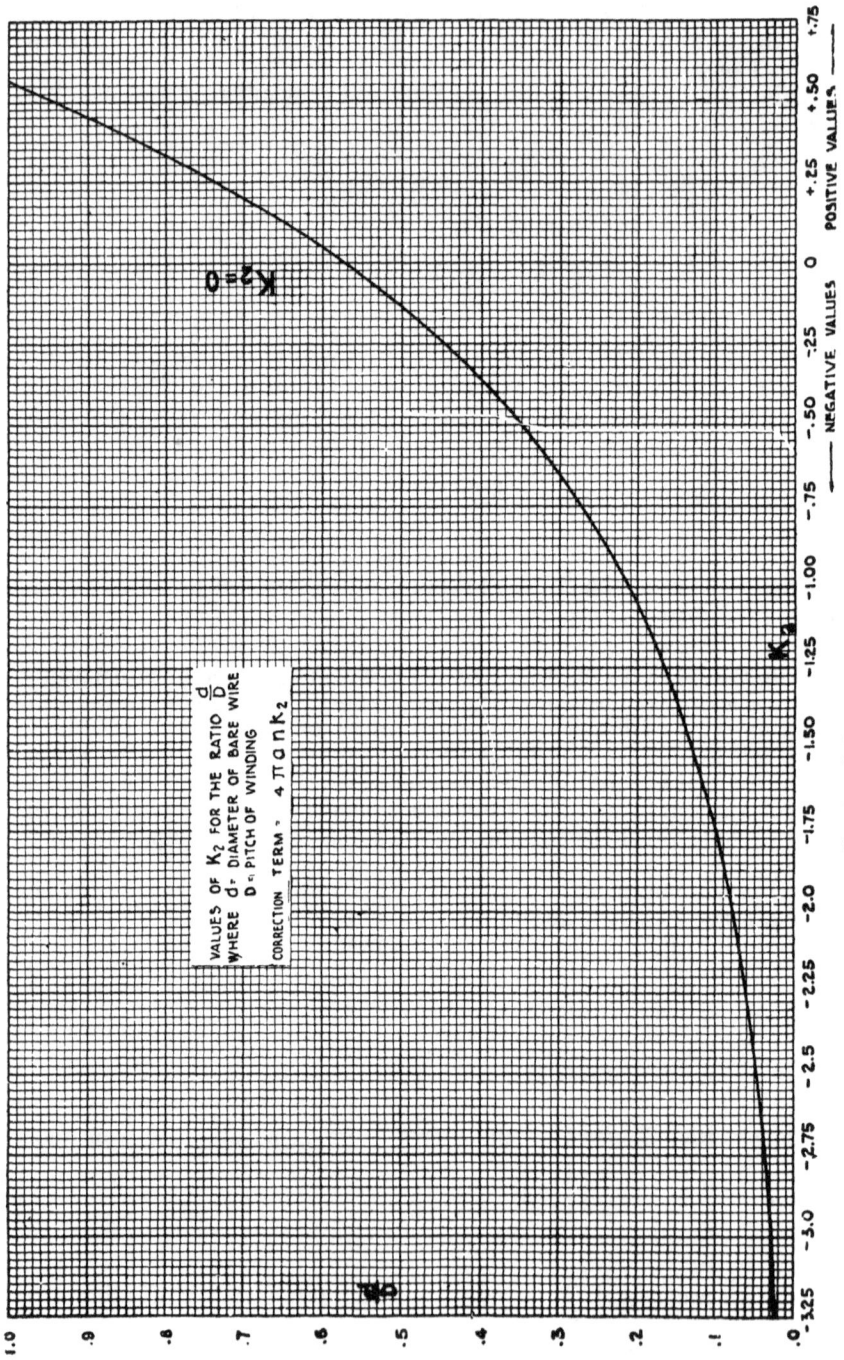

FIG 29 Values of K_2 for the ratio d/D

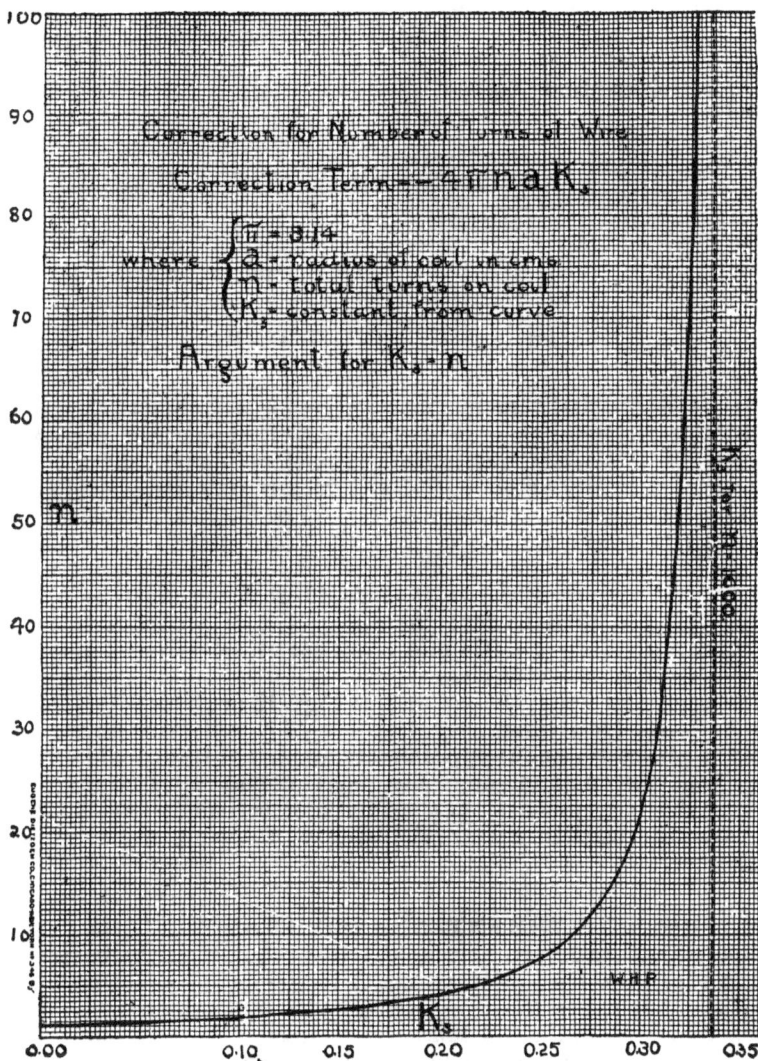

Fig. 30. Values of K_3 for various numbers of turns.

ance of a turn of wire and that of a turn of the current sheet. K_3 is another constant, which depends upon the difference in the mutual inductance of the turns of a practical coil from that of the current sheet inductance. In Fig. 29, K_2 is plotted against the ratio of $\dfrac{d}{D}$ and in Fig. 30, K_3 against the number of turns. It should be noted that K_2

may be positive, negative, or zero.   To the left of the zero line Fig. 29 it is negative, to the right, positive*.

Applying formula (41) to the problem just worked out we have first $\dfrac{d}{D} = 0.25$ and from Fig. 29,

$K_2 = -0.8$ (note negative sign)
$n = 3$ and, therefore, from Fig. 30, $K_3 = 0.175$ (approx.)
$4\pi = 12.566$

Hence

$L_c = 12.566 \times 8.89 \times 3 \ (-0.8 + 0.175)$
$L_c = 12.566 \times 8.89 \times 3 \ (-0.625)$
$L_c = (-208)$ cms.
$L = L_t - L_c = 1805 - (-208) = 2013$ cms.

The value, 2013 centimeters, is too large, for according to the previous calculations the inductance of the primary L should not exceed 1400 centimeters.   Moreover, 400 to 500 centimeters must be allowed for the connecting leads leaving about 900 centimeters (0.9 microhenry) as the actual inductance of the primary L, Fig. 13 or Fig. 26.   This will mean in practice that not quite two turns of the primary of the dimensions we have selected can be employed, for $\lambda = 200$ meters.   But it is well to have an additional turn or more, as it is doubtful if the amateur by using formula (34) can predetermine the dimensions of a condenser to have the exact capacitance of 0.008 mfd., because of the different inductivity values of various grades of glass.

The exact position of the tap on L in Figs. 13 or 26 for 200 meters can easily be found by a wavemeter, but it is of considerable advantage to be able to predetermine the approximate number of turns required before going about the construction.   The amateur may use formula (40) to calculate the inductance of any primary or secondary coil he has at hand, whether used in the transmitting or receiving radio frequency circuits.

DETERMINING THE ANTENNA INDUCTANCE.—We have shown that approximately 32,000 centimeters were required at the base of the aerial in Fig. 26 to raise its wave length to 200 meters.   We may assign part of the inductance to the *secondary of the oscillation transformer* and the remainder to the *aerial tuning inductance*.   We will allow 10,000 centimeters for the secondary and 22,000 centimeters for the aerial tuning inductance.

The primary coil as already determined consists of three turns, spaced 1″ from center to center and with a mean diameter of 7″.   The secondary may be larger in diameter, or smaller, or the same size.   It is rarely necessary to build the oscillation transformer so that these two coils are telescopic.   Sufficient coupling is generally provided when the coils are placed "end-on."

We will arbitrarily make the secondary coil of the same mean diameter, with a smaller conductor and less spacing between turns.   We will

---

* Values of K_2 and K_3 may be interpolated from page 284 "Radio Instruments and Measurements "

use a secondary coil of 10 turns of $\frac{3}{16}''$ copper tubing, with a pitch of $\frac{3}{4}''$ and a mean diameter of 7″.

Using formulae (**40**) and (**41**),

$$\text{equivalent length} = n \times D = 10 \times \tfrac{3}{4} = 7\tfrac{1}{2}''$$
$$n = 10 \text{ turns}$$
$$a = (3\tfrac{1}{2} \times 2.54) = 8.89 \text{ cms.}$$
$$b = (7\tfrac{1}{2} \times 2.54) = 19.05 \text{ cms.}$$
$$\frac{2a}{b} = \frac{7}{7\ 5} = 0.933$$
$$K = 0.703$$
$$L = 39.47 \times \frac{8\ 89^2 \times 10^2}{19\ 05} \times 0.703 = 11{,}509 \text{ cms.}$$

For the correction term using formula (**41**)

$$4\pi = 12.566$$
$$a = 8.89 \text{ cms.}$$
$$n = 10 \text{ turns}$$
$$\frac{d}{D} = \frac{0\ 1875}{0.75} = 0.25. \quad \text{Therefore } K_2 = -0.8$$
$$K_3 = 0.266$$
$$L_c = 12.566 \times 8.89 \times 10 \ (-0.83+0.266)$$
$$= 12.566 \times 8.89 \times 10(-0.564)$$
$$= -628 \text{ cms.}$$
$$L = 11509 - (-628) = 12{,}137 \text{ cms.}$$

This value is too high but it is well to have an extra turn or more to make up for inaccuracies in other computations.

The primary and secondary coils of the amateur transmitter need not have the overall dimensions given, viz., primary 3″x7″ and the secondary 7½″x7″. Other diameters, spacings, and lengths are satisfactory. These dimensions were selected principally to indicate the method of computation. They are, however, correct for an amateur's transmitter. The spacing between turns should not be reduced unless insulated wire is employed, or the secondary voltage of the power transformer is less than 10,000 volts.

INDUCTANCE OF THE ANTENNA LOADING COIL.—It is now necessary to determine the correct dimensions of a coil to have inductance of 22,000 centimeters. We will arbitrarily use a loading coil of the same dimensions as the secondary but twice the length. The experimenter must understand that a coil twice the length of our first calculation will give more than twice the inductance. But as explained in connection with the previous problem, a few additional turns are desirable.

Using the previous dimensions in regard to spacing and diameter, we will make the aerial tuning inductance of 20 turns of $\frac{3}{16}''$ copper tubing with a winding pitch of $\frac{3}{4}''$ and a mean diameter of 7″.

Using formula (**40**),

$$\text{equivalent length} = 20 \times \tfrac{3}{4} = 15''$$
$$n = 20 \text{ turns}$$
$$a = (3\tfrac{1}{2}'' \times 2.54) = 8\ 89 \text{ cms.}$$
$$b = (15 \times 2.54) = 38.1 \text{ cms.}$$
$$\frac{2a}{b} = \frac{7}{15} = 0.466$$
$$K = 0.828$$
$$L = 39.47 \times \frac{8\ 89^2 \times 20^2}{38\ 1} \times 0.828 = 27{,}117 \text{ cms.}$$

For the correction factor using formula (**41**)

$$4\pi = 12.566$$
$$a = 8.89$$
$$n = 20$$
$$\frac{d}{D} = \frac{0\ 1875}{0\ 75} = 0.25. \quad \text{Therefore } K_2 = -0.8$$
$$K_3 = 0.3$$
$$L_c = 12.566 \times 8.89 \times 20\ (-0.8+0.3)$$
$$= 12.566 \times 8.89 \times 20(-0.5)$$
$$= -1117 \text{ cms.}$$
$$L = 27{,}117 - (-1117) = 28{,}234 \text{ cms.}$$

The few extra turns will make up for inaccuracies due to factors surrounding the antenna which the wave length computations do not take into account.

Should the dimensions of the amateur's aerial differ from those given in the tables of Figs. 20 and 21, and should he desire to calculate the dimensions of a secondary coil to raise the wave length of 200 meters, he should employ the inductance and capacitance of the aerial in the tables that correspond closely to the dimensions of his aerial, and compute as explained in connection with formula (**39**). Simple methods of measuring the inductance and capacitance of any aerial will be described on pages 297 to 300. After the values have been determined by measurement, the necessary secondary inductance for a given wave length may be computed with considerable accuracy.

FINDING THE NUMBER OF TURNS FOR A GIVEN INDUCT-ANCE.—Suppose it is found that for a given wave length a secondary coil of 35,000 centimeters is required. By transposing formula (**40**) we may determine the required number of turns provided we first decide on the mean diameter and the length of the coil.

From (**40**),

$$n^2 = \frac{L\ b}{4\pi^2 a^2 K} \tag{42}$$

Where

n = number of turns
a = mean radius in cms.
b = length of "current sheet"

$$K = \frac{2a}{b}$$

L = inductance in cms.

Since there are n turns in the length b, the coil must be wound with a pitch $\frac{b}{n}$. If the pitch proves to be less than the diameter of the wire, or gives a winding with the turns too close, the designer must assume a new length and try again.

It is clear that if the coil is to be used as the secondary of the oscillation transformer in a transmitting set, it must withstand high voltages, and the turns must therefore be well spaced.

Let us arbitrarily take a coil of $\frac{3}{16}''$ copper tubing, wound on a form 24 $''$ in length with a mean diameter of 7 $''$, and determine the number of turns required for an inductance of 35,000 centimeters.

Using (**42**) above,

$$a = 3\tfrac{1}{2} \times 2.54 = 8.89 \text{ cms.}$$
$$b = 24 \times 2.54 = 60.96 \text{ cms.}$$

$$\frac{2a}{b} = \frac{7}{24} = 0.29. \quad \text{(From curve Fig. 27)}$$

$$K = 0.887$$
$$L = 35,000$$
$$4\pi^2 = 39.47$$

$$n^2 = \frac{35,000 \times 60.96}{39.47 \times (8.89)^2 \times 0.887} = 771 \text{ approx.}$$

$$n = \sqrt{771} = 27.8 = 28 \text{ turns (approx.)}$$

$$\frac{b}{n} = \frac{60.96}{28} = 2.17 \text{ cms.} = 0.85'' \text{ or,} \quad \frac{b}{n} = \frac{24}{28} = 0.85''.$$

That is, for an inductance of 35,000 centimeters on the above form we must wind 28 turns of $\frac{3}{16}''$ tubing, spaced 0.85 $''$ from center to center. This gives ample spacing between turns for moderate voltages.

INDUCTANCE OF A FLAT SPIRAL.—The pancake type of oscillation transformer is widely used. It consists of a few turns of copper strip or ribbon mounted on a bakelite or hard rubber base. It offers the convenience that a sliding contact may be mounted on a revolving arm, by which the inductance may be varied by any small fraction of a turn. The construction is slightly more difficult than the plain helix type and the calculation of the self-inductance rather tedious.

The following formula is applicable. For extreme accuracy a correction term must be applied. It is purposely left out here as the error* will not harm in the practical amateur set.

---

*See pp 260–261 "Radio Instruments and Measurements"

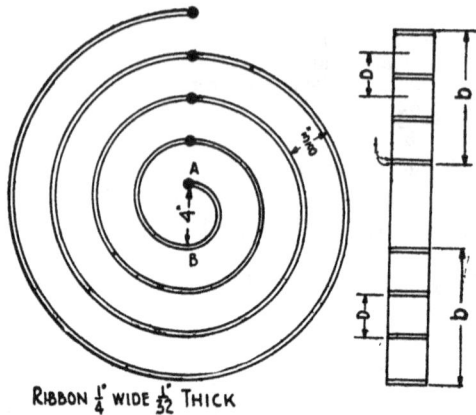

RIBBON ¼" WIDE 1/32" THICK

FIG. 31. Showing how the dimensions of a "pancake" coil are related to the inductance formula.

TABLE VI

$\dfrac{b}{c}$ or $\dfrac{c}{b}$	Y_1	$\dfrac{b}{c}$	Y_2
0.	0.5000	0.	0.597
0.025	.5253	.025	.598
.05	.5490	0.05	.599
.10	.5924	.10	.602
0.15	0.6310	0.15	0.608
.20	.6652	.20	.615
.25	.6953	.25	.624
.30	.7217	.30	.633
0.35	0.7447	0.35	0.643
.40	.7645	.40	.654
.45	.7816	.45	.665
.50	.7960	.50	.677
0.55	0.8081	0.55	0.690
.60	.8182	.60	.702
.65	.8265	.65	.715
.70	.8331	.70	.729
0.75	0.8383	0.75	0.742
.80	.8422	.80	.756
.85	.8451	.85	.771
.90	.8470	.90	.786
0.95	0.8480	0.95	0.801
1.00	.8483	1.00	.816

FIG. 32. Table of constants for use with formula (43) for spiral pancakes.

The following expression applies to pancake coils where b is less than c.

$$L = .01257 \, a \, n^2 \left[2.303 \left(1 + \frac{b^2}{32a^2} + \frac{c^2}{96a^2} \right) \log_{10} \frac{8a}{d} - y_1 + \frac{c^2}{16a^2} \, y_3 \right] \tag{43}$$

Where

L = inductance in microhenries
n = number of turns
b = width of ribbon (equivalent length)
$c = n \times D$, where D = pitch of the winding = center of cross section of one turn to center of next turn
$d = \sqrt{b^2 + c^2}$ = diagonal of the cross section
a = mean radius = $a_1 + \frac{1}{2}(n-1)D$, where $a_1 = \frac{1}{2}$ of the distance AB, in Fig. 31

$$\left.\begin{array}{c} \dfrac{c}{b} \text{ or } \dfrac{b}{c} = y_1 \\[2ex] \dfrac{b}{c} = y_3 \end{array}\right\} \text{ from table Fig. 32.}$$

Assume the coil in Fig. 31 consists of 5 turns of copper ribbon $\frac{1}{4}''$ in width, $\frac{1}{32}''$ in thickness wound with a pitch of $\frac{3}{8}''$. The inner diameter A to B is $4''$.

Remembering 1 inch = 2.54 cms. the following data obtains:

$n = 5$
$b = \frac{1}{4} \times 2.54 = 0.635$ cm.

$a_1 = \dfrac{4 \times 2.54}{2} = 5.08$ cm.

$D = \frac{3}{8} \times 2.54 = 0.9525$ cm.

$a = \dfrac{10.16}{2} + \frac{1}{2} (4 \times 0.9525) = 6.985$ cms.

$c = n \times D = 5 \times 0.9525 = 4.7625$ cms.

$\dfrac{b}{c} = \dfrac{0.635}{4.7625} = 0.1331$ and from table Fig. 32

$y_1 = 0.6180$ and $y_3 = 0.6059$
$a^2 = 48.79 \qquad b^2 = 0.4032 \qquad c^2 = 22.6814$
$d = \sqrt{b^2 + c^2} = \sqrt{0.4032 + 22.7214} = 4.808$

$\dfrac{8a}{d} = \dfrac{58\,88}{4.804} = 11.63.$ $\text{Log}_{10}\, 11.63 = 1.0652$

Therefore,

$$L = 0.01257 \times 6.985 \times 25 \left[2.303 \left(1 + \frac{0\,4032}{32 \times 48\,79} + \frac{22.6814}{96 \times 48.79} \right) 1.0652 - 0.6180 \right.$$

$$\left. + \frac{22\,6814}{16 \times 48.79} 0.60597 \right] = 4.3128 = 4.3 \text{ microhenries (approx.)} = 4300 \text{ centi-} \atop \text{(meters.}$$

A couple of turns of this coil would be sufficient with a condenser of 0.008 mfd. for the wave length of 200 meters.

TRANSFORMER DESIGN.—The following discussion deals in generalities. The author's aim is to give the amateur an insight into the relations between factors in transformer building. The phenomena of resonance and magnetic leakage are not taken into account.

It is recommended that the amateur not skilled in electrical construction or with limited time on his hands, purchase an assembled transformer. It will probably prove more satisfactory and more effective than the home-made instrument, principally because the manufacturer has the facilities for impregnating the secondary coils with insulating compound which is done by the aid of a vacuum pump. Moreover, through a series of experiments the manufacturer is enabled to obtain the correct operating characteristics for a particular service, which the general formulae for transformer design perhaps do not take into consideration.

FIG 33   Showing the construction of a commercial type of closed core transformer   This is a transformer of the shell type the primary and secondary coils being wound on the middle leg of the transformer core.

Actual details of construction and instructions for the assembly of the transformer designs here given are treated in the chapter following.

Advance knowledge of certain quantities is absolutely essential in transformer design. Take, for example, the diagrammatical sketch of the transformer in Fig. 34. For a given primary input, frequency, and e.m.f. we wish to determine the *length* and *cross section* of the *core*, the *turns* in the *primary* and *secondary*, the sizes of the primary and secondary wires, and the over-all efficiency.

There are two principal energy losses in transformation, viz., the *copper losses* and the *core losses*. The former are due to the cross section of the primary and secondary turns, that is, their resistance, and the latter to *eddy currents* and *hysteresis*. All of these losses are manifested in the form of heat.

The eddy currents are those which are induced in the iron laminations by magnetic induction. The losses due to eddy currents may be

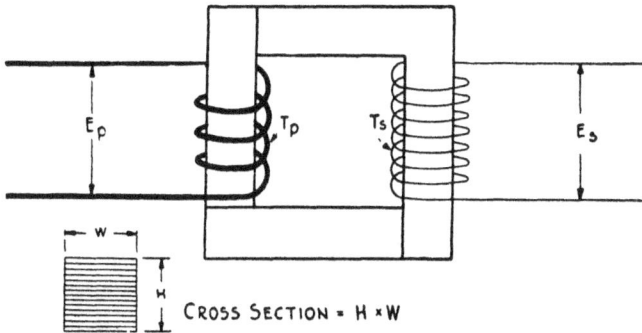

CROSS SECTION = H × W

FIG. 34   General construction of high voltage transformers suitable for amateur stations.

kept at a minimum by insulating the sheets of iron from one another by dipping them in varnish or some other insulating compound.   Rusty sheets of iron have sufficient resistance to reduce eddy currents.

The *hysteresis losses* are produced by the rapid reversals of flux through the core, which cause molecular friction and result in the generation of heat.

The quality of the iron, in so far as it pertains to its saturation point, has a marked effect upon the efficiency of the transformer.   It is to be thoroughly understood that in the designs to follow, the use of the best *transformer silicon steel* is assumed, such as the "Apollo special extra," or "Apollo special electrical."   The latter is a standard in the manufacture of 60-cycle transformers.   The standard gauge for 60-cycle work is usually No. 29, the thickness of which is 14.1 mils = 0.014".   Steel of a slightly heavier gauge may be employed if desired.

The hysteresis loss in transformer cores may be calculated approximately by the following formula:

$$W_h = \frac{K \times N \times B^{1.6}}{10^7} \tag{44}$$

W_h = loss in watts per cub. cm. of core
N = frequency in cycles per second
B = flux density per sq. cm.
1 sq. inch = 6.45 sq. cms.
10^7 = 10,000,000
K = hysteresis coefficient varying from 0.0006 to 0.006
= 0.0021 for plain transformer steel = 0.00093 for the best grades of silicon steel.

The eddy current loss may be determined by

$$W_e = \frac{K \times (t \times N \times B)^2}{10^{11}} \tag{45}$$

Where

W_e = loss in watts per cub. cm of core
K = 1.65 for plain transformer steel
= 0.57 for silicon steel
t = thickness in centimeters
N = frequency of applied e.m.f.
B = flux density = lines of force per sq. cm. of cross section.
10^{11} = 100,000,000,000.   1 inch = 2.54 cms.   1 sq. inch = 6.45 sq. cms.

TABLE VII

WATT LOSS PER LB. IN SILICON STEEL TRANSFORMER CORES

$K = 0.00092$ approximately

$n = 60$ cycles

$t = 14$ mils

Kilolines per sq. cm.	Lines per sq. inch	Total hysteresis and eddy current loss in watts per lb.	Hysteresis loss in watts per lb.	Eddy current losses in watts per lb.
2	12903	0.15	0.1	. . . .
3	19354	0.20	0.15	. . . .
4	25806	0.25	0.20	0.05
5	32258	0.30	0.25	0.05
6	38709	0.50	0.40	0.10
7	45161	0.60	0.48	0.12
8	51612	0.75	0.60	0.15
9	58064	0.88	0.70	0.18
10	64516	1.00	0.80	0.20

FIG. 35. Eddy current and hysteresis losses for the best grades of silicon transformer steel expressed in watts per lb.

It is convenient in transformer design to consider the two core losses as one. If the *total core loss* be expressed in watts per pound and the loss to be expected in a given transformer can be predetermined, it is clear that

$$\frac{\text{core loss}}{\text{loss per lb.}} = \text{total weight of transformer core.} \qquad (46)$$

And since 1 cubic inch of transformer steel weighs approximately 0.278 lb. then,

$$\text{volume of core in cubic inches} = \frac{\text{weight in lbs.}}{0.278} \qquad (47)$$

The third column of the table in Fig. 35 shows the *combined hysteresis and eddy current* loss for various flux densities per square inch (from 12,000 lines to 64,000 lines) for the best grades of *silicon transformer steel*. The fourth column shows the *hysteresis* loss and the fifth column the *eddy current* loss. This table is correct for the frequency of 60 cycles.

For the best grades of silicon steel at 60 cycles, it is safe to allow a flux density of *60,000 lines per square inch*. For higher frequencies the density should be proportionally less. At *500 cycles*, for example, *15,000 to 20,000 lines per square inch* is the maximum permissible saturation.

For operation at 60 cycles much lower flux densities must be employed for the poorer grades of iron often used by amateurs—not over 30,000 lines per square inch. This calls for an expensive design, i.e., increased number of primary and secondary turns with increased copper losses.

At this point the builder should understand that maximum efficiency is generally obtained when the copper losses = the core losses. The design of a ½ kw. 60-cycle transformer follows.

THE DESIGN OF A ½ KW. 60-CYCLE, 15,000 VOLT TRANSFORMER.

—We will determine the dimensions of a ½ kw. 60 cycle transformer built after the sketch of Fig. 34, the primary to operate off 110 volts and the secondary to deliver 15,000 volts.

If there were no magnetic leakage the secondary voltage would bear the relation to the primary voltage as the ratio of the secondary and primary turns, that is, if E_s = secondary voltage, T_s = secondary turns, E_p primary voltage, T_p = primary turns, then

$$\frac{E_s}{E_p} = \frac{T_s}{T_p} \qquad (48)$$

For a preliminary computation no great error results in considering the leakage to be zero.

From (48) we may obtain the *ratio of transformation* for if $E_p = 110$ volts, $E_s = 15,000$ volts, then

$$\frac{E_s}{E_p} = \frac{15000}{110} = 136.3$$

This means that the secondary must have 136.3 times the number of turns in the primary in order that the secondary e.m.f. may equal 15,000 volts.

The *primary turns* may be obtained from the formula following, which is the fundamental equation for the transformer:

$$T_p = \frac{E_p \times 10^8}{4.44 \times a \times B \times N} \qquad (49)$$

Where

T_p = primary turns
E_p = primary voltage
a = cross sectional area of the core in sq. inches
B = flux density per sq. inch of core cross section
N = primary frequency

It is to be noted that the number of turns depends solely upon the e.m.f., flux and frequency. The size of the wire is determined by the capacity of the transformer.

In the problem under consideration E and N are known, but a, the cross sectional area, is preferably decided upon from knowledge gained through experience, and B, the flux density, by the permeability of the transformer steel to be used.

With the best grades of transformer steel it is safe to allow 60,000 lines of force per square inch. We will make the height and width of the transformer core 1¾".

Hence,

$E = 110.$ $10^8 = 100,000,000.$ $a = 1.75 \times 1.75 = 3.0625''.$ $B = 60,000.$ $N = 60.$

$$T_p = \frac{110 \times 100,000,000}{4.44 \times 3.0625 \times 60,000 \times 60} = 225 \text{ turns}$$

For the secondary turns, transposing (**48**),

$$T_s = \frac{E_s}{E_p} \times T_p$$

$$T_s = \frac{15,000}{110} \times 225 = 30,660 \text{ turns.}$$

The procedure from this point on is to determine the *resistance* of the primary and secondary turns, the primary and secondary *currents*, and the length of the *cores* upon which the primary and secondary coils are mounted. The pieces of iron which close the magnetic circuit between the two cores will be called the *yokes*.

When these data are obtained, we may determine the *copper losses* in the primary and secondary; the sum of these losses in turn giving the core losses from which the volume of the complete core is obtained.

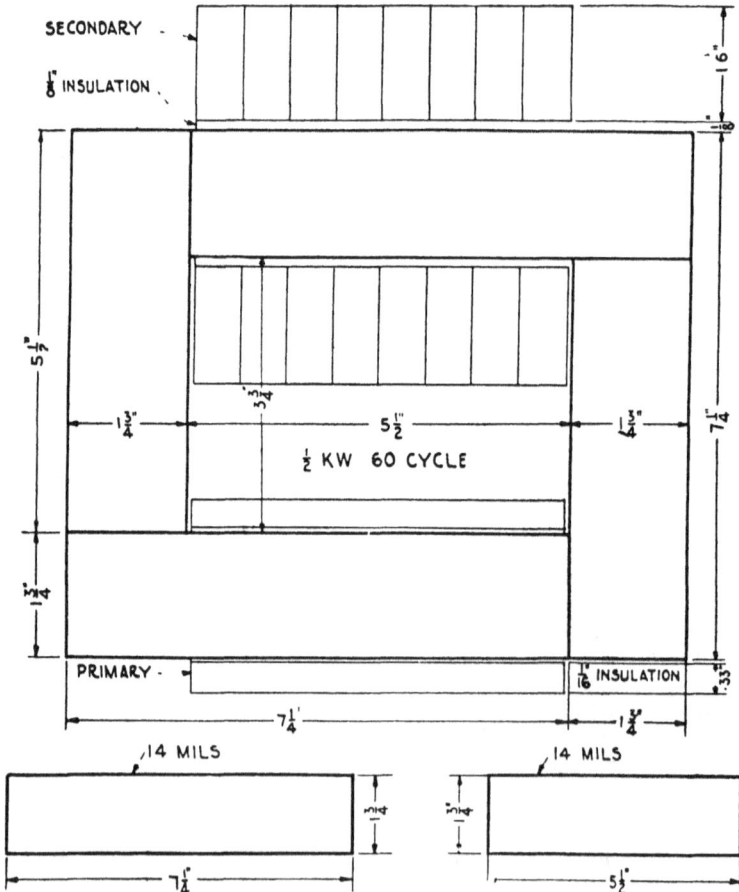

Fig. 36.   General core dimensions of a ½ kw , 60-cycle, 15,000 volt transformer.

The builder must bear in mind that the maximum efficiency is generally obtained when the *total copper loss = the total core loss*.

To keep the voltage between layers at a safe value the secondary should be made up of "pies" $\frac{1}{4}''$ in width. After the pies are wound they should be thoroughly impregnated with molten paraffine and separated from one another by discs of empire cloth.

If the secondary is divided into 18 pies $\frac{1}{4}''$ in width, the space required by the pies will be $\frac{18}{0.25} = 4.5''$. *Micanite* or *fibre washers* (or similar insulating material) should be placed at each end of the secondary to prevent the high voltage currents from discharging through the core. If $\frac{3}{4}''$ is allowed for the taping of the secondary pies and the insulating discs between, and $\frac{1}{8}''$ for the end washers, it is clear that the length of the secondary will be $4.5 + 0.75 + 0.25 = 5.5''$, which is the height of the transformer window Fig. 36.

The length of the core obviously is $5.5 + 1.75 + 1.75 = 9''$; and the length of the two cores $= 9 \times 2 = 18''$.

The volume of the two cores in cubic inches $= 1.75 \times 1.75 \times 18 = 55.1$.

If the copper losses in the primary and secondary are now determined, the core losses may be fixed and the length of the yokes determined accordingly.

To obtain the *mean length* of a secondary turn, and hence the total number of feet in the winding, it will be assumed that the coil is rectangular in shape, although even if the winding is begun on a square form it will gradually become a circle. It is necessary first to determine the sizes of the wires. The sizes of the primary and secondary wires are determined from knowledge of the primary and secondary currents. The currents should be calculated from the k v.a. readings, which as explained before involve the power factor.

The power factor in transformer circuits is usually rather low, not much over 0.8 and since for the primary circuit,

$$W_p = E_p \times I_p \times P\,F. \tag{50}$$

Then the primary current,

$$I_p = \frac{W}{110 \times P\,F}$$

$$= \frac{500}{110 \times 0\ 8} = 5 \text{ amperes (approx.)}$$

The secondary current,

$$I_s = \frac{E_p \times I_p}{E_s} \tag{51}$$

Hence

$$I_s = \frac{110 \times 5}{15.000} = 0.036 \text{ ampere.}$$

TABLE VIII
WIRE TABLE
B & S GAUGE

Gauge No. B. & S.	At 20°C. Dia. in mils	Dia. in inches	Area Cir. mils	Square inches	Ohms per 1000 ft. at 77°F. or 25°C.	Pounds per 1000 ft.	Max. Dia. S.C.C. in mils.	Turns per Linear inch S.C.C.	Max. Dia. D.C.C. in mils	Turns per Linear inch D.C.C.	Dia. in Mils S.S.C. (G.E.Co.)	Dia. in mils D.S.C. (G.E. Co.)	Turns per S.S.C. (G.E. Co.)	Turns per inch (G.E. Co.)	Approx. Turns per inch Enameled
6	162	.1620	26250	.02062	.4028	79.46	172	5.60	189.	5.44					
7	144.3	.1443	20820	.01635	.5080	63.02	154.3	6.23	173.30	6.08					
8	128.5	.1285	16510	.01297	.6405	49.98	137.5	6.94	142.5	6.80					
9	114.4	.1144	13090	.01028	.8077	39.63	122.4	7.68	127.4	7.64					
10	101.9	.1019	10380	.008155	1.018	31.43	117.9	8.55	112.9	8.51					
11	90.74	.0974	8234	.006467	1.284	24.92	96.74	9.60	101.7	9.58					
12	80.81	.08081	6530	.005129	1.619	19.77	86.81	10.80	91.8	10.62					
13	71.96	.07196	5178	.004067	2.042	15.68	77.96	12.06	83.0	11.88					
14	64.08	.06408	4107	.003225	2.575	12.43	70.08	13.45	75.1	13.10					14
15	57.07	.05707	3257	.002558	3.247	9.858	63.07	14.90	68.1	14.68					16
16	50.82	.05082	2583	.002028	4.094	7.818	56.82	16.60	60.8	16.35					18
17	45.26	.04526	2048	.001609	5.163	6.200	51.26	18.20	55.3	18.08					21
18	40.30	.0403	1624	.001276	6.510	4.917	46.30	20.20	50.3	19.90					23
19	35.89	.03589	1288	.001012	8.210	3.899	41.89	22.60	45.9	21.83					27
20	31.96	.03196	1022	.0008023	10.35	3.092	37.96	25.30	42.0	23.91					29
21	28.46	.02846	810.1	.0006363	13.05	2.452	34.46	28.60	38.5	26.20					32
22	25.35	.02535	642.4	.0005046	16.46	1.945	31.35	31.00	33.3	28.58					36
23	22.57	.02257	509.5	.0004002	20.76	1.542	28.57	34.30	30.60	31.12	26.00	29.00	38	34	40
24	20.10	.02010	404.0	.0003173	26.17	1.2223	26.10	37.70	28.10	33.60	23.00	26.00	43	38	45
25	17.90	.01790	320.4	.0002517	33.00	.9699	23.90	41.50	25.90	36.20	21.00	24.00	47	41	50
26	15.94	.01594	254.1	.0001996	41.62	.7692	21.94	45.30	23.04	39.90	19.00	22.00	52	45	57
27	14.20	.01420	201.5	.0001583	52.48	.6100	20.20	49.40	22.20	42.60	17.00	20.00	58	50	64
28	12.64	.01264	159.8	.0001255	66.17	.4837	18.64	54.00	20.64	45.50	15.60	18.60	64	53	71
29	11.26	.01126	126.7	.00009953	83.44	.3836	17.26	58.80	19.36	48.00	14.00	17.00	71	58	81
30	10.03	.01003	100.5	.00007894	105.20	.3042	16.03	64.40	18.03	51.10	12.50	15.00	80	66	88
31	8.928	.00892	79.70	.00006260	132.70	.2413	14.93	69.00	16.93	56.80	11.4	13.90	87	71	104
32	7.950	.00795	63.21	.00004964	167.30	.1913	13.93	75.00	15.95	60.20	10.50	13.00	95	76	120
33	7.080	.00708	50.13	.00003937	211.00	.1517	13.08	81.00	15.08	64.30	9.50	12.00	105	83	130
34	6.305	.006305	39.75	.00003122	266.00	.1203	12.31	87.60	14.31	68.60	8.80	11.30	110	88	140
35	5.615	.005615	31.52	.00002476	335.50	.09542	11.62	94.20	13.61	73.00	7.60	9.6	130	104	160
36	5.000	.005	25.00	.00001964	423.00	.07568	11.00	101.00	13.	78.50	7.00	9.00	140	110	
37	4.453	.004453	19.83	.00001557	533.40	.06001	10.45	108.00	12.45	84.00	6.00	8.00	160	120	190
38	3.965	.003965	15.72	.00001235	672.60	.04759	9.965	115.00	11.96	89.00					
39	3.531	.003531	12.47	.000009793	848.10	.03774	9.531	122.50	11.53	95.00					
40	3.145	.00314	9.888	.000007766	1069.00	.02993	9.145	130.00	11.15	102.50	5.00	7.00	200	140	230

Fig. 37.   Wire table for use in inductance calculations and transformer design.

TABLE IX

WEIGHTS OF SMALL SIZES OF MAGNET WIRE
(G. E. Co.)

Size	Weight in Pounds per 1000 Feet				
B. & S.	S. C. C.	D. C. C.	S. S. C.	D. S. C.	Enamel
10	....	31.9	....	....	....
13	....	16.0	....	....	....
14	12.684	12.918	....	....	12.684
15	10.082	10.274	....	....	10.053
16	8.012	8.176	....	....	7.973
17	6.375	6.510	....	....	6.322
18	5.081	5.188	....	....	5.009
19	4.043	4.130	....	....	3.966
20	3.215	3.289	....	....	3.136
21	2.569	2.628	....	....	2.475
22	2.055	2.106	....	....	1.970
23	1.630	1.676	1.57	1.604	1.555
24	1.297	1.344	1.241	1.298	1.232
25	1.036	1.082	.991	1.040	.980
26	.828	.873	.791	.833	.777
27	.661	.703	.631	.666	.616
28	.524	.562	.499	.521	.485
29	.421	.457	.397	.416	.384
30	.336	.372	.315	.332	.303
31	.271	.307	.254	.267	.242
32	.215	.248	.203	.214	.192
33	.174	.201	.161	.172	.152
34	.141	.161	.130	.140	.121
35	.12	.137	.110	.119	.101
36	.099	.112	.089	.096	.081
38	....	....	.058	.065	.051
40	....	....	.037	.040	.031

FIG. 37a. Useful table giving the weights of small sizes of magnet wire.

For small transformers, the cross section of the wire should be such that the current density for the primary does not exceed *1300 amperes per square inch and for the secondary 1100 amperes per square inch.* We will allow 1200 for the primary and 1000 for the secondary.

The diameter of the primary and secondary conductors may then be determined from the formulae following:

$$\text{Current density} = \frac{I}{A} \tag{52}$$

where

I = current in amperes
A = area of the conductor in square inches.

Transposing,

$$A = \frac{I}{\text{current density}}$$

For the primary

$$A = \frac{5}{1200} = 0.0041 \text{ square inches.}$$

TABLE X

CLOSED CORE TRANSFORMERS

Power rating	Frequency	Primary Voltage	Secondary voltage	Turns primary	Turns secondary	Number of primary layers	Number of secondary pies	Turns per pie	Length of primary in ft.	Length of secondary in ft.	Approx. weight of primary in lbs.	Approx. weight of secondary in lbs.	Core dimensions inside	Core dimensions outside
¼ kw., 250 watts	60	110	15000	260	35438	5	16 (¼")	2214	178	40444	2 3	5 5	4½"x3¾"	7¾"x7"
½ kw., 500 watts	60	110	15000	225	30660	4	18 (¼")	1703	161	35412	2 5	7 2	5½"x3¾"	9"x7¼"
1 kw., 1000 watts	60	110	18000	172	28139	3	30 (¼")	938	147	33710	5 8	12 6	9" x3 7"	13"x7 7"
½ kw., 500 watts	500	110	10000	112	11200	4	10 (.35")	1120	89	9900	2 0	4 1	4" x4"	7½"x7½"

OPEN CORE TRANSFORMERS

Power rating	Frequency	Primary Voltage	Secondary voltage	Turns primary	Turns secondary	Number of primary layers	Number of secondary pies	Turns per pie	Length of primary in ft.	Length of secondary in ft.	Approx. weight of primary in lbs.	Approx. weight of secondary in lbs.	Core dimensions inside
*1 k·w., 1000 watts	60	110	18000	440	42750	2	38 (.6")	1125	513	53437	16 3	13¼	3" in dia. x 25" long
½ kw., 500 watts	60	110	18000	340	34500	2	20 (.6")	1700	328	45425	4 3	8 5	2½" in dia.x14" long

Power rating	Dimensions of long core pieces	Dimensions of short core pieces	Size Primary wire	Size Secondary wire	Resistance of Primary at 77° Fahr.	Resistance of secondary at 77° Fahr.	Cross section of core	Flux lines per sq. inch	Primary current approx.	Secondary current approx.
¼ kw., 250 watts	6⅛"x1⅝"	5⅜"x1⅝"	No.14 d.c.c.	No.35 d.c.c.	0.458 ohm	13457 ohms	1⅝" sq.	60000	2.84 amps.	0.0208 amp.
½ kw., 500 watts	7¾"x1¾"	5½"x1¾"	No.13 d c.c.	No.33 d.c.c.	0.328 ohm	7472 ohms	1¾" sq.	60000	5 amps.	0.036 amp.
1 kw., 1000 watts	11"x2"	5.7"x2"	No. 9 d.c.c.	No.30 d.c.c.	0.1187 ohm	3539 ohms	2" sq.	60000	11 amps.	0.067 amp.
½ kw., 500 watts	5.9"x1.7"	5 9"x1 7"	No.11 d.c.c.	No.31 d.c.c.	0.1 ohm	1257 ohms	1¾" sq.	15000	7.5 amps.	0.075 amp.

OPEN CORE TRANSFORMERS

Power rating	Dimensions of long core pieces	Dimensions of short core pieces	Size Primary wire	Size Secondary wire	Resistance of Primary at 77° Fahr.	Resistance of secondary at 77° Fahr.	Cross section of core	Flux lines per sq. inch	Primary current approx.	Secondary current approx.
1 kw., 1000 watts	......	....	No.10 d.c.c.	No.32 d.c.c.	0.552 ohm	8992 ohms			13 amps.	?
½ kw., 500 watts	....	....	No.14 d.c.c.	No.32 d.c.c.	0.844 ohm	7585 ohms			7 amps.	?

*Cores made of a bundle of fine iron wires—No. 24 to No. 30

FIG. 38. Data for open and closed core transformers of various power inputs.

For the secondary

$$A = \frac{0\ 036}{1000} = 0.000036 \text{ square inches.}$$

As will be noted from the fifth column, Fig. 37, giving the area of B&S wire sizes in square inches we find that the nearest size corresponding to 0.000036 is No. 33, and to 0.0041, No. 13.

To determine the *mean length* of the primary and secondary turns, the *depth* of the windings, their *resistances*, etc., the procedure is as follows: If 30,660 turns are to be split between 18 pies, then,

turns per pie	$= \frac{30660}{18} = 1703$ turns.
From the table Fig. 37, turns per inch, No. 33 d.c.c.	$= 64.$
Turns per ¼″ layer	$= \frac{64}{4} = 16.$
Layers per pancake	$= \frac{1703}{16} = 106.$
From the table Fig. 37, diameter of No. 33 d c c.	$= 0.01508″.$
Depth of secondary winding	$= 106 \times 0.01508 = 1.59″.$

If the secondary pie is rectangular in shape as the core, and the core is covered with insulation ⅛″ in thickness, then

the mean length of a secondary turn	$= (1.75 + 1.59 + 0.125)\ 4 = 13.86″.$
Number of feet in secondary winding	$= 13.86 \times \frac{30660}{12} = 35{,}412.3$ ft.
From the table Fig. 37, resistance of No. 33 B&S wire per 1000 ft.	$= 211$ ohms.
Resistance of secondary	$= 35.4123 \times 211 = 7472$ ohms.

Allowing 5″ winding space for the primary we find from the table Fig. 37 that No. 13 d.c.c. wire permits 11.88 turns per inch, hence

turns per layer	$= 5 \times 11.88 = 59.40 = 60.$
The number of layers	$= \frac{225}{60} = 3.75 = 4$
From Fig. 37, diameter of No. 13 d.c.c.	$= 0.083″$
Depth of winding	$= 0.083″ \times 4 = 0.332″$
Allowing 1/16″ for insulation between primary and core, mean length of turn	$= (1.75 + 0.0625 + 0.332)\ 4 = 8.57″$
Number of feet in primary winding	$= \frac{225 \times 8\ 57}{12} = 160.6 = 161$ ft.
From table Fig. 37 resistance of No. B&S per 1000 ft.	$= 2\ 04$ ohms
Resistance of primary	$= 161 \times 2.04 = 0.328$ ohm

Having determined the resistances and current for the primary and secondary windings, the primary loss may be expressed:

$$W_p = I_p^2 \ R_p \tag{53}$$

The secondary loss:

$$W_s = I_s^2 \ R_s \tag{54}$$

Hence

$$W_p = 5 \times 5 \times 0.328 = 8.2 \text{ watts}$$
$$W_s = 0.036 \times 0.036 \times 7472 = 9.68 \text{ watts}$$

Total copper losses $= 9.68 + 8.2 = 17.88$ watts.

As already noted the core and copper losses should be approximately equal. Hence, core loss $= 17.88$ watts. Assume that the transformer steel (at a flux density of 60,000 lines) has a core loss of 0.82 watt per lb. at 60 cycles (which is a fair average for silicon steel), then it will require $\dfrac{17.88}{0.82} = 21.7$ lbs. of steel for the complete cores and yokes. But since 1 cubic inch of transformer steel weighs 0.278 lbs., the core must have a volume of $\dfrac{21.7}{0.278} = 78$ cubic inches.

We know that the length of each core is $5.5 + 1.75 + 1.75 = 9''$, and two cores $= 2 \times 9 = 18''$. Hence the volume of the cores (as previously shown) in cubic inches $= 1.75 \times 1.75 \times 18 = 55.1''$. This leaves $78 - 55.1 = 22.9$ cubic inches for the yokes. Dividing the volume of the yokes by the cross section we obtain the length of the two yokes, i.e., $\dfrac{22.9}{3.0625} = 7.47''$. Dividing this by 2 gives $3.73''$ ($3.75''$ approx.) as the length of each yoke.

The dimensions of the completed transformer are given in Fig. 36 and in the table Fig. 38. The long pieces of steel are $5.5 + 1.75 = 7\frac{1}{4}''$ in length. The short pieces are $3.75 + 1.75 = 5\frac{1}{2}''$ in length.

If the steel is $0.014''$ in thickness there will be required, $\dfrac{1.75}{0.014} = 125$ sheets for each core and yoke. That is, the builder requires 250 of the long sheets and 250 of the short sheets mentioned in the paragraph preceding.

It is now in order to determine the over-all efficiency of the transformer.

$$\text{Efficiency} = \frac{\text{output}}{\text{input}} \tag{55}$$

$$= \frac{\text{output}}{\text{output} + \text{iron losses} + \text{copper losses}}$$

$$= \frac{500}{500 + 17.88 + 17.88} = 93.3\%.$$

The table in Fig. 38 gives the dimension of ¼, ½, and 1 kw. transformers to operate off 60 cycles at 110 volts. The secondary voltage is approximately 15,000.

Sketches of the dimensions of the 60-cycle transformers appear in Figs. 36, 39 and 40.

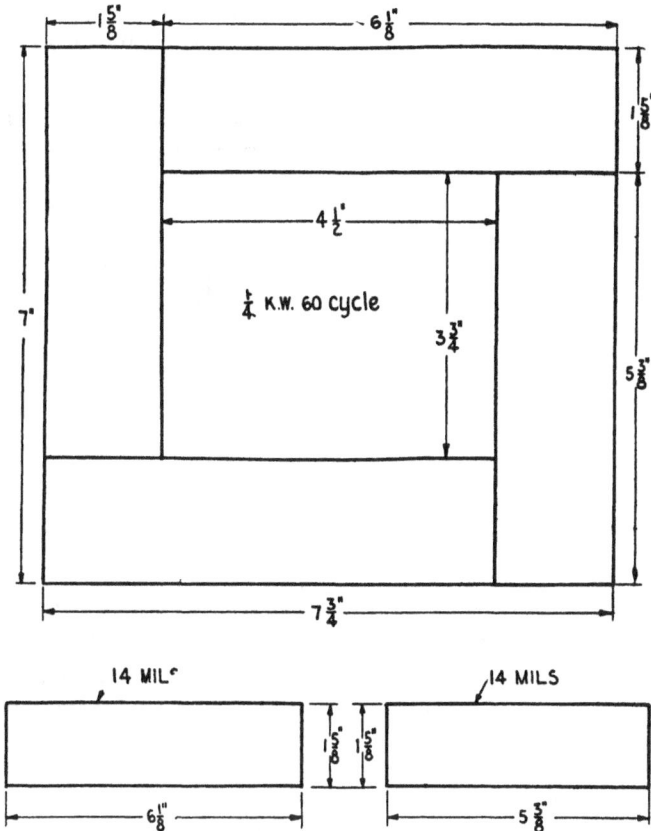

Fig. 39. General core dimensions for a ¼ kw., 60-cycle, 15,000 volt transformer.

Data is given also for the design of a ½ kw. 500-cycle transformer. The total core loss per pound at this frequency is 1.2 watts. Data and constructional details for 500 cycle transformers of other capacities are not given for the reason that the amateur usually does not possess a 500-cycle source. The dimensions for other inputs may be obtained by the line of reasoning presented in connection with the 60-cycle transformers. The results will be approximate but a series of subsequent experiments with the desired secondary capacity in shunt will reveal the correct operating characteristics. For example, a lesser or greater number of primary and secondary turns may be tried with a wattmeter and ammeter connected in the primary circuit to determine the most satisfactory adjustments. Usually a primary reactance or choke coil external to the transformer will permit the desired regulation of current and watts input.

The theory of the open core transformer has not been touched upon because the formulae for their design are for the most part empirical. The designs for the two sizes given, however, have been tried out in practice and have proven satisfactory. The closed core transformer for a given output requires less material than the open core and is, therefore, less expensive to build.

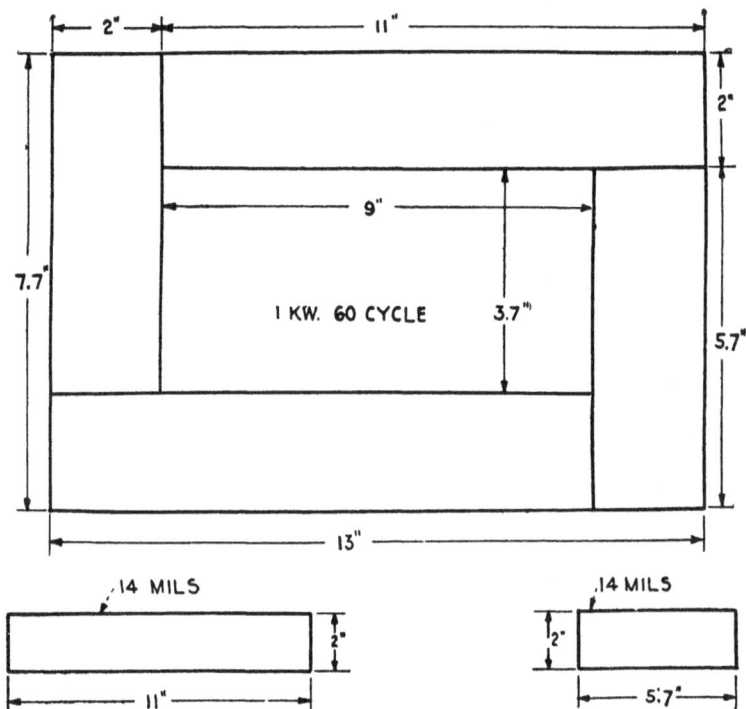

Fig 40. General core dimensions for a 1 kw., 60-cycle, 18,000 volt transformer.

SIXTY-CYCLE TRANSFORMERS AT LOWER FLUX DENSITIES.

—The builder should note that the table for 60-cycle transformers calls for the use of *high grade transformer steel*. If poorer grades of steel, i.e., plain steel or Russia iron, are employed the core losses are greater, requiring many additional pounds of wire to give the same voltage as the transformers with the better grades of steel.

To illustrate the point, assume that the transformer steel will only permit 30,000 flux lines per square inch; then the 1 kw. 60 cycle transformer, operated at the voltages and frequencies given in the preceding table will require a core 2″ square, with inside dimensions 11″x 6″ and outside dimensions 15″x 10″. The primary will require 320 turns of No. 10 d.c.c (295′), and the secondary 44,000 turns of No. 31 d.c.c (64,336′). This amounts to 9.4 lbs. of wire for the primary and 19 lbs.

for the secondary. The secondary may be split into 24 pies, $\frac{1}{4}''$ wide, with 1833 turns per pie.

The $\frac{1}{2}$ kw. transformer will require a core $1\frac{3}{4}''$ square, with inside dimensions $5\frac{1}{2}'' \times 7\frac{1}{4}''$ and outside dimensions $9'' \times 10\frac{3}{4}''$. The primary will require 450 turns of No. 13 d.c.c. (375') and the secondary 61,335 turns of No. 34 d.c.c. (83,376'). This amounts to 6 lbs. of wire for the primary and 13.3 lbs. for the secondary. The secondary turns may be divided between 24 pies, $\frac{1}{4}''$ wide, with 2555 turns per pie.

TABLE XI
DATA FOR SPARK COILS

Size	Dia. core	Length of core	Insulation over core in layers	Size wire primary	Primary turns
1″	$\frac{1}{2}''$	$5\frac{3}{4}''$	2 layers emp. cloth	No. 18	170
2″	$\frac{5}{8}''$	7″	2 layers emp. cloth	No. 16	184
3″	$\frac{3}{4}''$	8″	2 layers emp. cloth	No. 16	208
4″	1″	$8\frac{3}{4}''$	3 layers emp. cloth	No. 16	232

Size	Thickness insulation over primary	Size sec. wire	No. lbs. sec.	No. sections sec.	Sq. inches of foil vibrator condenser
1″	6 layers emp. cloth	No. 38	$\frac{3}{4}$ lb.	2 sect.	800
2″	6 layers emp. cloth	No. 36	1 lb.	2 sect.	1400
3″	8 layers emp. cloth	No. 36	$1\frac{1}{2}$ lb.	2 sect.	2000
4″	8 ayers emp. cloth	No. 36	2 lbs.	3 sect.	2500

FIG 41. Winding data and general dimensions of spark coils suitable for low-power amateur transmitters

The $\frac{1}{4}$ kw. transformer will require a core $1\frac{1}{2}''$ square, with inside dimensions $2\frac{1}{2}'' \times 4\frac{1}{2}''$, and outside dimensions $5\frac{1}{2}'' \times 7\frac{1}{2}''$. The primary is wound with 611 turns of No. 16 d.c.c. (478') and the secondary with 83,279 turns of No. 34 enamel (83,265'). This amounts to 4 lbs. of wire for the primary and 10 lbs. for the secondary. The secondary may be split into 15 pies, $\frac{1}{4}''$ wide, with 5551 turns per pie.

A table of the dimensions and general data for *spark coils* is given in Fig. 41. General dimensions are given for 1″, 2″, 3″ and 4″ coils to be operated off 6 to 12 volt storage batteries.

A fundamental wiring diagram of an induction coil with a magnetic interrupter is shown in Fig. 41a. P, the primary coil, is wound over the iron core C. The core is composed of a bundle of fine iron wires (No. 22 or 24) which are bound together with tape and then covered with one or two layers of empire cloth. The secondary S is split into sections. Each section is wound with many layers of wire separated by thin sheets of paraffined paper. The condenser K-1, prevents arcing at the interrupter contacts. The dimensions of the condenser for various coil sizes are given in the table. They are made up of several sheets of tinfoil separated by paraffined paper.

PRIMARY IMPEDANCES OR "CHOKES."—Primary chokes or reactance coils must be employed with the 60-cycle transformers, designed in accordance with the data in Fig. 38. The flux leakage in these designs is comparatively low, and when the spark discharges across the gap, the transformer is on short circuit. This may cause an excessive rise of the primary current and may burn out the transformer.

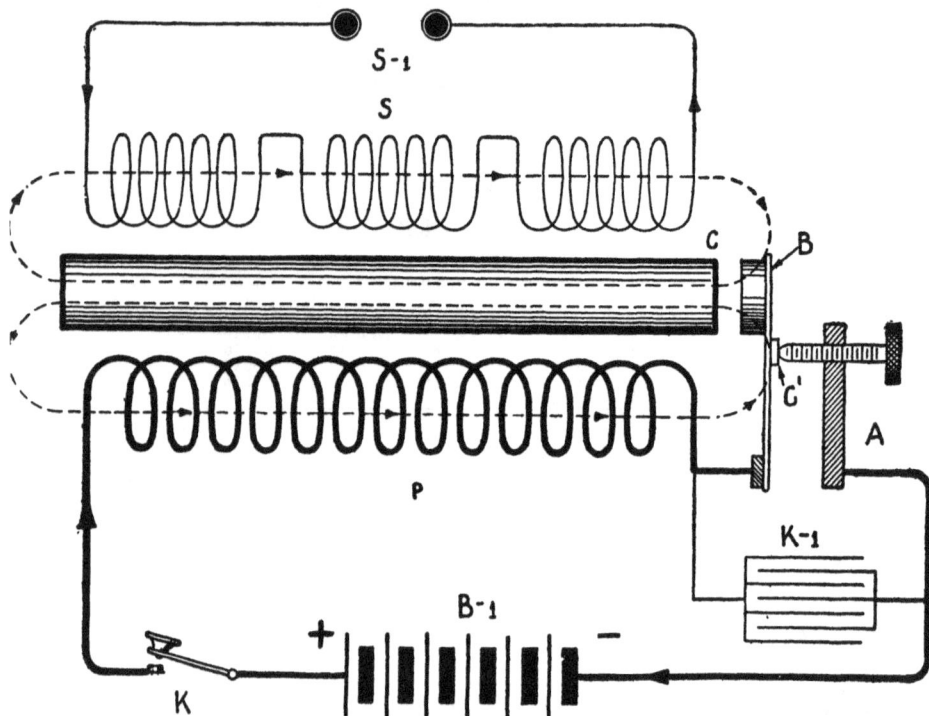

FIG. 41a. Fundamental wiring diagram of an induction coil for the production of high voltage currents from a d c. source.

An experimental primary choke for the 1 kw. transformer may be made on a core 2″ square, 15″ in length wound with 4 layers of No. 9 d.c.c. wire. The coil may be tapped at the middle and ends of each

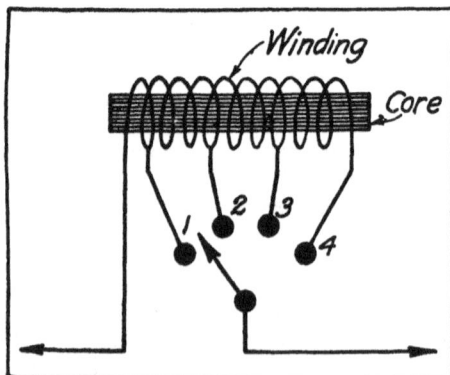

FIG. 42. Reactance regulator of the open core type. Such chokes are useful for regulating the power input

layer, and leads brought out to a multipoint switch. As an alternative the coil may be wound on a tube so that the core can be moved in and out of the coil. Very close regulation of the primary current may be obtained in this way and the necessity for a multipoint switch avoided.

The choke for the $\frac{1}{2}$ kw. transformer may be wound on a core $1\frac{3}{4}''$ square, $15''$ in length wound with four layers of No. 13 d.c.c. wire, tapped in the middle of each layer.

The choke for the $\frac{1}{4}$ kw. transformer may be wound on a form $1\frac{1}{2}''$ square, $14''$ in length wound with four layers of No. 14 d.c.c. wire. The coil is tapped in the middle of each layer.

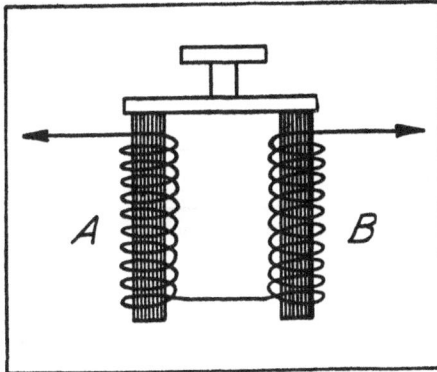

Fig. 43. "U" shaped reactance coil.

These primary chokes will be of considerable value in obtaining resonance between the transformer and the frequency of the source. Such regulation is particularly desirable if a series multiplate spark gap is employed. The chokes must be employed, in any event, to cut down the primary current.

The chokes may be wound on a straight iron core as in Fig. 42, or on a "U" shaped core as in Fig. 43. The reactance of the choke in Fig. 43 is varied by drawing the iron core in and out of the coils A and B.

MAGNETIC LEAKAGE.—With all types of spark transmitters, the power circuits should be designed to have a certain amount of magnetic leakage. The reason for this is that the discharge at the spark gap places the secondary of the transformer on short circuit. This not only causes an excessive rise of current which may burn out the transformer, but the arc formed at the gap prevents *quenching of the primary oscilla-tions.* Lack of quenching permits the antenna circuit to re-transfer a part of its energy to the spark gap circuit and this interchange of energy raises and lowers the impedance of the oscillation transformer so that oscillations of two frequencies flow in the antenna circuit resulting in *double wave emission.*

To secure *single wave emission* under such conditions, the coupling of the oscillation transformer must be reduced until quenching takes place. This, of course, decreases the antenna current and, accordingly, the range of the set.

On the other hand if the transformer circuits have sufficient magnetic leakage the primary current is limited to a definite maximum; there is then less tendency towards arcing at the spark gap, the spark note is clearer and a closer coupling may be employed at the oscillation transformer.

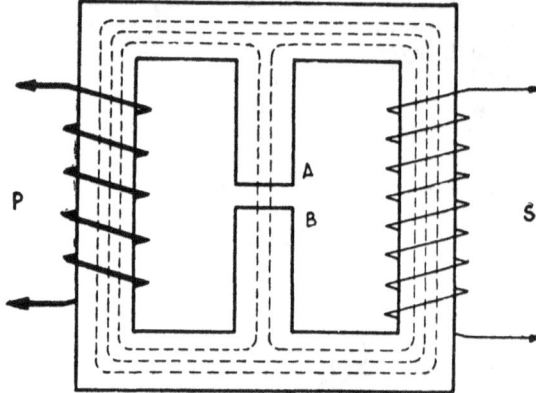

Fig. 44. A magnetic leakage gap placed between the primary and secondary cores of a closed core transformer for regulation of load.

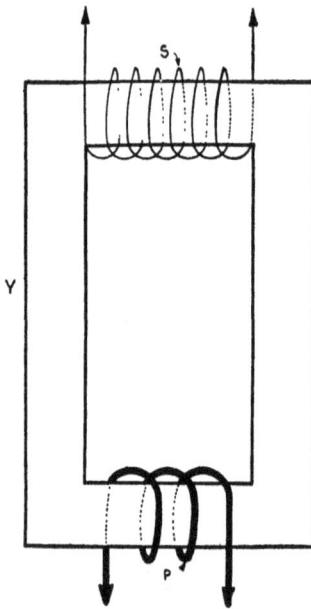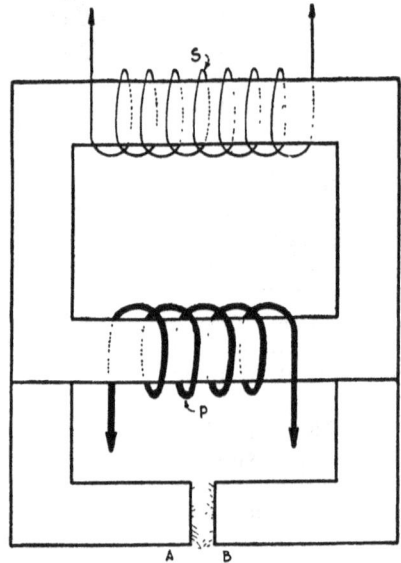

FIGURE 45   FIGURE 46

Fig. 45. Showing how magnetic leakage may be obtained by mounting the primary and secondary coils on the short legs of a transformer core
Fig. 46. An external magnetic leakage gap suitable for high voltage transformers.

The requisite flux leakage does not necessarily take place in the transformer. Special design of the alternator gives flux leakage at the armature when the transformer is short circuited. The primary chokes described in the preceding paragraph give practically the same effect.

Figs. 44, 45, and 46 show how magnetic leakage may be secured. In Fig. 44 the tongues *A B*, form an air gap across which leakage takes place. In Fig. 45, the primary *P* and secondary *S* are wound on the short legs of the core thus giving leakage through the yokes *Y*. In Fig. 46 the transformer core has the adjustable air gap *A B*, through which the proper regulation may be obtained. Experimentation with these various methods permits the builder to obtain any desired operating characteristic. The power factor may thus be brought up to 0.8 or 0.9 and wastage of the primary power may be thereby prevented.

RESONANCE TRANSFORMER.—The principle of resonance has been utilized in transformer circuits to assist quenching of the primary oscillations by preventing arcing at the spark gap. As an illustration, take the circuit of Fig. 47. Since the primary coil and the secondary coil of the high voltage transformer are closely coupled, any change of capacity in the secondary condenser will tune the primary circuit, including the armature coils of the alternator, to the frequency of the source *N* as well.

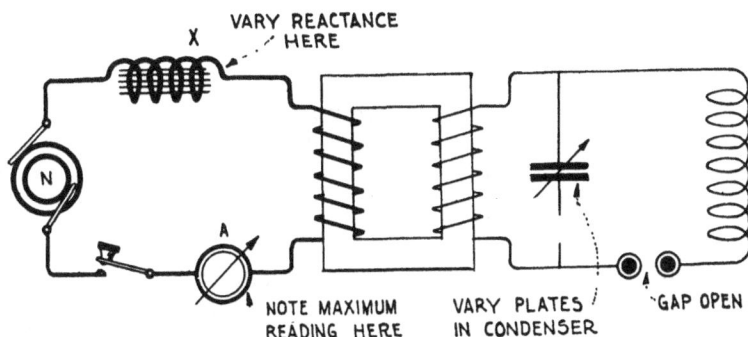

Fig 47   Showing how resonance may be obtained in transformer circuits such as are used in radio transmitters.

Resonance may be established by varying the capacity of the high voltage condenser, noting the reading of the ammeter *A*, until a maximum is secured. Or, the capacity of the condenser being fixed resonance may be secured by variation of the reactance *X*. During this test, the spark gap should be opened sufficiently to prevent sparking.

It then will be found that some value of secondary capacity or a combination of some value of secondary capacity and primary reactance will give a maximum as read at the primary ammeter.

The beneficial effects of resonance are then as follows: When the spark discharges at the gap, the secondary of the transformer is short circuited, the resonance condition between the source *N* and the transformer circuits is destroyed, and the gap potential falls off rapidly reducing the tendency to arc. Due to the destruction of the transformer arc the primary oscillations are more effectively quenched and single wave emission results.

It has been found difficult with resonance transformers to maintain a uniform spark discharge; that is, a clear spark tone. If the transformer is worked from 15 to 20 per cent off resonance, the spark tone is more readily maintained without sacrificing greatly the good effects of resonance.

The natural frequency of the transformer circuits is generally 15 to 20 per cent greater than that of the alternator. Just how much greater it must be for satisfactory operation must be found by experiment for a great variety of transformer designs are encountered in practice. The resonance condition is dependent upon many things, such as the inductance of the dynamo armature, the mutual inductance between primary and secondary and the capacity of the secondary condenser.

The secondary voltage of a resonance transformer may be considerably greater than the turn ratio $\dfrac{T_s}{T_p}$. A transformer not constructed with particular regard to insulation may burn out as the resonance point is approached. It may develop with some designs that the capacity required for resonance will exceed the maximum possible for 200 meters, viz., 0.01 mfd. Only the experimenter who possesses a motor generator will be able to take full advantage of the resonance transformer phenomenon. The *varying power factor* of transmission lines would introduce a constantly changing set of conditions which could not be compensated for rapidly enough to be effective.

THEORY OF SPARK DISCHARGERS.—When the closed oscillation circuit of a wireless telegraph transmitter acts inductively upon the antenna circuit, and the spark gap is one in which the insulating qualities of the included air are not quickly restored between sparks, part of the energy of the antenna oscillations is retransferred to the closed circuit. The interchange of energy between the open and closed circuit results in a complex action which modifies the frequency of the free oscillations which would exist in the closed circuit if it were not coupled to the antenna. In fact, it results in the production of oscillations of two different frequencies in both the spark gap and the antenna circuits, and the aerial radiates two waves.

Double wave emission tends to decrease the range of radio transmitters and beyond this interferes with stations operating on other wave lengths. The receiving apparatus can generally be tuned to one wave length only, as the energy of the other wave is of no use. But if the radiated energy be confined to one wave length, the maximum effect will be obtained at the receiver and the liability to "jam" another station is farther removed.

It is customary to say that "tight coupling" between the closed and open circuits results in "double wave emission," whereas "loose coupling" results in single wave emission. The truth of either statement depends upon the construction of the spark gap, in fact upon the design of the whole transmitter. For if a so-called *quenched gap* be employed, single wave emission is secured with relatively close couplings at the oscillation transformer. But, if a plain gap or any type of gap in which there is tendency to arc, be employed, single wave emission is possible only with loose coupling.

The quenched gap transmitter allows the primary circuit to oscillate through but a few cycles, whereupon the primary oscillations cease.

By this time the antenna oscillations have attained their maximum amplitude but because the primary circuit is then silent, the antenna circuit continues to oscillate at its own frequency and decrement until

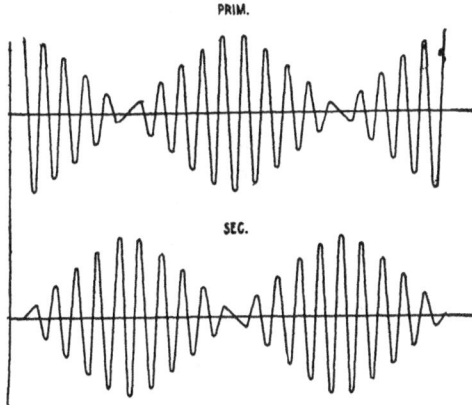

PRIM.

SEC.

FIG. 47a. Graphs showing the complex oscillations obtained in coupled radio frequency circuits with imperfect quenching

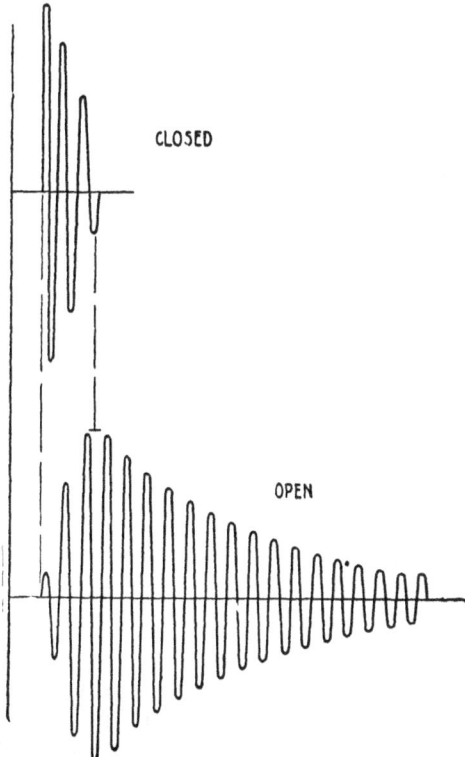

CLOSED

OPEN

FIG. 47b. Graphs showing the primary and secondary oscillations in coupled circuits with good quenching

the energy originally imparted to it is dissipated. With gaps which do not quench readily, and with tight coupling at the oscillation transformer, oscillations of two frequencies are generated in both the closed and open circuits. Single wave emission can only be obtained by reducing the transformer coupling until a wavemeter shows but one maximum. The antenna current will, on this account, be less than that obtained with a gap giving better quenching.

The spark gap phenomena just outlined are shown by the graphs of Figs. 47a and 47b. Graph 47a shows the complex oscillations obtained with *poor quenching* and graph 47b indicates the results of *good quenching*. In the latter case the primary oscillations stop after a very few swings even when the oscillation transformer is closely coupled.

We take the liberty of quoting a summary of the phenomena of spark dischargers in radio frequency circuits from the author's "Practical Wireless Telegraphy:"

"It should be understood that a transmitter in proper adjustment for practical use never radiates a double wave, but, in fact, should always radiate a single wave, of wave length and damping normal to the antenna circuit as adjusted, but with any transmitter two waves will appear if the spark gap is not in proper condition. The remedy, in event of the latter, is to restore the gap to its proper working qualities or to loosen the coupling at the oscillation transformer.

"When the spark discharges across the gap it acts as a trigger to start the primary circuit into oscillation and the stored energy of the condenser will be transferred to the antenna circuit until in the course of a few oscillations (the number decreasing as the coupling is closer), the voltage in the primary circuit becomes so low that the spark will no longer discharge across the gap; the primary oscillations will then cease. The exact value of the minimum voltage for non-sparking will depend upon the resistance of the gap.

"The resistance of the spark gap always increases as the oscillating current decreases and if it were not for the burnt gases which exist in the immediate vicinity of the discharge gap the original resistance would be restored at the end of the first half oscillation. Since there is a lag in the cooling and dissipation of the hot gases this does not occur, but if the gap is properly cooled, that is—the electrodes do not get too hot, the resistance becomes so high after a few oscillations that the reduced voltage of the condenser cannot maintain the spark. In other words the spark is quenched and the oscillations of the two circuits will take the form shown in Fig. 47b, that is—after a few swings of the primary circuit, the primary oscillations will cease and the antenna circuit will oscillate at its natural frequency and decrement.

"These are the precise actions taking place in a properly adjusted radio transmitter, no matter what type of spark gap is employed, if the electrodes are clean and smooth, the ventilation adequate, and the coupling of the oscillation transformer is not too close.

"The rate at which the gases in the gap are dissipated or the non-conducting qualities of the gap restored determines how close the coupling can be made without interfering with the quenching of the spark. If the coupling is so close that the reaction of the secondary upon the primary not only does not extinguish the spark but transfers energy back to the primary, then the spark will not be quenched until the energy of the entire system has fallen to a low enough value to allow the high resistance of the gap to be restored. The complex oscillations shown in Fig. 47a will then result and double wave emission will result. If the spark quenches after one or two of the "beats" shown in Fig. 47a, and the antenna still has energy to be radiated, then investigation with a wavemeter may show three apparently different wave lengths radiated.

"In all spark gaps there is a tendency toward "arcing," that is—for the spark to be followed by passage of the power current across the gap. This will prevent the restoration of the high resistance—the spark will not quench.

"The plain, open spark gap without artificial means of cooling requires very careful adjustment and reduction of coupling to give proper operation, that is—freedom from double wave emission. Unless the spark voltage is carefully adjusted, the tendency toward arcing is difficult to control and the action tends to be irregular. The use of special cooling means, such as a series of gaps or air blasts, enables good quenching to be obtai.ed with sufficient regularity to give a clear spark tone."

DAMPING DECREMENT.—The U. S. laws require that the decrement of the antenna oscillations shall not exceed 0.2 per complete cycle. This means that each spark discharge in the closed circuit must set up no less than 23 complete oscillations in the antenna circuit. The oscillations in the wave train are considered to have stopped when one of the successive cycles has fallen to 0.01 of the amplitude of the first oscillation.

If the inductance, capacitance, and resistance of a radio frequency circuit are fixed, the free oscillations generated therein decay at a fixed rate in a way similar to the decaying oscillations of the pendulum. Each successive cycle is the same fraction of the preceding cycle throughout the train. Thus the amplitude of a cycle may be 0.8 of the amplitude of the preceding one and so on. Instead of expressing the amplitudes of successive cycles by a numerical ratio it is more convenient to express the decay in terms of the logarithm of the ratio. Hence we have the term, *the logarithmic decrement.*

The upper graph in Fig. 47c shows a group of damped oscillations.

Let the ratio of $\frac{A\ B}{C\ D} = 1.105$ and $\frac{C\ D}{E\ F} = 1.105$. Referring to a table

of Naperian logarithms, we find that the logarithm of 1.105=0.1, which is the logarithmic decrement of that particular wave train. The lower graph indicates a wave train the decrement of which = 0.69, i.e., a case of excessive damping.

To find the number of oscilla.ions in a wave train, when one of the subsequent cycles has fallen to 0.01 (1 per cent) of their initial amplitude, we need only divide the natural logarithm of 100 by the decrement.

Since $\log^e 100 = 4.6$ then $\frac{4\ 6}{0\ 1} = 46$ complete oscillations for a decrement

of 0.1. The smaller the decrement the "sharper" will be the radiated wave. More accurate tuning at the receiver is then possible with a consequent increase in the strength of signals.

A wave whose decrement exceeds 0.2 per complete oscillation will interfere with stations not sharply tuned to it.   That is such a wave will set up currents in a receiver over a considerable range of wave lengths. The decrement of the radiated wave may be kept at a minimum by keeping the resistance of the oscillatory circuits low, by providing a good earth connection, and by reducing the coupling at the oscillation transformer until single wave emission results.   Methods of measuring the logarithmic decrement will be described in chapter XI.

If the inductance, capacitance and resistance of an oscillation circuit are known the decrement may be calculated by the formula following:

$$\delta = 3.1416 \; R \sqrt{\frac{C}{L}} \qquad\qquad (56)$$

Where

$$\delta = \text{decrement per complete cycle}$$
$$R = \text{resistance in ohms}$$
$$C = \text{capacitance in farads}$$
$$L = \text{inductance in henries}$$

If $R = 1$ ohm, $L = 10,000$ cms. and $C = 0.01$ mfd., then

$$\delta = 3.1416 \times 1 \sqrt{\frac{0.0000001}{0.00001}}$$
$$= 3.1416 \times 1 \times 0.1$$
$$= 0.31416$$

PURE AND SHARP WAVES.—A sharp wave, according to the U. S. definition, has a decrement of 0.2 or less.   A pure wave emission may consist of two waves, provided the amplitude of the lesser does not exceed 0.1 of the amplitude of the stronger wave.   The lesser wave in such circuits is usually of negligible importance.   Hence we may say that a pure and sharp wave is one of single frequency whose decrement is 0.2 or less.

Fig. 47d shows a graph of pure wave according to the U. S definition Fig. 47e indicates a sharp wave whose decrement is 0.12   Fig. 47f shows a very broad wave which the U. S. laws will not permit.   The lesser wave is 175 meters, the stronger wave 245 meters.   The closed and open circuits were individually tuned to 200 meters.   This is substantially the form of wave motion that will be obtained with the average amateur set, if the coupling be too close and the quenching poor.

As we have said, single wave emission in the ordinary amateur set is obtained by reducing the coupling at the oscillation transformer until a wave meter shows but one wave.   The decrement of the wave can be measured with any wavemeter including in its circuit a "current square" meter.   The process will be described in chapter XI, page 301.

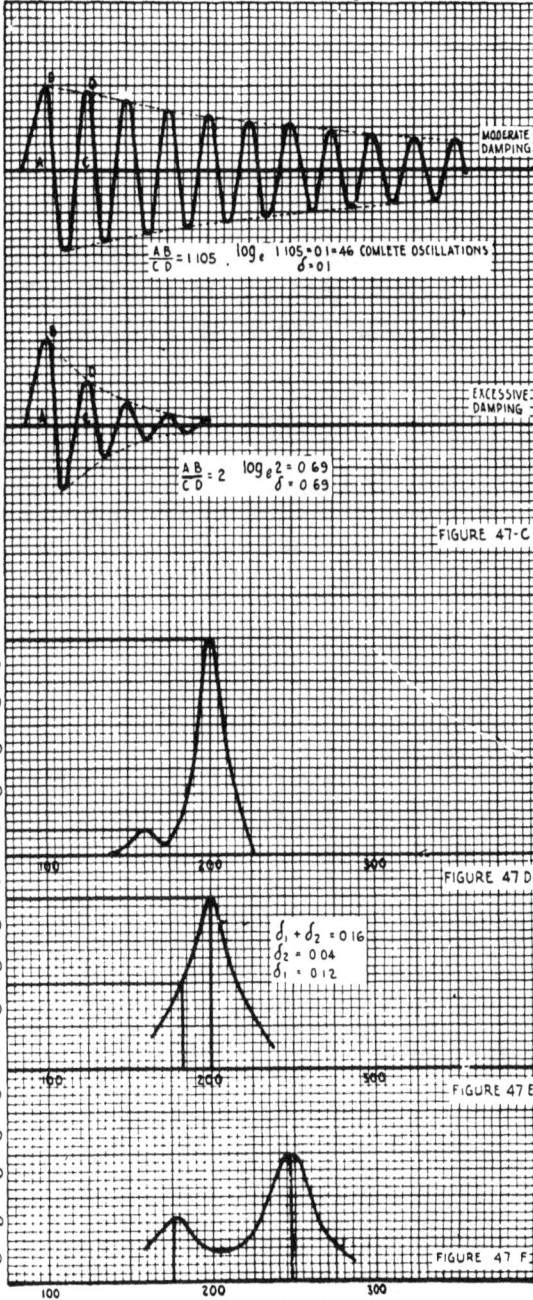

AB/CD = 1105   log_e 1105 = 0 1 = 46 COMLETE OSCILLATIONS
δ = 0 1

MODERATE
DAMPING

EXCESSIVE
DAMPING

AB/CD = 2   log_e 2 = 0 69
δ = 0 69

FIGURE 47-C

FIGURE 47 D

δ_1 + δ_2 = 0 16
δ_2 = 0 04
δ_1 = 0 12

FIGURE 47 E

FIGURE 47 F

FIG. 47c. A group of damped oscillations where δ=0.1, and another group where δ=0.69. The former is a case of moderate damping and the latter a case of excessive damping.
FIG. 47d. Graph of a 'pure' wave according to the U. S. definition. The amplitude of the lesser wave must not exceed 0.1 of the amplitude of the greater wave.
FIG. 47e. Resonance curve of the antenna oscillations for a decrement of 0.12 per cycle.
FIG. 47f. Resonance curve of a double wave emission which the U. S. law will not permit. This is a type of wave emission which the amateur must avoid.

CHAPTER IV

CONSTRUCTIONAL DETAILS OF AMATEUR WIRE-LESS TRANSMITTERS—PANEL AND ISOLATED TYPES—BUZZER TRANSMITTERS

CONSTRUCTION OF A HIGH VOLTAGE TRANSFORMER.— Assume that the steel strips for the transformer cores and yokes are cut to shape in accordance with the designs given in the preceding chapter. If, in the cutting process, the iron is burred or bent, the strips may be placed between two smooth slabs of wood and straightened out with a hammer. All burrs should be removed with emery paper or with a file and afterwards both sides of each strip should be coated with *shellac varnish* and set up on edge to dry.

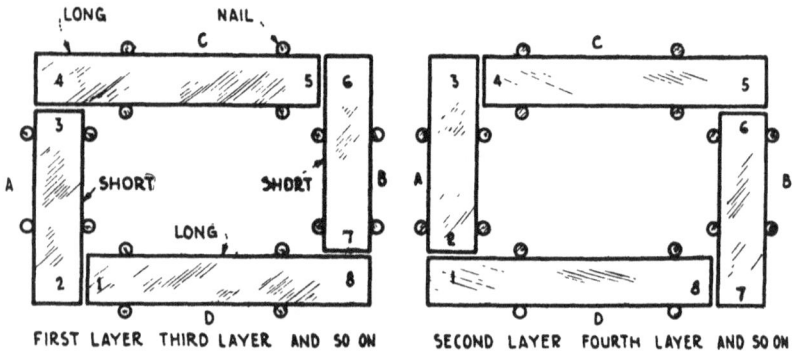

FIG. 48. Showing the core assembly of a closed core transformer.

The core should be assembled as shown in Fig. 48, where the short strips, i.e., the strips for the *yokes* of the core, are denoted by A and B, and the long strips by C and D To be sure that the pile will be even after it is stacked up drive nails in the base board as a guide, as shown in the sketches of this figure The end 3 of strip A is placed against the side 4 of strip C; the end 5 of strip C against the side 6 of strip D; the end 7 of strip B against the side 8 of the strip D, and the end 1 of strip D against the side 2 of strip A.

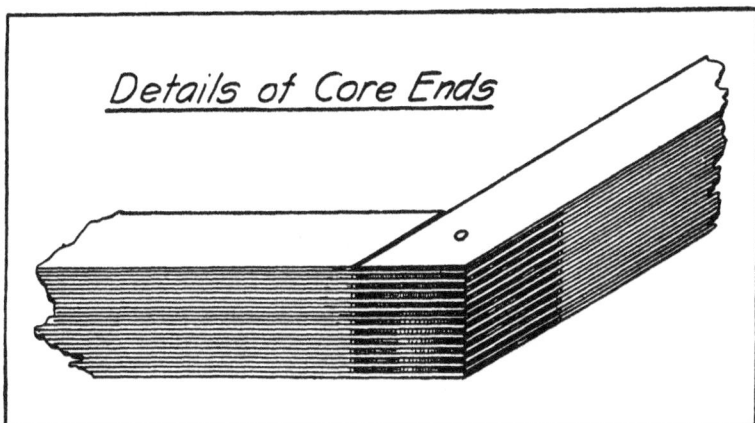

FIG. 49. Showing the overlapping laminations at the corners of the transformer core.

As shown in the right-hand of drawing the next layer over-laps the bottom layer so that the ends of the core when completed will look like that in Fig. 49. Holes may be bored in the corners of the core through which bolts are passed to compress the sheets into a compact bundle.

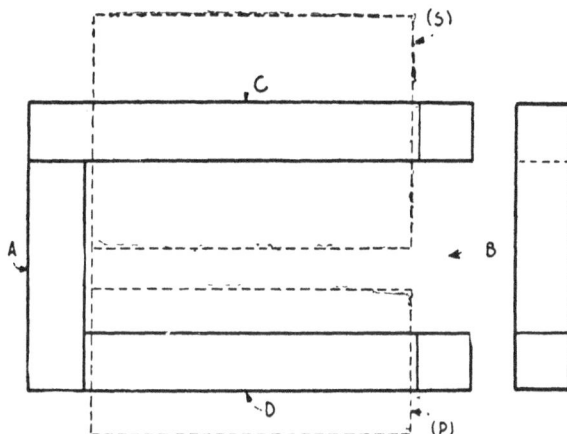

FIG. 50. Core with one yoke removed to permit the primary and secondary coils to be placed in position.

If the builder decides to mount the primary and secondary coils on the long legs he may remove one yoke of the completed core as in Fig. 50 and place in position the pieces of steel for that yoke after the primary and secondary coils are in place. The author has found it more convenient to first stack up the core completely, winding it loosely with tape so that afterwards the yoke B can be removed for placing the primary and secondary windings. After these coils are placed the yoke B may be slid back into place without tearing it down. If this proves too difficult place the pieces of steel for the last yoke in position one at a time.

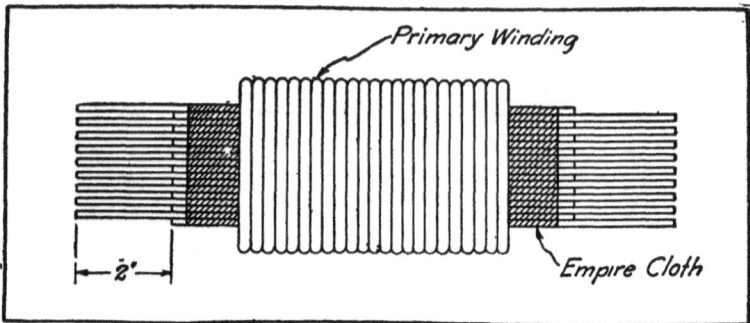

FIG 51.　Showing the appearance of the transformer core with the primary coil in place.

The appearance of the core and coil after the primary is in position is shown in Fig. 51, which shows a few layers of empire cloth, each layer about 7 mils in thickness, wound over the core.　The primary turns are then wound over the empire cloth as indicated.

In the transformer designs already presented the primary and secondary coils are approximately square, but some experimenters prefer to place round insulating tubes over the cores.　The tubes may be of *bakelite, dilecto,* or *micanite.*　If circular tubes are used the dimensions of the windings will be greater than those given in the table Fig. 38 and the transformer window may have to be enlarged to prevent the primary and secondary coming in contact.

Sufficient insulation is however provided by winding several layers of empire cloth over the primary and secondary cores.　The coils will then be approximately square.　As an alternative, wind several layers of *sheet mica* around the primary and secondary cores and hold them in place by tape.　Take care that the mica does not break at the corners permitting the high voltage currents generated in the secondary to discharge through the iron core.　Usually, insulation $\frac{1}{16}''$ in thickness between the core and primary, and $\frac{1}{8}''$ between the secondary and the core is sufficient, although for the 1 kw. transformer (in Fig. 40), the insulation between the secondary coils and the core should be $\frac{3}{16}''$ in thickness.

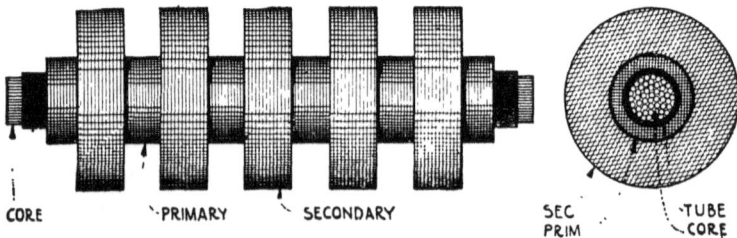

FIG 51a　Assembly of the open core transformer

After the primary and secondary coils are placed on the core place the remaining yoke in position, then immerse the transformer in transil oil or in a beeswax compound.　A formula for this will be given in a following paragraph.

A general idea of the assembly of an open core transformer may be obtained from Fig. 51a. The core is a bundle of fine iron wire covered with six or seven layers of empire cloth. The primary turns are then wound on, given a coating of shellac and allowed to dry. The primary is then slipped in the insulating tube. The secondary pies are slid over the tube and connected in series.

In the commercial types of open core transformers the secondary pies, after being connected in series, are placed in a box and covered with insulating wax; but if the pies are carefully insulated from one another this will not be necessary.

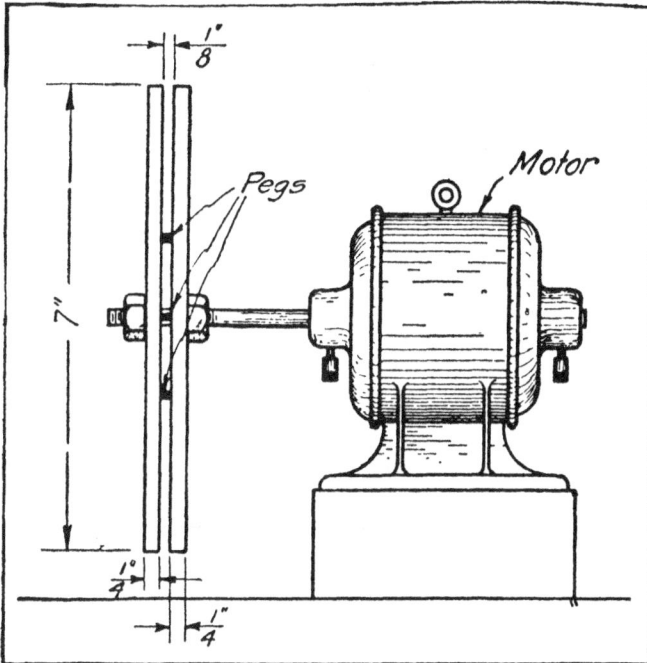

FIG. 52. Apparatus for winding the secondary "pies" of high voltage transformers.

A MACHINE FOR WINDING SECONDARY PIES.—A simple winder for the secondary pies may be constructed in accordance with the drawing in Fig. 52. Two wooden discs 7″ or 8″ in diameter are mounted on the shaft of a small driving motor. The discs are about ¼″ in thickness. The space between them is gauged by the two nuts shown in the drawing. Four pegs are placed in the disc to form a square of dimensions slightly greater than the dimensions of the secondary core with its insulation. Two or three layers of empire cloth or paraffined paper are wound over these pegs, the ends of which are glued together. The wire is then given one or two turns around the winding form and the motor kept in rotation until the desired number of turns are obtained.

The use of *double cotton covered* wire is assumed in the transformer designs given in Fig. 38. The wire should be run through melted paraffine during the winding process.

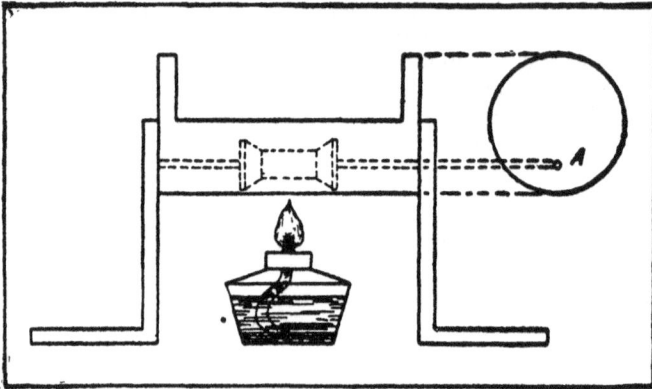

Fɪɢ. 53.   Device for impregnating the secondary wire of a transformer with paraffine.

A convenient device for impregnating the secondary wire with paraffine is shown in Fig. 53, where a baking powder can is cut away and an empty spool rigidly attached to a brass shaft *A* placed at the bottom of the tank. The shaft is fastened in place so the spool cannot turn. A quantity of melted paraffine is kept hot by a small alcohol lamp immediately underneath. The spool from which the secondary wire is drawn is mounted on a spindle immediately below, the wire being passed around the spool in the paraffine tank and then on to the pie winder of Fig. 52.

Do not allow the spool in the tank to turn or it will throw the paraffine badly.

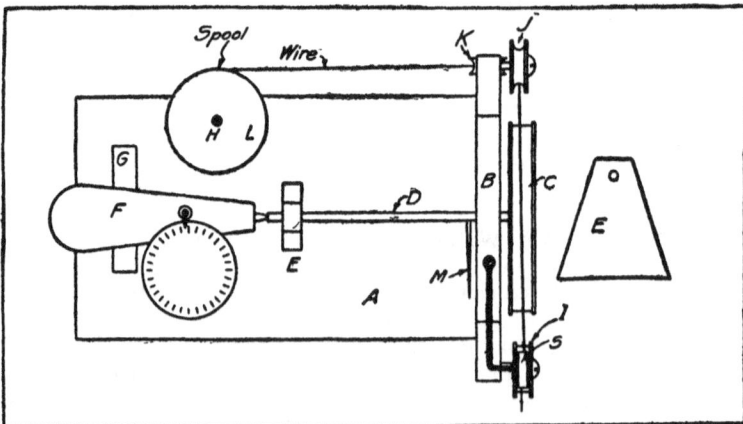

Fɪɢ 54.   Wire measuring device for the experimental workshop

A WIRE MEASURING DEVICE FOR USE IN COIL WINDING.—

Melvin Wallace of Oregon designed the wire measuring instrument shown in Figs. 54, 55 and 56. This instrument will be useful to the amateur experimenter who wishes to measure the actual number of feet of wire placed on a transformer pie, or on tuning coils to be wound for receiving apparatus. The construction is as follows: First cut from a 1″ x 6″ piece of hard wood stock, the piece A in Fig. 54, which is 8″ in length. At one end attach the piece B which is to carry the friction wheels I and J. Next cut a measuring wheel as at C. It should be 4½″ in diameter cut from the same stock. A flat groove should now be cut in the rim until it has the exact length of 12″. Then place the wheel on the shaft D which projects through the end piece B, through the bearing piece E and is attached to the revolution recorder F which, in turn, is securely clamped to the block G.

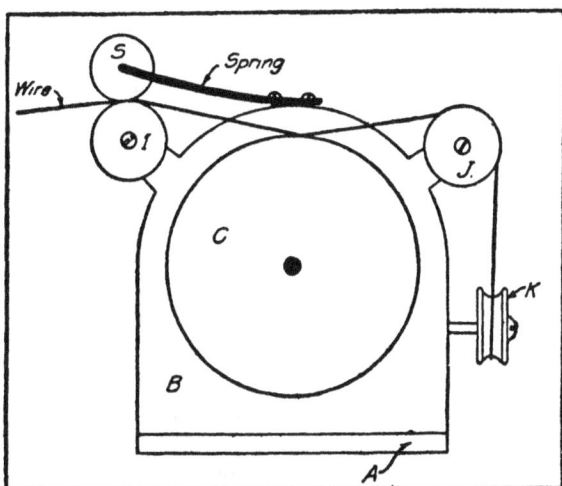

Fig 55.   Side view of wire measuring device.

Thread a piece of iron rod H, size optional, at one end, and place it in the board A at H to carry the spool of wire.

The friction wheels I, J and K in Figs 54 and 55 must now be constructed and attached as in the drawings. Now construct the wheel S (Fig. 55) counter to the wheel I, and fit the former to the flat groove of the latter. Mount the tension wheel S on a spring which is preferably adjustable so that in case of a break or stop the wire will not slip.

Place the pointer M indicated in Fig. 54 on the shaft D close to the piece B. Next cut a piece of white paper the size of the wheel C and calibrate it into 12 equal parts; then glue it to the inside of the piece B to form a scale. The pointer will then indicate the number of inches and the revolution counter the number of feet which have passed the spot on the top of the wheel C. When large sizes of wire are being measured the effective diameter of the wheel S is increased so that the indicated length will be somewhat smaller than the actual length.

To operate the recorder, run the wire from the spool L, as in Figs. 54 and 55, under friction guide K, and over J, continue it over and around the measuring wheel C, and on through the tension holders I and S. From this point lead the wire directly to the tube being wound. Take care that all bearings and friction surfaces fit snugly yet freely, that the resistance and friction may be reduced to a minimum. Too much tension might break or strain the insulation of very small wire. Fig. 56 gives drilling dimensions of the supporting upright.

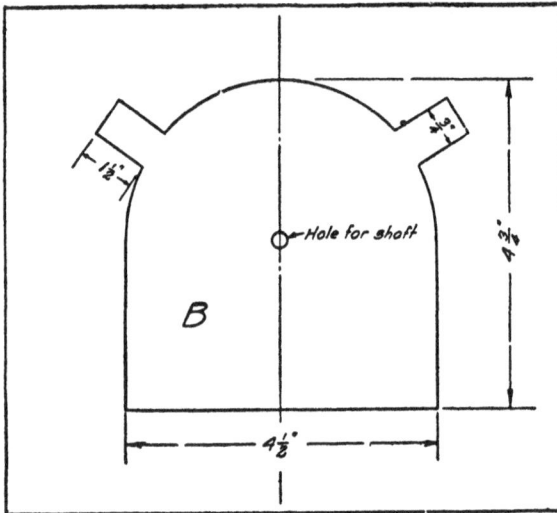

FIG 56. Elevation of support for wire measuring instrument.

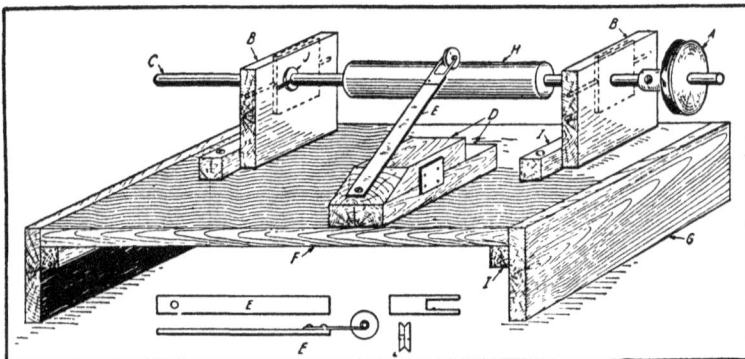

FIG. 57. Winding machine for tuning coils and secondary "pies."

COIL WINDER.—We will now describe a winding machine devised by Lattimer Reader of New York City, suitable for winding *tuning coils* or *secondary "pies"* A perspective view of the machine is shown in Fig. 57 and a side view in Fig 58 The sketches for the most part explain themselves. The winder is designed to be operated from the driving wheel and pedal of a sewing machine. The designer declares that with the aid of the guide mounted on the block immediately in front of the machine, he was able to wind one section of a 3″ spark coil with No. 38 wire in two hours. These are the dimensions for the various parts: The wheel A (Fig. 57) may be 1½″ or 2″ in diameter. The bearing B may be a piece of brass 5″ long, 2½″ wide and ¾″ thick. The shaft C should be 10″ in length and ½″ to ⅝″ in diameter.

The block D carrying the rider may be made of two pieces 2½″ wide by 1½″ thick and 7″ long. They should be cut as shown and fastened to permit the shifting of the piece holding the guide   The arm of the guide E should be 6″ long, ¾″ wide, about 1⁄16″ thick and made out of sheet iron. A thin piece of brass or tin turned over to receive a nail will form a support for the small guide pulley.

The base F may be 15″ long, 18″ wide and ¼″ in thickness; the sides G, 8″x4″. The form H is a piece of broom handle to fit over the shaft. The blocks I should be ¾″ square.

The side view, Fig. 58, shows the coil winder belted to a sewing machine.

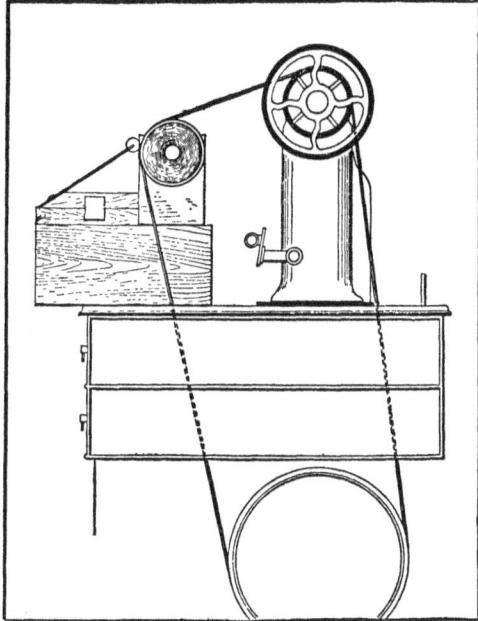

FIG 58.   Side view of coil winding machine.

CONNECTING THE TRANSFORMER PIES.—When the secondary pancakes are p'aced on the transformer core, take great care to connect them so that the current circulates throughout the whole secondary in the same direction. The correct method of connecting these pies is shown in Fig. 58a. It will be seen from these that each pair of adjacent pies must be placed on the core so that the inside terminal of one pie is connected to the inside terminal of the adjacent pie. The outside terminal of one of these pies must be connected to the outside terminal of the next pie and so on through the series.

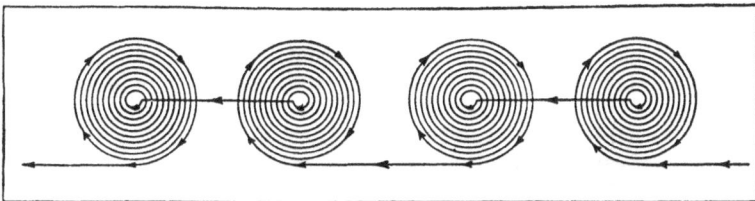

FIG 58a.   Showing the correct method of connecting the "pies' in a transformer secondary   The coils are connected so that the current throughout the whole secondary flows in the same direction

In other words, if all pies are wound in the same direction they must be stacked up so that the outside terminals of two adjacent pies come out of the coils in opposite directions: likewise the inside terminals of the next two adjacent pies.

INSULATING COMPOUND FOR TRANSFORMERS.—A wax preparation for imbedding the completed transformer, for insulation purposes, may be made of 4 lbs. of rosin, 1 lb. of beeswax, and 1½ lbs. of paraffine. These ingredients should be put in a galvanized iron dish and then heated and thoroughly mixed. If the wax is too brittle use less rosin and if too soft, use less paraffine. The experimenter may omit the beeswax if it proves too expensive.

If the transformers are to be immersed in oil, a galvanized iron box is generally used. If *transil* oil is not available use *double boiled linseed oil.*

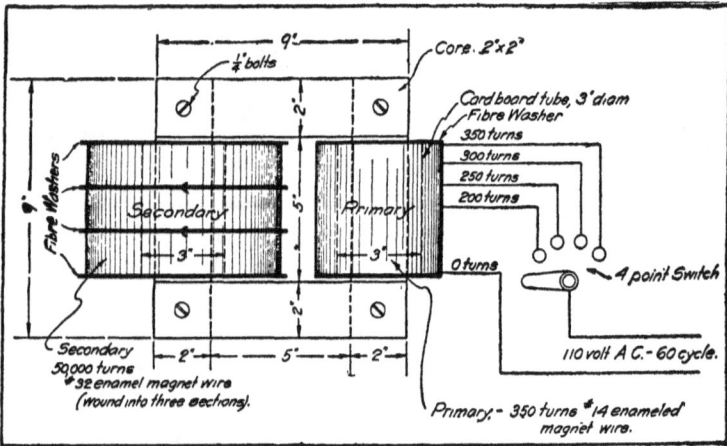

Fig. 59.   Closed core transformer with a magnetic leakage gap as designed by an amateur experimenter.

TRANSFORMER WITH MAGNETIC LEAKAGE GAP.—As the amateur experimenter cannot always purchase the best grades of silicon transformer steel he is apt to employ whatever material he may have on hand. An example of a transformer which an amateur has constructed at low cost is shown in Figs. 59 and 60.

This transformer was designed by Mark Biser of Maryland. The constructional details and over-all dimensions shown in Fig. 59 are for the most part self-explanatory. The cross section of the core is 2″ square, made of sheet steel strips 7″ in length, 2″ in width and with an average thickness of 3/32″. The primary and secondary coils are wound on opposite legs of the core and held in place by cardboard forms 3″ in diameter which are slipped over and secured to the core. Fibre discs are inserted between the sections of the secondary and at other points as indicated in the drawing. The primary consists of 350 turns of No. 14 d.c.c. magnet wire and the secondary of 50,000 turns of No. 32 enamel magnet wire.

The secondary is divided into three sections. A layer of thin oiled paper is placed between each layer of the secondary.

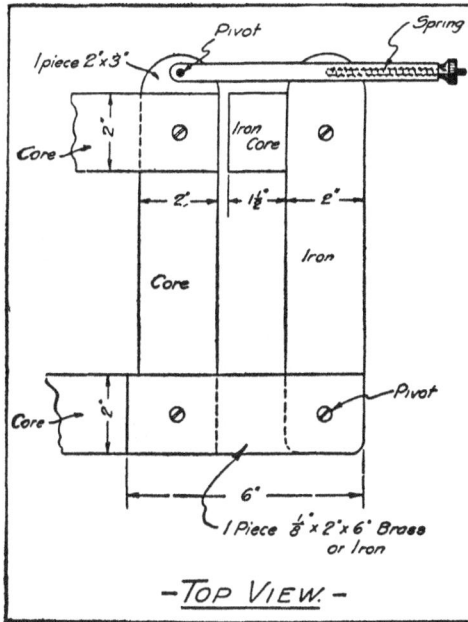

Fig. 60. Details of the leakage gap.

To give this transformer satisfactory operating characteristics and also to permit variation of the power input over a considerable range, Mr. Biser fitted it with a *magnetic leakage gap* which is controlled from the outside of the case by means of a suitable handle. A general idea of the construction of the leakage gap can be obtained from Fig. 60. An end elevation is shown in Fig. 61. By turning the knurled knob on the adjusting shaft, the length of the leakage gap, which is shunted around the primary core, can be varied very considerably, permitting the

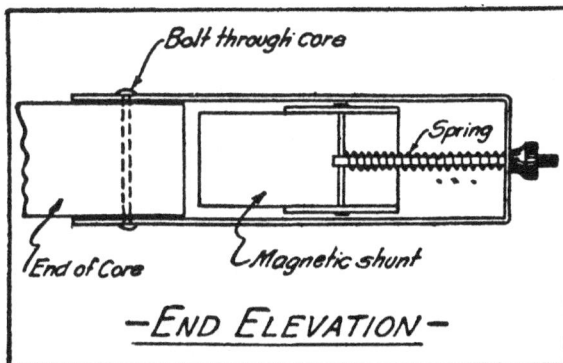

Fig. 61 End elevation of the leakage gap

secondary voltage to be changed between 10,000 and 20,000 volts. As mentioned in the preceding chapter, this gap is of particular advantage for increasing the power factor in transformer circuits.

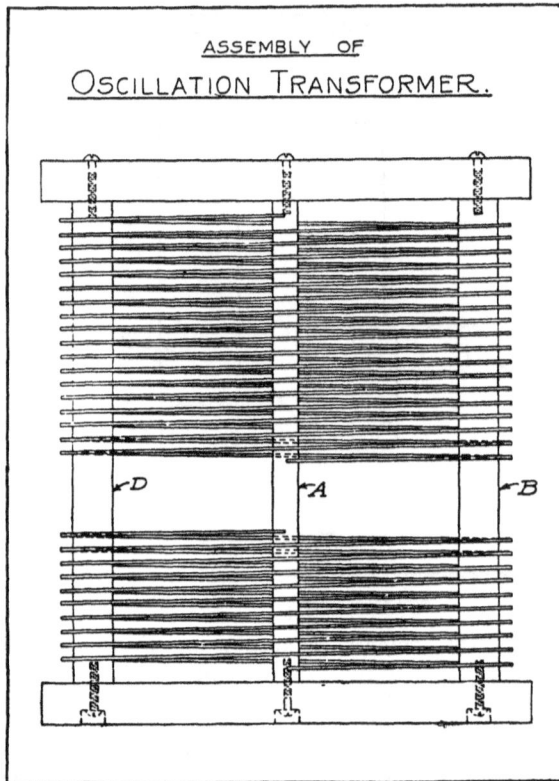

FIG. 62   Helical type of oscillation transformer with turns wound on edgewise, as constructed by an amateur experimenter.

OSCILLATION TRANSFORMERS.—The "helical" or pancake types are generally preferred. The amateur often errs in building an oscillation transformer of too many turns. Not infrequently stations are fitted with an oscillation transformer of sufficient inductance for waves up to 1000 meters in length, although the set is to be operated at 200 meters.

It is not even necessary to adopt the expensive designs following, when a few turns of heavily insulated wire will do for the primary and the secondary. The conductor should be of low resistance, at least ⅜″ in diameter for the 1 kw. set, and ¼″ in diameter for the ½ kw. set.

The reason that *copper tubing* is as suitable as a solid conductor is that the "skin effects" of high frequency currents are very marked and since high frequency currents penetrate only a small depth into conductors only the surface is of importance.

If the *pancake type* of oscillation transformer is adopted, use copper strip $\frac{1}{4}''$, $\frac{3}{8}''$ or $\frac{1}{2}''$ wide and about $\frac{1}{16}$ of an inch in thickness. It is better to use a bare conductor so that for tuning purposes contact may be made with a fraction of a turn. Some experimenters prefer the helical type of transformer with the turns wound edgewise.

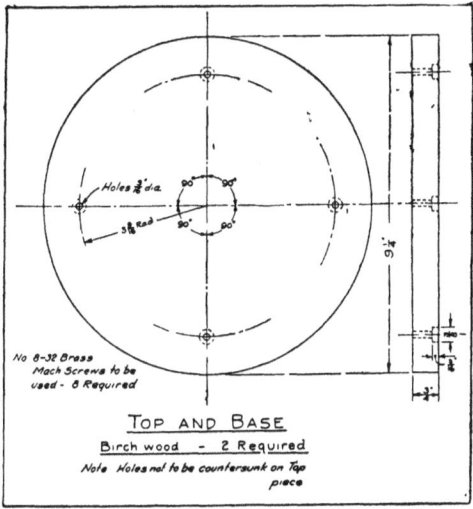

FIG 63.   Details of the top and base

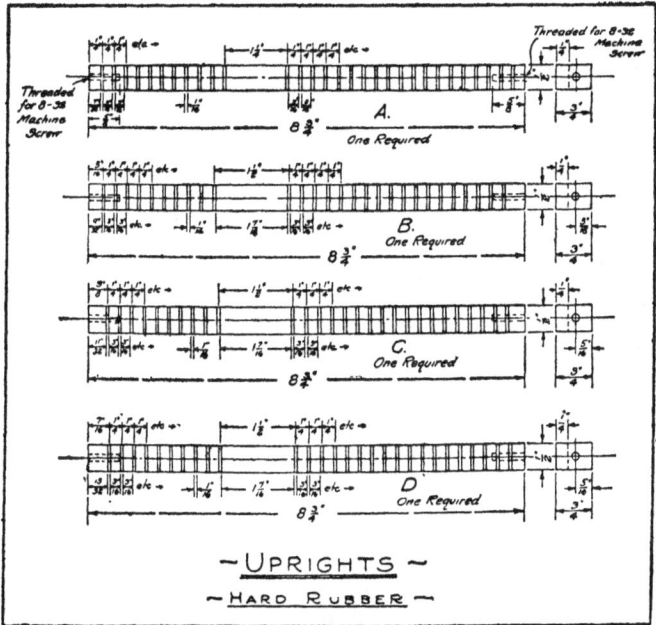

FIG 64   Showing the dimensions of the uprights for the oscillation transformer

A transformer of the latter type has been designed by Ralph Hoaglund of Massachusetts. The assembled instrument is shown in Fig. 62, a plan of the top and base in Fig. 63, and the dimensions of the uprights in Fig. 64.

The primary has 10 turns and the secondary 18 turns. The inside diameter of the coil is 7½″, the strip being ½″ wide and 1⁄16″ thick. The dimensions of the uprights are 8¾″ x ¾″ x ½″. The base and top are each 9¼″ in diameter and ¾″ in thickness. Exact measurements for the positions of the slots in the strips are shown in Fig. 64. The slots

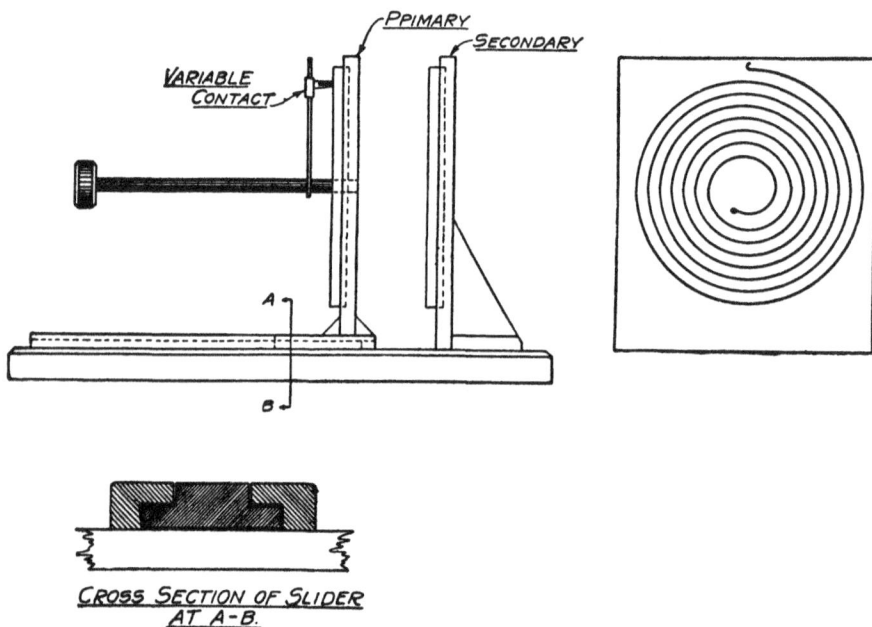

FIG. 65. Design for the "pancake" type of oscillation transformer.

for upright A begin ¼″ from the bottom; for upright B, 5⁄16″ from the bottom; for upright C, ⅜″ from the bottom; and for the upright D, 7⁄16″ from the bottom. The slots should be cut on a slight slant.

For 200 meters a part of the smaller coil may be employed as the *primary* and the larger coil as the *secondary*. The coupling is, mechanically speaking, fixed, but may be varied electrically by placing two taps on the secondary, through which the *used* secondary turns may be brought nearer to or placed further away from the primary, according to requirements.

A suggested design for the *pancake type* of oscillation transformer is shown in Fig. 65. The primary inductance is varied by a sliding contact mounted on an arm attached to the control handle. The contact slides on the coil and as it is rotated, any fraction of a turn can be cut in the circuit. The primary coil is mounted on a *wooden slider* which is shown in cross-section. This permits the coupling between the primary and secondary to be varied. The conductor of pancake coils is usually mounted in slots cut into strips of hard rubber or bakelite which are supported on wooden uprights.

By using copper strip $\frac{3}{8}''$ wide and $\frac{1}{16}''$ thick, the following dimensions will be found satisfactory. The primary has 8 turns spaced $\frac{1}{4}''$. The

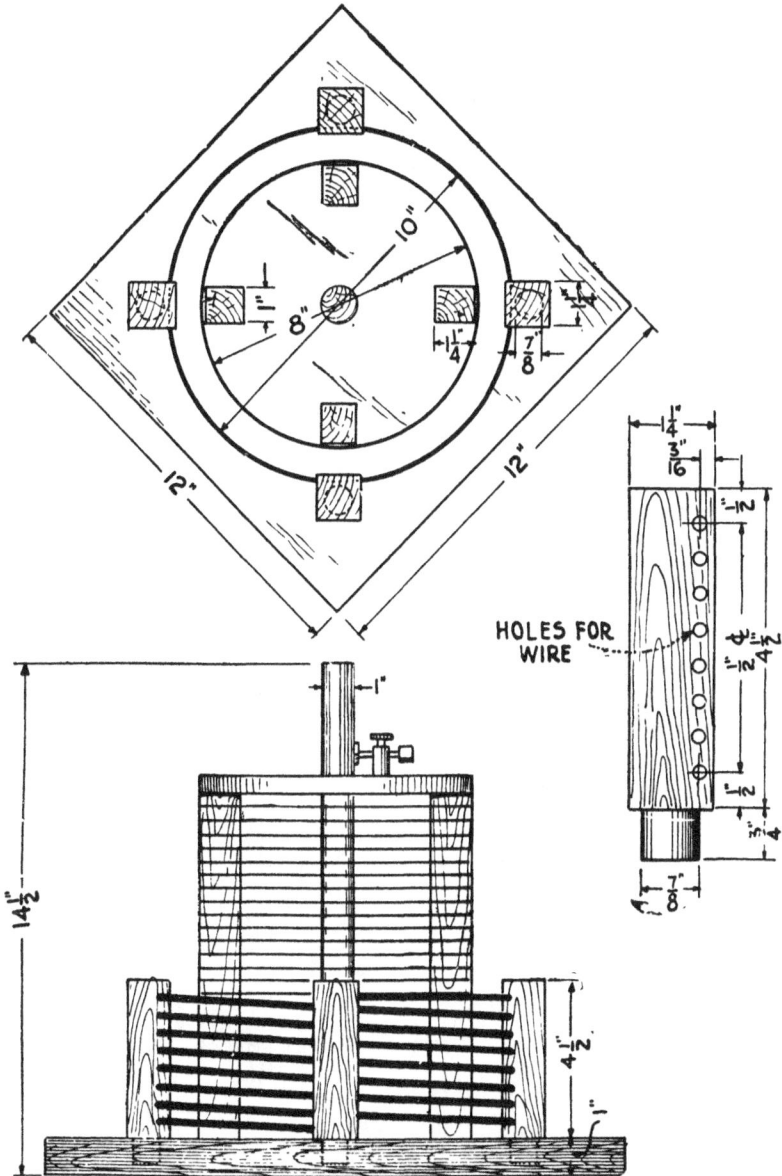

Fig. 66. Helical oscillation transformer providing variable coupling.

inside diameter is $4\frac{3}{4}''$ and the outside diameter $10''$. The secondary has 20 turns spaced $\frac{1}{4}''$, the outside diameter being about $14''$ and the inside diameter about $4\frac{1}{2}''$.

A *helical oscillation transformer* designed for variable coupling is shown in Fig. 66. The primary is wound with No. 6 spring brass wire and the secondary with No. 8. The primary turns are threaded through holes cut in the wooden uprights; the secondary is mounted in the same way. The primary coil is $10''$ in diameter and the secondary $8''$ in diameter. Four turns are sufficient for the primary and 12 to 15 turns for the secondary, although a greater number are shown. The spacing between the primary turns should be $\frac{3}{8}''$, and about $\frac{1}{4}''$ or $\frac{5}{16}''$ between the secondary turns. *Wooden uprights* will do in dry climates, but if the apparatus is to be operated in a wet zone use hard rubber or some similar material.

A small binding post with a set screw mounted on the top of the secondary locks the secondary in position for any desired coupling.

The dimensions shown can be varied somewhat to suit the builder. It is in fact preferable, that the necessary number of turns be predetermined by the inductance formula given in Chapter III.

HIGH VOLTAGE CONDENSERS.—We usually distinguish high voltage condensers from low voltage types by the material of the dielectric. *Glass* and *mica* are generally employed for high voltages. *Paraffined paper, thin sheets of hard rubber, mica* or *air* at ordinary pressures are ordinarily used as the insulating medium for low voltages.

The most satisfactory condensers for the amateur's transmitter are the *copper plated Leyden jar* and the *mica condenser*. The oil-immersed *glass plate condenser*, however, is widely used principally because the experimenter has less difficulty in obtaining the materials for construction.

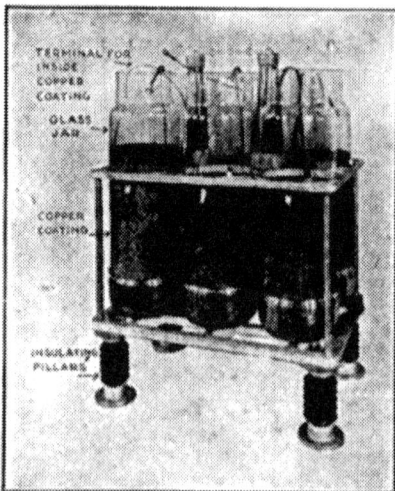

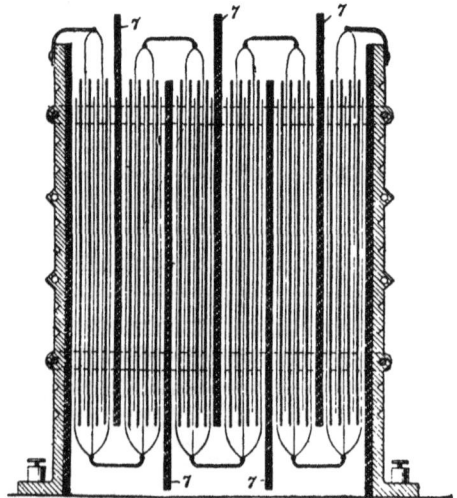

Fig 67   Commercial type of high voltage condenser using copper plated Leyden jars mounted in a metal jar rack
Fig 67a.   Showing in a general way the construction and assembly of the mica insulated condenser.

A *commercial type* of *high voltage condenser* is shown in Fig. 67. Six copper plated glass jars of 0.002 mfd each are mounted in a metal rack and connected in parallel. The outside coatings of the jars are connected through the metal cups at the bottom and the frame. The inside coatings are connected to a common terminal through braided conductors, which are soldered to the inside coatings.

Mica condensers are generally employed in government radio sets. They have the advantages of compactness, fairly small losses, and little danger of breakdown. The reason for the last is that the breakdown voltage of very thin sheets of mica is high (60,000 volts per centimeter of thickness), consequently even if used with high voltages very thin sheets may be placed between the conducting surfaces of the condenser giving a relatively high value of capacity. Several units may then be connected in series so that the voltage across each unit is comparatively low. A fundamental idea of the assembly of the mica condenser may be obtained from Fig. 67a.

It is a common practice to connect 10 or more units in series, the units being stacked up in a metal container and thoroughly impregnated with a good insulating compound by a vacuum process. With 10 units in series on a 15,000 volt transformer the voltage per unit is only 1500 volts. The liability to rupture is thus far removed. On the other hand, if ten banks of Leyden jars were connected in series, the result would be a very expensive and bulky condenser. Both mica condensers and Leyden jars may be purchased ready for use.

Commercial types of Leyden jars will not stand potentials in excess of 15,000 volts. If the transformer voltage exceeds this value a *series-parallel* connection must be used. This connection requires *four times the number of jars* used in a simple parallel connection.

It is clear from formula (35) page 46, that the resultant capacity of two equal condenser banks in series is one-half that of one bank. Assume, then, that standard jars of 0.002 mfd. each are available. Four jars in parallel will have a capacitance of $4 \times 0.002 = 0.008$ mfd.—the correct value for the amateur's transmitter operated from a 60-cycle source. Sixteen of these jars will be required if the voltage of the transformer is 20,000 and two banks are to be connected in series. Eight jars in parallel will constitute a bank, and the capacity of each bank will be $8 \times 0.002 = 0.016$ mfd. With two such banks in series, the resultant capacity will be $\dfrac{0.016}{2} = 0.008$ mfd. The voltage across each bank will be $\dfrac{20,000}{2} = 10,000$ volts.

The majority of amateur experimenters use the *foil-coated plate glass condenser*. The capacitance of such a condenser may be calculated with a fair degree of accuracy by formula (**34**). Glass free from lead and "bubbles" should be selected and photographic plates, which may be obtained from a local photographer, are often used. *Flint glass* has a small heat loss and is preferred.

A plate of glass 8″ x 8″ and about ⅛″ in thickness, covered on both sides with tin foil, has a capacitance of about 0.0005 mfd. Sixteen such plates in parallel will give 16 x 0.0005 = 0.008 mfd.

A plate of common glass 14″ x 14″ x 3/32″ covered with foil 12″ x 12″ will have a capacitance of about 0.0016 mfd. A plate with a conducting surface of twice the area will, of course, have $2 \times 0.0016 = 0.0032$ mfd. A

plate of one-quarter that area will have $\dfrac{0.0016}{4} = 0.0004$ mfd.

The difficulty of calculating the capacity of a glass plate condenser lies in the varying value of the *dielectric constant* (K). If it is definitely known formula (34) is very accurate.

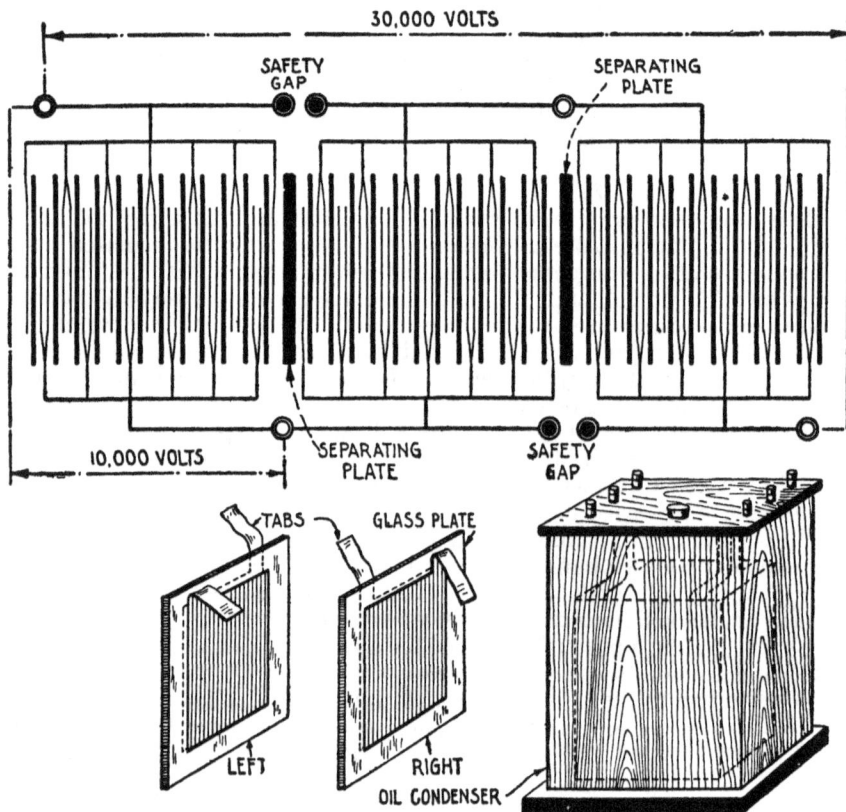

FIG. 68. Giving a general idea of the assembly of the glass plate type of high voltage condenser.

ASSEMBLY OF A GLASS PLATE CONDENSER.—The assembly of an oil-immersed condenser is shown in Fig. 68. The foil is cut with a lug projecting from one corner. The lugs issue alternately from the *right* and *left hand* corners of adjacent plates.

The foil should not come too close to the edge lest sparking over the edges occur. Allow a space of at least one and one-half inches.

To attach the tin foil to the plates: First, cover the surface of the glass with a good grade of *thin fish glue*. Place in position the sheets of tin foil, which have been properly cut and rolled to smoothness by a "squeegee" roller Next place the plates in a rack and allow them to dry slowly, after which they should be covered with one coat of orange shellac. Then pile them together, *first a left plate, then a right plate* and so on throughout the series until the unit is complete. Bind the entire unit with *insulating tape* or a *canvas strip* and immerse it in a *tank of insulating oil.* Swan & Finch's Special Atlas AAA is recommended as a good grade of oil.

Another method for placing the tin foil on the glass is: When the sheets of tin foil have been properly cut place the glass plates in an oven and heat them for five minutes. Then remove and rub them with a good grade of *beeswax.* While the plates are still warm, rub the sheets of tin foil into place by a "squeegee" and paint the edges with hot beeswax. Beeswax has in many instances been found superior to shellac because it does not blister.

While in the sketch of Fig. 68 three banks are connected in series two banks are really sufficient, unless the dielectric strength of the glass em-

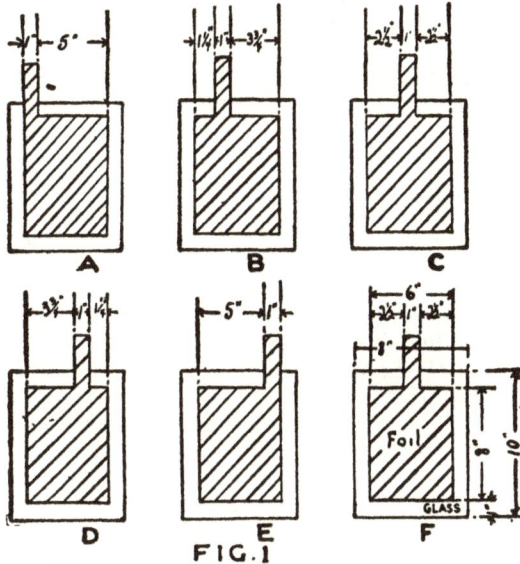

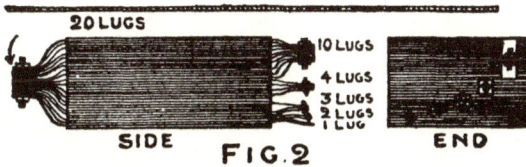

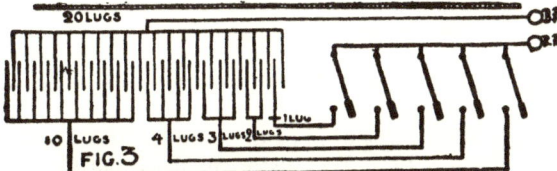

FIG. 69. The construction of a high voltage condenser of variable capacity.

ployed happens to be very low. The safety gaps shown in the drawing are not strictly essential. A protective gap connected across the transformer secondary (to be described later) affords all the protection necessary.

It is desirable to have a *high voltage condenser of variable capacity* in the amateur station. Such a condenser permits the operator to experiment with different capacities for varying conditions of service. Sometimes greater efficiency will be obtained with a capacity quite different from that originally intended, due to the difference in the designs of high voltage transformers and the possibility of the set being operated at frequencies above 60 cycles. At the frequency of 500 cycles, for example, a condenser of 0.004 mfd. is sufficient for a ½ kw. set; for a ¼ kw. set, 0.002 mfd. will do.

E. C. Eriksen of California designed the variable high voltage condenser shown in Fig. 69. The condenser is constructed from 8″ x 10″ photographic plates which are generally free from flaws.

The film may be removed from the plates by immersing them in two solutions. The first solution is made up in the proportion of ¼ ounce of sodium floride to 16 ounces of water. The second solution consists of ¼ ounce of sulphuric acid and 16 ounces of water. After the plates have been placed in the first solution for a couple of minutes and then in the second solution the film should come off easily.

Now coming to the construction of Eriksen's condenser it should be noted that all of the lugs leading from the foil at one end are joined together, while those at the opposite end are staggered so that they can be connected together in groups. This construction allows the condenser to be mounted in a wooden box with knife blade switches placed on the side to cut in and out the various sections.

The foil is cut 6″ x 8″ with a lug 1″ wide, 3″ long. The lugs are alternated so that on adjacent plates they come out at opposite ends of the pile. The first lug should be placed as at *A;* the next two as at *B;* the next three as at *C;* the next four as at *D;* the next ten as at *E* and the remaining twenty as at *F.*

The plates are then covered with vaseline, the tin foil placed in position and rolled with a "squeegee," and afterwards the foil is coated on the outside with vaseline. Next, the plates are piled up in the order described before so that 20 lugs appear at the center at one end, while at the other end there will be five bunches of 1, 2, 3, 4, and 10 lugs each. These lugs must come out from consecutive plates and be placed in the same position on each plate. The pile may then be wound tight with tape.

The plates may then be placed in a wooden box with outside dimensions 10″ x 14″ x 5″. The corners of the box are mitered, the bottom is glued on, and the top is fastened with nickel-plated, round-head screws. The box should be shellaced or boiled in melted paraffine. When the plates are in place it should be filled with a good grade of transformer oil. The oil and vaseline both serve to reduce the brush discharge.

The knife blade switches should be mounted on a hard rubber slab placed on one end of the box. Wood will not withstand high voltages for any considerable period

By this arrangement any number of plates from one to twenty may be connected in the circuit. For instance, if 17 plates are desired the switches connected to the 10-lug, 4-lug, and 3-lug bunches should be closed

Forty plates will do for a ½ kw. set. The required number for any wave length with a fixed primary inductance can easily be found by a wavemeter. If the transformer voltage exceeds 20,000, a series parallel connection must be used.

Some wireless experimenters prefer to mount their condenser plates in a wooden rack. This construction certainly has the advantage of neatness and is simpler in that a punctured plate may be removed without taking down the whole condenser.

The condenser rack shown in Fig. 70 was designed by a contributor to The Wireless Age. It may be built to take any desired number of plates. It consists mainly of the slotted pieces A cut to take the plates which are separated, say ½ inch. Two copper busbars C are mounted on the front of the rack from which connection is made to the plates. No lugs are brought from the plates, but connections between adjacent plates and to the busbars are made by the brass contact clips B constructed in conformity with the detail Fig. 70. The remainder of the construction is self-explanatory.

If voltages below 15,000 are used the condenser may be operated in the open air, but for higher voltages the rack, plates and all may be immersed in oil. If the condenser is operated in the open air, brushing at the edges of the plates can be reduced by coating them with beeswax.

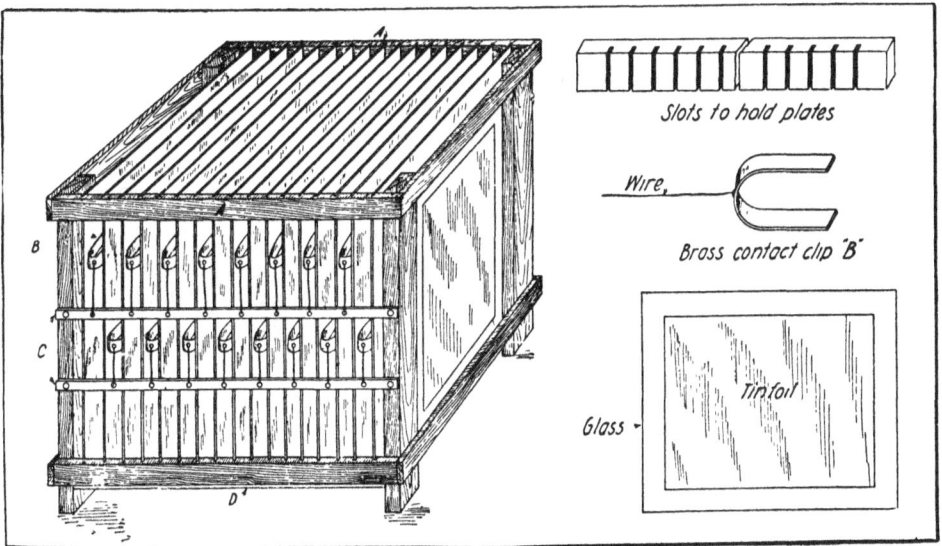

Fig 70   Condenser rack for the glass plate type of high voltage condenser.

SPARK DISCHARGERS.—The theory of spark dischargers has been given in the preceding chapter. It should be evident that the design of the spark gap has a marked effect upon the efficiency of a wireless transmitter. It is not sufficient to merely supply a discharge path for the energy in the condenser. Means must be provided for rapidly quenching out the oscillations in the closed circuit before the antenna oscillations have the opportunity to retransfer their energy to the closed circuit; otherwise oscillations of two frequencies will flow in the antenna circuit and the set will radiate two waves.

Single wave emission, with tight coupling at the oscillation transformer, can only be obtained with a gap and associated power circuits designed so that between individual sparks the insulating qualities of the gap are rapidly restored. If the gap action is prolonged, single wave emission can only be secured with very loose coupling. A reduction of coupling cuts down the antenna current resulting in a decreased transmitting range.

There are four types of spark dischargers in commercial use, viz:

   (1) **The plain gap.**
   (2) **The non-synchronous rotary gap.**
   (3) **The synchronous rotary gap.**
   (4) **The multiplate or series gap** (commonly called the "quenched" gap).

The plain gap is generally used with spark coil transmitters. The non-synchronous gap is employed to give a musical tone when the power source is 60 cycles or less. The synchronous gap is rarely used by experimenters, unless the frequency of the power supply exceeds 120 cycles. The multiplate gap is seldom encountered in the amateur station although it will increase the transmitting range if the complete transmitter is properly designed.

If the frequency of the power supply exceeds 120 cycles the *synchronous gap* has some advantages—the principal one being the uniformity of the spark note. But the amateur, in nine cases out of ten, will use the *non-synchronous* rotary because his power supply is usually 60 cycles. A *synchronous gap* at this frequency should give 120 sparks per second, and the pitch of the note would be much lower than that obtained with the non-synchronous gap. The latter can be designed to give from 200 to 400 sparks per second. The note has musical characteristics composed of overtones, undertones, and the fundamental, and it is far from displeasing to the average operator's ear. In fact many operators prefer it.

The experimenter cannot use a synchronous gap where the power supply is furnished by a local company, for uniform sparking can only be obtained by mounting the gap on the shaft of the alternator of the power supply.

The synchronous gap has been mounted on the shaft of an a.c. synchronous motor connected to the power supply, but it is difficult to maintain an accurate synchronous adjustment over an extended period as the motor is apt to get slightly out of step with the frequency of the source.

PLAIN SPARK GAP.—The fundamental idea underlying the construction of this gap is shown in Fig. 71. Two zinc electrodes S-1 and S-2, are mounted on brass rods supported rigidly by two binding posts, so that the sparking surfaces are strictly parallel. The cooling flanges are cast on the rod or placed on afterward.

The base should be of hard rubber, bakelite, or glass. The binding posts should be mounted so that they do not come in contact with wood or with any other material liable to leak high voltage currents.

For spark coils the zinc electrodes should be approximately ¼" in diameter. For more powerful transmitters up to 1 kw., the electrodes need not exceed ⅜" to ⁷⁄₁₆". Blunt electrodes give a rough spark note. Sharp electrodes give notes of higher pitch but they do not cool so rapidly, tending to prolong the spark discharge with consequent *lack of quenching*. *Blunt electrodes on stationary gaps aid quenching* and would be preferable did they not tend to destroy the pitch of the note.

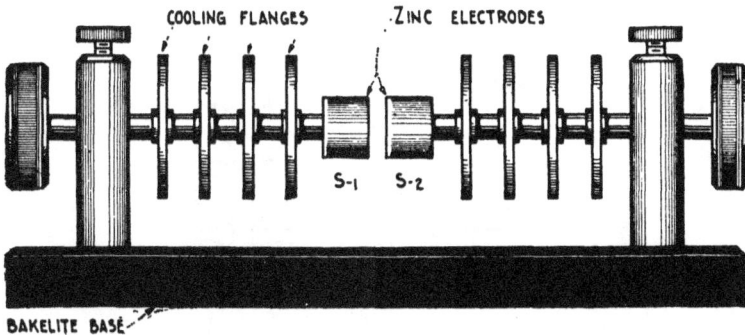

FIG. 71. Showing the construction of a plain type of spark gap utilizing two stationary electrodes.

The crash of the spark of a 1 kw. transmitter is exceedingly loud. The gap should be placed in a *muffling box*, the leads to the gap brought into the box through *bushings* possessing the best insulating qualities. A few holes must be left in the box for ventilation; two at the bottom and two at the top are generally satisfactory. Placed this way they tend to create a draft and assist the ventilation. The muffling box may be made of wood or metal. A wooden box should be lined with asbestos ¼" in thickness.

Brass or *copper spark electrodes* are nearly as satisfactory as zinc but zinc electrodes give very uniform spark notes and are preferable for that reason. The ingenious experimenter can easily design a plain gap fitted with an adjusting handle which will permit the length of the gap to be regulated while the spark is discharging.

THE NON-SYNCHRONOUS ROTARY GAP.—The rotor of this type is mounted on the shaft of a d.c. or a.c. motor. A ⅛ h.p. motor has sufficient power to drive the average gap. The speed of the motor is unimportant, provided a sufficient number of spark electrodes are supplied to give the desired spark frequency.

Three different designs are shown in Figs. 72, 73 and 74. In Fig. 72 the driving motor M carries the disc D, which may be made of micanite, bakelite, dilecto or micarta. Around the circumference of the disc, near the edge, holes are drilled to take the rotating electrodes E which are equally spaced.

The electrodes E need not be more than ⅛" or ³⁄₁₆" in diameter and may be made of zinc or copper. Copper is more commonly used. The stationary electrodes D^1 and D^2 are the same diameter as the electrodes on the disc. They may be fitted with *cooling flanges* not shown in the drawing.

The motor armature must have *end-thrust washers* of the correct thickness so that the end play does not exceed $\frac{1}{64}$ of an inch. Otherwise the electrodes on the disc will jam against the stationary points and wreck the gap. The electrodes D^1 and D^2, are then set up as close as possible to electrodes E, without making actual contact, *for the shorter the gap the clearer will be the spark note.* A high pitched note is the goal of all amateurs.

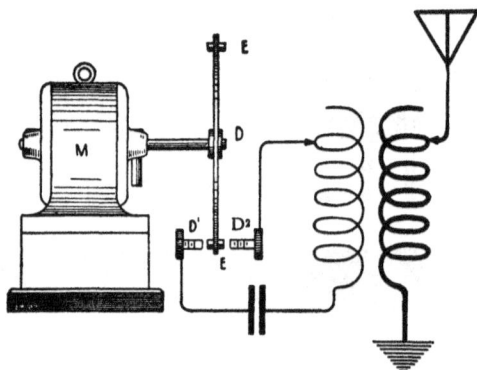

Ƒɪɢ 72   A type of the non-synchronous rotary gap first developed by the Marconi Company. This gap is favored by the amateur experimenter.

The electrodes E must all be lined up equally in respect to electrodes D^1 and D^2. Do this by: setting up the ends of D^1 and D^2 at equal distances from the disc by means of a micrometer gauge. The total distance between them must be just slightly greater than the length of the lugs E. The latter can then be centered in the disc and fastened tightly by nuts on both sides of the disc. The lugs E are, of course, threaded throughout their length.

For general amateur working a gap giving 240 sparks per second is satisfactory. Assume that the motor M rotates 1800 r.p.m., or 30 revolutions per second. Then, if the disc has 8 spark electrodes, there will be produced in one second $8 \times 30 = 240$ sparks.

A disc $6\frac{1}{2}''$ in diameter will do for a gap with 8 electrodes. The holes for the electrodes on the disc should be drilled around a circle drawn $3''$ from the center with a scriber. The electrodes will then be about $1.6''$ apart. If the motor runs only 900 r.p.m. the disc should have 16 electrodes and should be say $10''$ in diameter.

The *non-synchronous gap* shown in Fig. 73 has been used in commercial and amateur stations and is one of the earlier designs developed by the Marconi Company—the pioneer in the use of rotary gaps. The construction will be understood from the drawing. Two nearly semicircular pieces of brass mounted on the frame of the driving motor are insulated from each other and from the motor frame by hard rubber posts extending outward from a metal spider (shown partially in the drawing).

The rotating member is an aluminum arm mounted on the armature shaft which carries two sparking electrodes. The stationary electrodes are slotted and held firmly, but not tightly, by machine screws, so that if the rotating electrodes should accidentally come in contact with the stationary electrodes the latter will turn on the machine screws and protect the gap from being wrecked.   One advantage of the gap con-

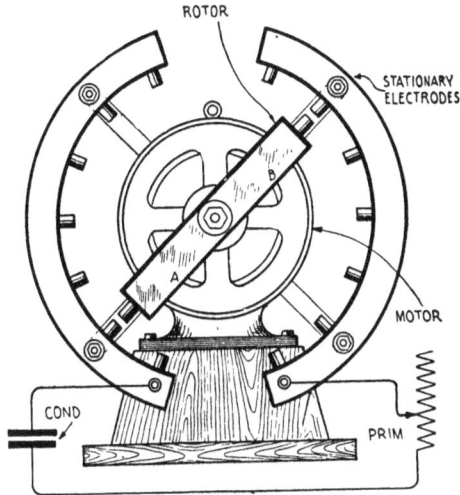

Fig. 73   Modified form of the non-synchronous gap as developed by the American Marconi Company.

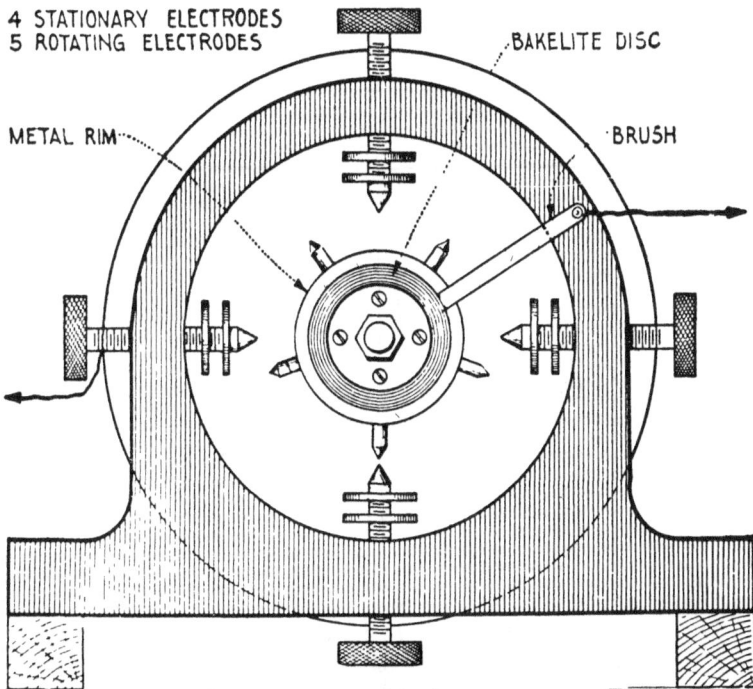

Fig 74   "Vernier" type of non-synchronous rotary gap.

struction of Fig. 73 is that while the rotating electrodes are self-cooled by rotation of the arm, the stationary electrodes are given the opportunity to cool between sparks; for, as is self-evident, the spark discharge travels around the circle from one pair of opposite stationary electrodes to the next opposite pair and so on. The rotating arm need not be more than 7″ in length and it should be insulated from the driving shaft by a hard rubber bushing. The bushing is placed over the shaft and the arm mounted on the end.

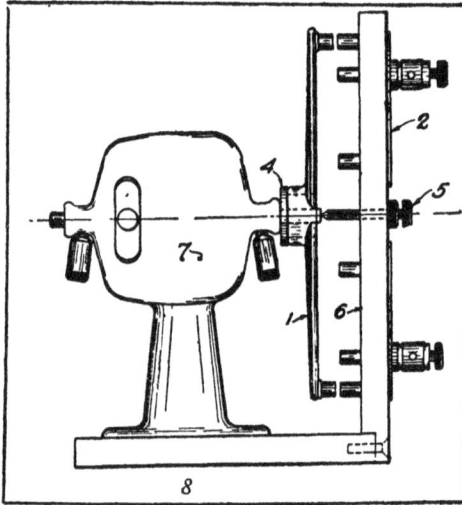

FIG. 75. Non-synchronous gap designed by an amateur experimenter.

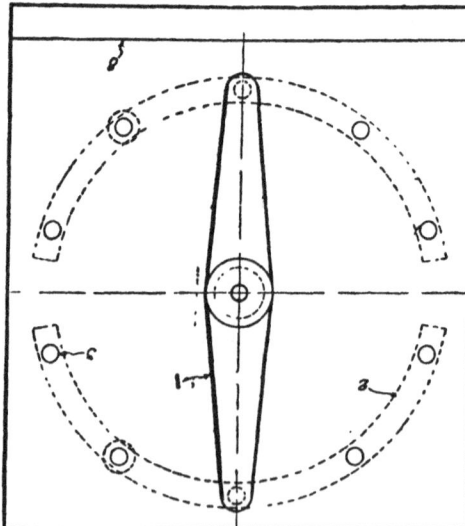

FIG. 76. Elevation of the support for the stationary electrodes.

The *"vernier"* gap of Fig. 74 has been used by Experimenters. Note that there are four stationary electrodes and five rotating ones. The disc which carries the rotating electrodes is insulated from the motor, but is connected to one terminal of the closed oscillation circuit by a brush. The other terminal of the oscillation circuit goes to the stationary electrodes which are connected together.

It is clear that in one revolution of the disc there will be $4 \times 5 = 20$ sparks; hence, if the disc revolves 30 r.p.s., i.e., 1800 r.p.m., there will be $30 \times 20 = 600$ sparks per second. High spark frequencies may thus be secured with low disc speeds. The note of this gap is somewhat superior to that obtained with the two types described before.

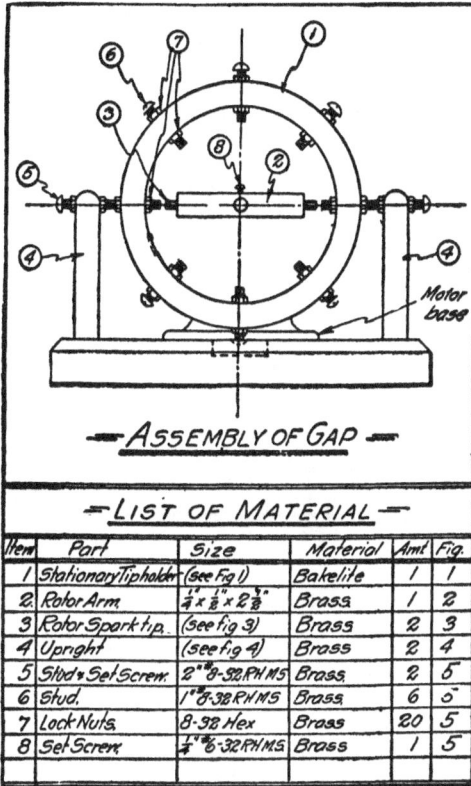

ASSEMBLY OF GAP

LIST OF MATERIAL

Item	Part	Size	Material	Amt	Fig.
1	Stationary Tip holder	(see Fig 1)	Bakelite	1	1
2	Rotor Arm	$4'' \times \frac{1}{8}'' \times 2\frac{3}{8}''$	Brass	1	2
3	Rotor Spark tip.	(see fig 3)	Brass	2	3
4	Upright	(see fig 4)	Brass	2	4
5	Slod & Set Screw.	$2''$ 8-32.RH MS	Brass	2	5
6	Stud.	$1''$ 8-32 RH MS	Brass	6	5
7	Lock Nuts.	8-32 Hex	Brass	20	5
8	Set Screw	$\frac{1}{4}''$ 6-32 PH MS	Brass	1	5

Fig. 7. Rotary gap for low powers as designed by an amateur experimenter.

Various amateurs have shown many designs for the non-synchronous gap. The constructional details may well be left to the builder as he will use the material he can most readily obtain. Those shown in the following drawings may suggest other designs which the amateur can work out for himself. In order to obtain the best quenching the electrodes should be wide and thin

A gap which is a variation of Fig. 73, designed by Ralph Hoaglund is shown in Figs. 75 and 76. The stationary electrodes, ten in number, are mounted on a wooden upright, parallel to the rotating arm. A fibre bushing for insulating the arm from the shaft is shown at 4. The pointed set screw 5 acts as an end-thrust to eliminate end-play in the armature shaft and also acts as a contact for connection with the rotating arm. There are ten stationary electrodes spaced evenly on a circle 6″ in diameter. Fig. 76 is an elevation of the support for the stationary electrodes.

E. Chester Stephen of New Jersey has presented the design shown in Fig. 77 which is self-explanatory.

O. Cote of Rhode Island built an aluminum rotary gap like that in Figs. 78 and 79. The rotor is cast or cut from aluminum. The driving motor is one of the small Ajax type operated off a step down auto-transformer energized from 110 volts a.c. source. The disc is $3\frac{1}{4}$″ in diameter and has twelve discharge points. Each spark electrode on the disc is $\frac{1}{4}$″ x $\frac{5}{16}$″. The disc is $\frac{1}{4}$″ thick.

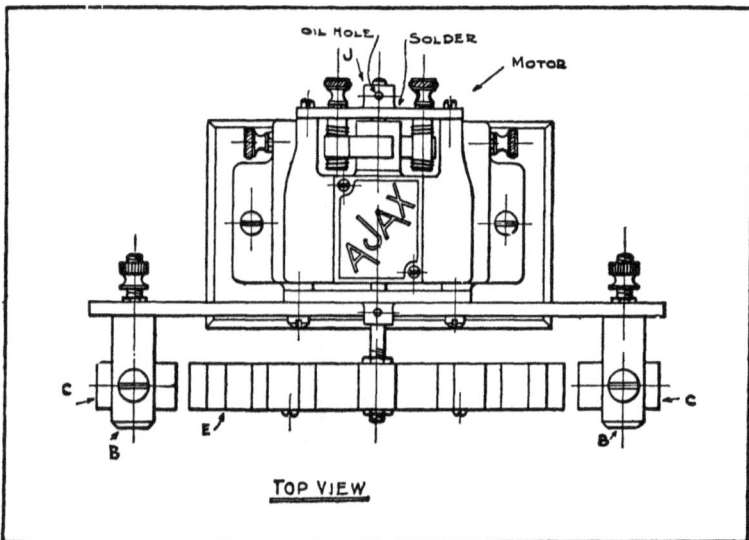

FIG. 78.   Rotary gap fitted with an aluminum rotor.

A sound muffled rotary gap designed by W. E. Wood of Missouri is shown in Fig. 80. The rotor is enclosed in a fibre tube large enough to permit it to rotate freely. The end of the tube is covered with a circular piece of glass as is shown in the illustration.

E. G. Mohn and Walter Maynes of California have constructed a gap rotated by a novel a.c. motor following the design suggested by Prof. A. S. Gordon of the Polytechnic High School, San Francisco. The motor shown in Fig. 81 consists of an electromagnet wound with $\frac{1}{2}$ lb. of No. 24 d.c.c. wire, and a steel bar magnet mounted on a shaft immediately in front of the pole pieces. The shaft carrying the bar magnet also carries the rotary disc. The only disadvantage of the design is that in order to start the bar magnet into rotation it must be turned by hand.

If the amateur generates a.c. within his station the non-synchronous gap may be mounted on the generator shaft. For example, a 4 pole generator driven 1800 r.p m. would require 4 rotating electrodes for synchronous discharges but if the disc is fitted with 8 or 12 electrodes much higher spark frequencies can be obtained; that is, several sparks per cycle may be produced.

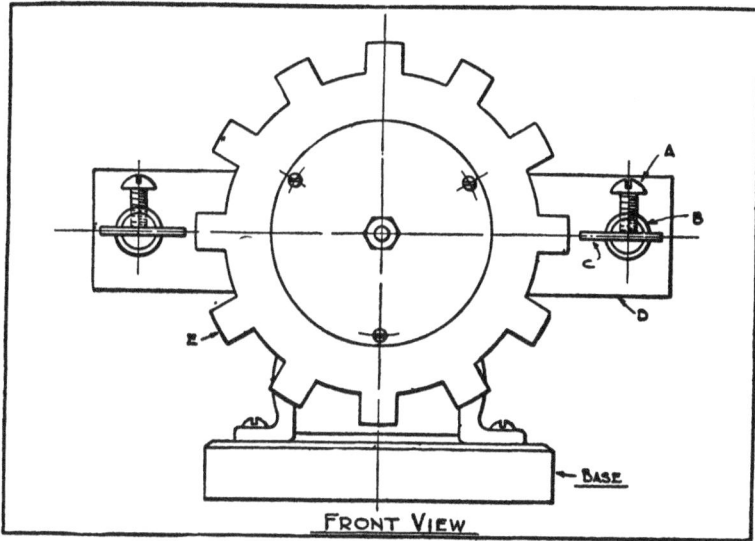

FIG 79. Details of the rotor.

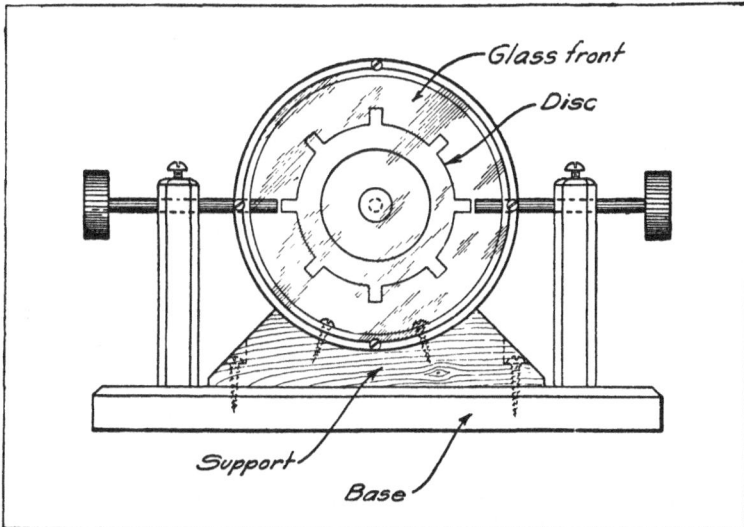

FIG. 80. Muffling box for a rotary gap

SYNCHRONOUS ROTARY GAP.—The experimenter is not apt to use a synchronous gap because he wants a higher spark frequency from that which the 60-cycle synchronous discharger gives.

The synchronous gap is the same as the non-synchronous type except that the rotor is mounted on the generator shaft and must have a *spark electrode for each field pole of the generator*. The stationary electrodes are mounted on a rocker arm so that they can be shifted through an arc of 25° to 40°. By shifting the rocker arm, the spark can be made to occur at the *peak* of the condenser voltage, giving a discharge for each half-cycle of the charging current.

With a 500-cycle source the synchronous gap is very desirable, but for frequencies below 120 cycles the non-synchronous gap gives the higher note.

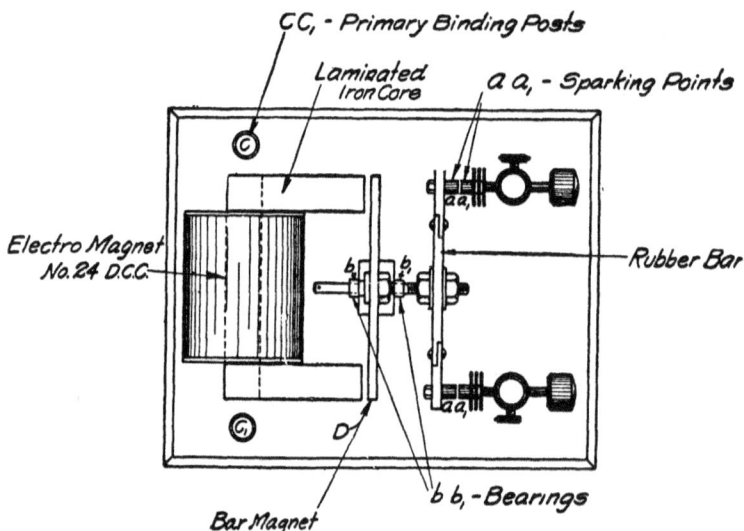

FIG 81. Non-synchronous rotary gap driven by a home-made alternating current motor.

THE MULTIPLATE OR "QUENCHED" GAP.—Although recognized as the most efficient gap this type is rarely used by wireless experimenters. This is due chiefly to the fact that the quenched gap gives a poor note with a 60-cycle source. The spark tone is not much higher in pitch than the plain gap The multiplate gap is expensive to build, for precise machining is essential for good results.

Although the quenched or multiplate gap quenches the primary oscillations more effectively than other types and gives a higher value of antenna current, unless certain factors of design are taken into account it may prove less efficient than an ordinary gap.

Before describing the construction of the "quenched" gap, certain basic considerations will be pointed out that must be given serious attention when this gap is employed with ordinary 60 cycle transmitters.

First, a resonance transformer is desirable as explained on page 83.

Second, a reactance coil must be placed in the primary circuit of the high voltage transformer.

Third, a rather smaller condenser capacity than that ordinarily used with other gaps must be employed.

Fourth, an oscillation transformer of the pancake type, permitting a continuous variation of inductance, is essential to obtain rapid adjustments of resonance between the spark gap and antenna circuits.

Fifth, a motor-generator is most desirable as the source of current in order that the phenomenon of resonance at audio frequencies may be utilized to advantage.

The author conducted numerous experiments with quenched gaps in connection with ordinary amateur transmitting apparatus and found that a marked increase of efficiency was obtained by observing the above precautions.

Regarding the construction of the quenched gap: The sparking plates are made of copper, carefully ground, and perfectly milled. They are provided with heat dissipating surface of sufficient area to take care of the required output. The individual plates are separated by *insulating washers* of the correct thickness, so that when the plates are compressed the sparking surfaces will be separated by no more than 1/100th of an inch. It is particularly important that the space between the sparking surfaces be *air-tight*. To this end the intervening discs or washers are treated with special insulating compounds, such as varnish, paraffine or boiled linseed oil. Under the heat of the gap, the compound will soften and if the plates are tightly pressed together an air-tight joint results.

As it is customary to allow 1,200 volts per gap the number of gaps to be employed in a given set can be readily calculated from the applied voltage. Not only the voltage of the transformer is taken for computation; but the available potential when the condenser is connected in

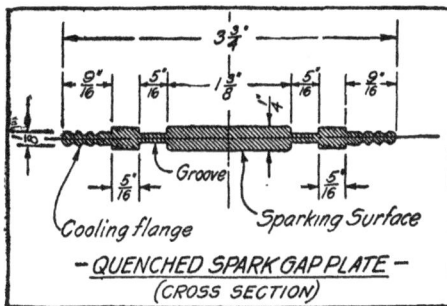

Fig. 82. Cross section and dimensions of the sparking plates of a quenched gap suitable for amateur transmitters up to ¾ kw.

shunt to the secondary winding must be taken into consideration. This value of potential can be determined by the sphere gap method, for which a table is given on page 313. Briefly, this method involves the use of two spheres or balls of a certain diameter, the spacing between which is gauged by a micrometer adjustment, with a corresponding scale. The discharge balls are connected in shunt to the source of high potential and gradually separated until the spark ceases to discharge. The length of the gap is noted and reference made to the table for the corresponding value of voltage. Remember that this table gives the *maximum voltage* per cycle and not the r.m.s. value.

The details of a gap suitable for ¼ and ½ kw. sets are given in Figs. 82, 83, 84 and 85. The cross section and the dimensions of the copper plate are given in Fig. 82. Fig. 83 shows the general appearance of the plate. Note that the copper surfaces on which the fibre disc rests has a number of circular grooves. As the compound on the discs softens, it flows into the grooves and makes an air-tight joint.

The plates are first cast to shape and then machined. This must be done very accurately for the sparking surfaces of adjacent plates must be strictly parallel.

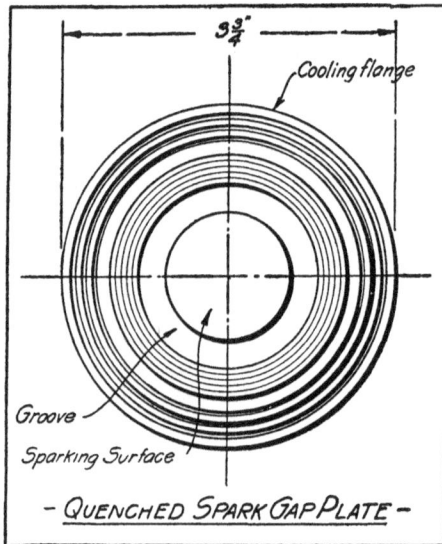

FIG. 83. Showing the finished spark plates of a quenched gap.

The sparking surface of each plate as noted in Fig. 82 is 1⅜″ in diameter. The groove surrounding it is 5⁄16″ wide and about ⅛″ deep. The shoulder which presses against the insulating discs is also 5⁄16″ wide. The entire plate is 3¾″ in diameter. The washers between plates, are 3⅞″ outside diameter, and 1¾″ inside diameter.

The assembled gap is shown in Fig. 85. The plates are mounted between two metal castings bolted to a cast iron base. A pressure bolt for compressing the plates is mounted on the right hand end It should be a ⅜″ bolt with a rather fine thread.

Although the drawing shows two insulating discs at either end of the quenched gap plates it is better to place a cast iron plate between the right hand disc and the frame as otherwise the pressure on the bolt necessary to prevent air leakage may puncture the disc The iron end-plate should be centered and countersunk to take the pointed end of the compression bolt.

If the insulating discs are made the same diameter as the plates, the experimenter can mount two parallel insulating rods between the end castings to form **a** rack upon which the plates rest. The plates and washers will then center themselves automatically when they are dropped into the rack. It is readily seen that, if the plates are stacked up unevenly, the inside edge of the insulating discs may not cover the parallel rims of two adjacent gaps; but if they are correctly centered, the inside edge of the washer will come to the center of the groove. In event of the former, sparking will take place at the rim and the insulating disc will be punctured, putting that particular gap out of commission.

The gap should be cooled by directing the blast of a 6-inch fan on the plates.

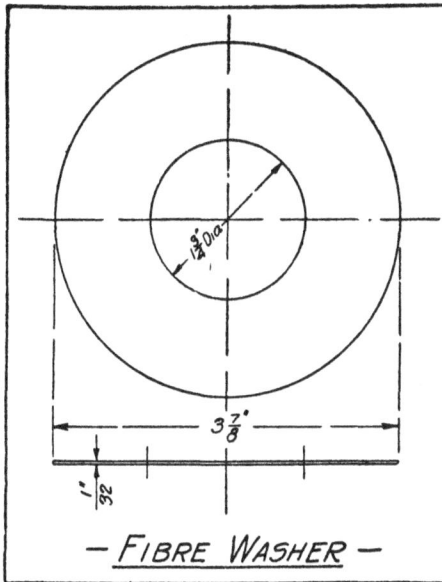

FIG 84   Dimensions of insulating gaskets.

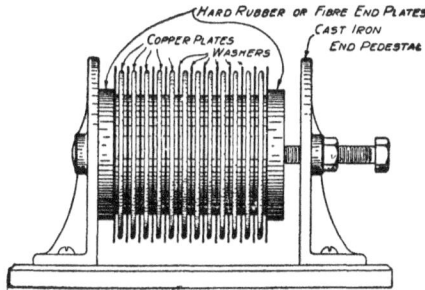

QUENCHED SPARK GAP (Assembled)

FIG. 85   An assembled quenched gap   The plates are mounted in a metal containing rack and compressed tightly by a bolt inserted in the end casting

ADJUSTMENT OF THE QUENCHED GAP TRANSMITTER.—

If the amateur had complete control over his power supply it would be possible to give him the design for a complete "quenched" transmitter which would give maximum efficiency without experimenting. But this is not the case, and moreover, a variety of transformer designs are met with in practice which make some experimenting necessary to get the best results.

Assuming that the complete transmitter is wired up as in Fig. 86 the method for adjustment is as follows. A wattmeter W and an ammeter A-2 are connected in the primary power circuit. The wattmeter gives a check on the primary power and the ammeter shows whether the current carrying capacity of the primary coil P is exceeded.

If the gap S-1 has 16 plates and the transformer potential is about 15,000 volts, cut in about 10 gaps for a preliminary trial. Then cut in about one-half of the primary reactance X, close the key, and note the primary power and current. Change the number of gaps and note the effect on the power. If the power reading is too low, reduce the reactance of X and bring it up to normal. These tests indicate that the circuits are working properly.

Now disconnect the spark gap S-1 and vary the capacitance of the condenser C (which is assumed to be a high voltage condenser of variable capacity), until ammeter A-2 reads a maximum. If the capacitance of C exceeds 0.008 mfd. for resonance, reduce C and increase the reactance of X. When resonance is found add a few more plates to C until the transformer circuits are 15 to 20% off resonance.

Then connect about 10 gaps in the circuit, close the key, and if the primary power does not exceed normal proceed to tune the set and afterward adjust the spark note.

Set the antenna circuit, A, L-1, A-1, E, to 200 meters by means of a wavemeter. Then vary the inductance of the primary L and watch the aerial ammeter A-1 for a maximum reading. If the resonance is sharp, that is if a slight change of L has a marked effect upon the antenna current, the gap is quenching. It may be well then to try various degrees of coupling and different numbers of gaps to determine if the antenna current can be increased.

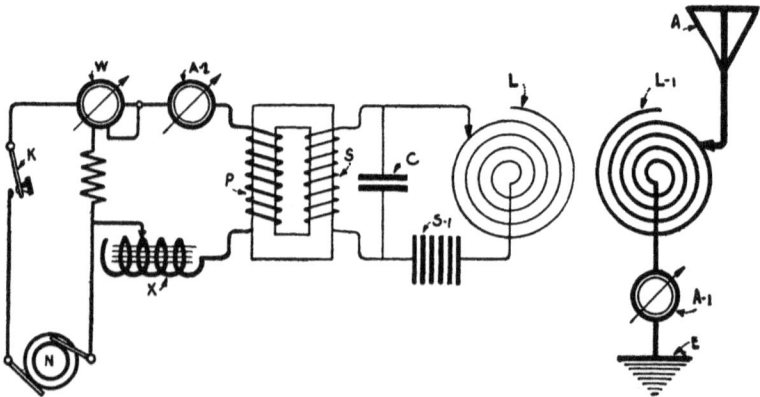

Fig. 86 Essentials of a transmitting circuit suitable for use with quenched spark dischargers. An oscillation transformer providing a continuous variation of inductance is preferred

A further change in the capacity of C or the reactance of X may help matters. By carrying on a few experiments along these lines, the experimenter will hit on a combination of values giving a strong antenna current of low decrement and a single wave emission. The author has proved this in practice.

Do not allow the gap to overheat. Avoid a temperature over 200 ° F.

If the transformer has a number of taps on the primary coil the reactance X is unnecessary. A secondary reactance may be used. It is made up of a number of pancakes like those of the transformer secondary, placed over an iron core. The primary reactance is, however, less expensive to construct. Very precise tuning of the spark gap and antenna circuits is required for maximum antenna current when using the quenched gap.

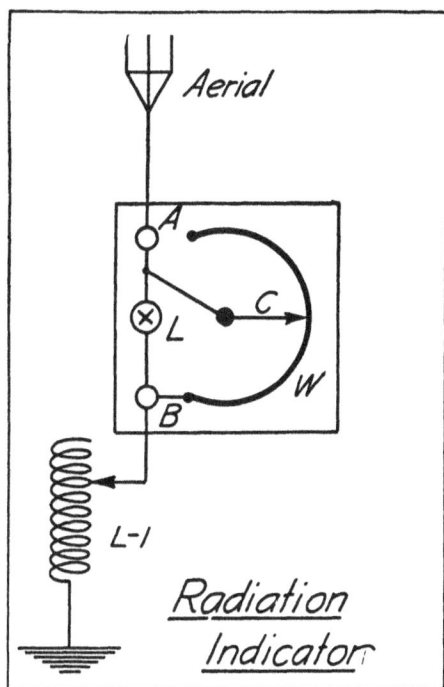

Fig. 87. A resonance indicator which may be substituted for a hot-wire ammeter in the antenna circuit of the amateur transmitter Resonance is indicated by the maximum glow of a small incandescent lamp which is shunted by a variable inductance.

RESONANCE INDICATOR.—To measure the antenna current and to determine whether the open and closed circuits are in substantial resonance, use a *hot wire ammeter* or an ammeter with a *thermo-couple*. These instruments indicate the strength of the current flowing through them by the heating of a wire through which the current is passed.

Magnetic measuring instruments cannot be used in radio frequency circuits because of their self-induction and because their coils will not withstand the high voltages of wireless transmitters.

Hot wire ammeters are rather expensive and are not commonly used in amateur stations. Any device that will indicate when the antenna current is maximum, although it is not calibrated in amperes, will serve as a substitute.

A *small battery lamp* shunted by a turn of wire and connected in series with the antenna circuit will indicate by its maximum incandescence that the closed and open circuits are in resonance. An indicator of this type is shown in Fig. 87, where a *two- or four-volt battery lamp* is connected in series with the antenna circuit through the binding posts *A* and *B*. The lamp *L* is shunted by a loop of copper wire or flat copper strip *W*, over which the arm *C* makes contact. The length of the loop will vary with the strength of the antenna current; usually its radius does not exceed 3″.

A few trial experiments with a given transmitter will reveal the correct position for the slider *C*. It is less difficult to judge a current maximum in the lamp if the shunt is adjusted for a cherry red glow as the point of resonance is approached.

ANTENNA AMMETERS.—Two types are in general use. One utilizes the expansion of a wire or strip heated by the current to move an indicating mechanism; the other type employs a *thermo-couple* mounted on a wire heated by the radio frequency current.

The *thermo-couple*, when heated by the radio frequency current, generates a direct e.m.f. which, in turn, actuates a magnetic instrument having the characteristics of a millivoltmeter. The meter may be calibrated directly in *amperes*.

In the design of hot wire meters coarse wires are to be avoided for their resistance changes markedly with the frequency of the current. On the other hand, if a wire of very small diameter is employed radio frequency currents will penetrate it to the center and its resistance will remain substantially constant over a large range of frequencies. If large currents are to be measured several such small wires must be connected in parallel and one of them selected to work the indicating mechanism or to heat the thermo-couple. The wires must be parallel, must be exactly of equal length, and the external leads must be attached at a point where the current will distribute itself equally through all wires. When several wires are employed they are usually strung on a cylinder.

An *ammeter* which any wireless man who is clever with his tools may construct at a small cost is shown in Fig. 88. *C D* is a piece of silk fibre about 3″ long. *G H* is another piece about 2″ long and is attached to the center of *C D* by a bit of beeswax. The lower end of *G H* is wound about the shaft which carries the spring *S* and the *pointer*. The spring tends to pull the pointer toward the maximum scale position, but it is resisted by the fibre thread and the wire *W-1*, *W-2*.

When no current flows through *W-1*, *W-2*, the pointer rests at zero, but when the wire is heated by the passage of radio frequency currents the tension on *C D* is reduced, relieving the tension opposing the spring *S*, which in turn, pulls the pointer across the scale.

In the diagram *B-1* and *B-2* are *heavy copper lugs*. *W-1*, *W-2* is a piece of No. 40 Therlo wire (0.003″ in diameter) about 4″ long. Its resistance is about 9.5 ohms. Using only one wire, the instrument will measure at its maximum scale, 0.1 ampere. With several wires in parallel the range may be increased to any desired value. The shunt wires must be of the same length and the same diameter as the wire which works the mechanism. All wires are preferably arranged around a cylinder, being spaced equi-distantly.

An instrument employing a *thermo-couple* has been designed by M. K. Zinn of Indiana, who says that it was patterned after the data given by Chas. Ballantine. Instruments of this type were first employed, commercially, by the Marconi Company and they are now common in radio sets. Fig. 89 shows the general details of construction. Fig. 90 is a sketch of the assembled instrument and Fig. 91 shows the method of calibration.

Six bare No. 40 copper wires are arranged around circular discs 1″

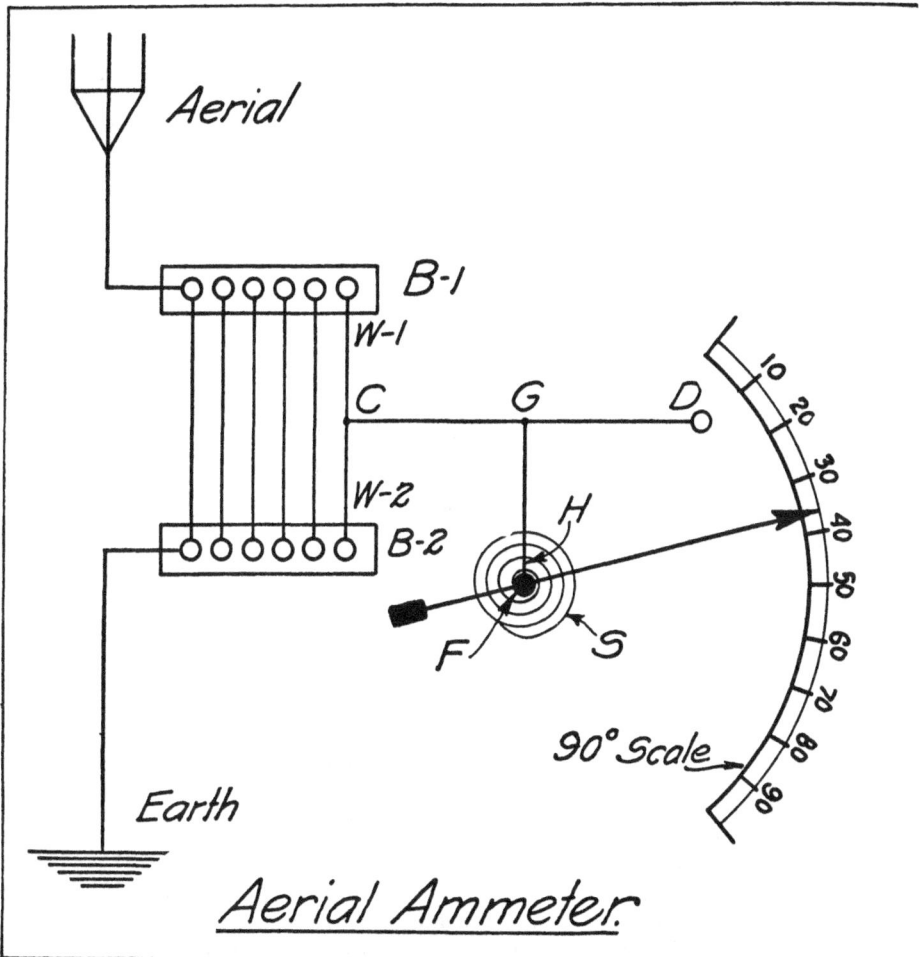

FIG. 88. Experimental hot-wire ammeter which may readily be constructed by the wireless experimenter.

in diameter, $\frac{1}{16}''$ in thickness. The centers of the discs are drilled for
8/32 machine screws, through which they are fastened to the posts
A and B.

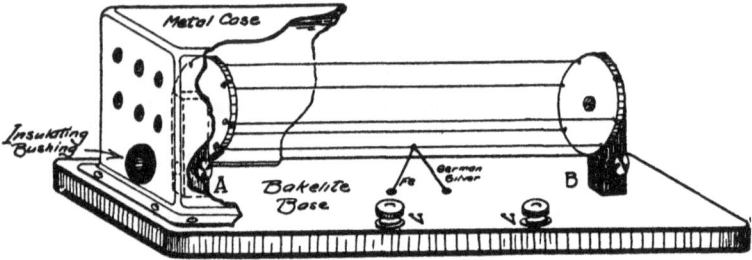

Fɪɢ. 89.   A thermo-couple suitable for radio frequency measurements.

The slots for the wires are cut at intervals of 60° with a fine hack-saw.
The terminal posts are made of brass stock $\frac{1}{4}''$ square, drilled and tapped
as shown in Fig. 90.   The posts are mounted on a $\frac{1}{4}''$ bakelite base,
$2''$ x $6''$ being separated $4\frac{3}{4}''$.   The six copper wires may then be soldered
in position.

The *thermo-couple* consists of a piece of No. 36 iron wire and another
piece of No. 36 German silver wire, each $3''$ long.   The ends of the two

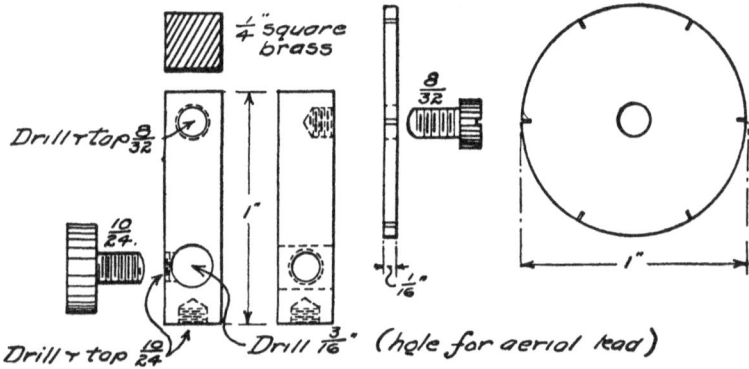

Fɪɢ 90.   Details of the terminal posts.

wires are carefully twisted together and then soldered to one of the
No. 40 heating wires.   The free ends of the thermo-couple extend to two
binding posts mounted on the base.   These posts are for connection to a
millivoltmeter which is to be calibrated in amperes by comparison with
a standard ammeter.

The method of calibrating the instrument is shown in Fig. 91, where
a battery B, a rheostat R, and a standard ammeter A_m are connected in
series with the heating wires.   The millivoltmeter is connected to the
iron element of the thermo-couple at Fe and to the German silver
element at Gs.

Various currents covering the whole range of the millivoltmeter scale are passed through A_m. The scale on the millivoltmeter should at any instant be given the value read from the meter A_m. More accurate methods of calibration will be found in standard electrical hand books.

Frank O'Neill of California designed the meter of Fig. 92. The pointer carrying mechanism was made from an old alarm clock. The hot wire, which is a piece of No. 36 wire about 4″ long, is strung between a hook and an adjusting screw as indicated. The pointer, which is not shown, is a piece of No. 28 copper wire soldered on the shaft carrying the wheel and spring.

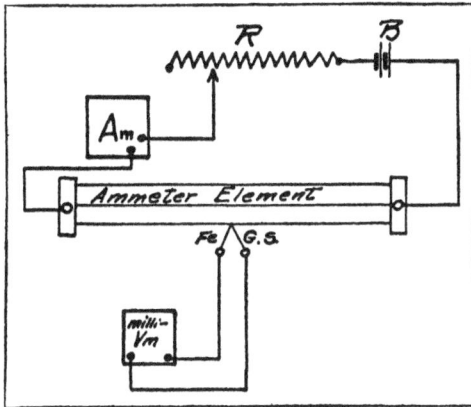

FIG. 91. Showing how an ammeter with a thermo-couple may be calibrated from a standard.

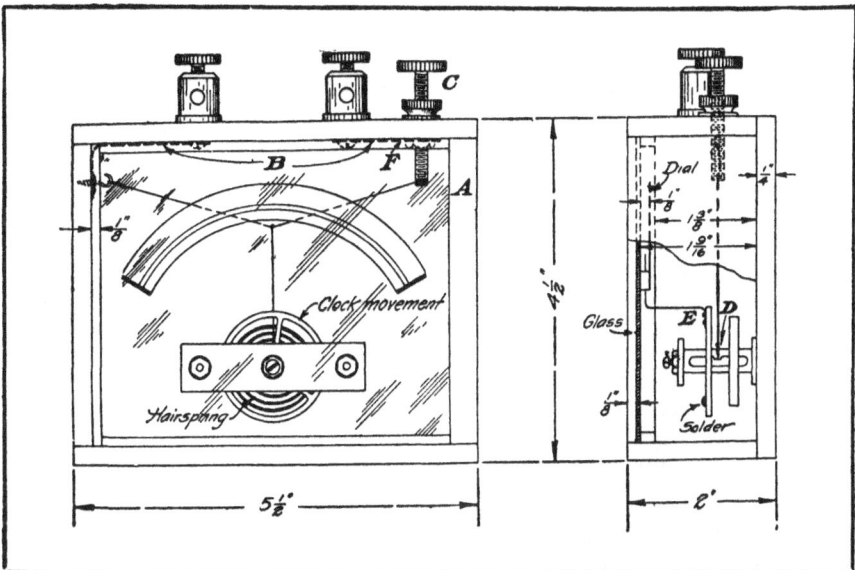

FIG. 92 A hot-wire ammeter which the amateur may construct from parts found around the workshop.

AERIAL TUNING INDUCTANCE.—Ordinarily the 200-meter station does not require an aerial coil separate from the secondary of the oscillation transformer. The secondary of the oscillation transformer usually has sufficient turns to raise the wave length of the average amateur aerial to 200 meters.

The aerial tuning coil may be of the *helical type* or a *flat spiral pancake;* the exact dimensions for any given wave length may be calculated from the formulae in Chapter III. If bare strip or bare copper tubing is employed, the turns should be spaced about $3/8''$ to prevent sparking between them. If heavily insulated copper conductor is used the turns may be wound as closely as possible. A few turns of No. 6 stranded wire will be satisfactory.

SHORT WAVE CONDENSER.—The short wave condenser is required only when the natural wave length of the antenna circuit exceeds 200 meters. For the average station operating at 200 meters, some capacitance between 0.0002 and 0.0005 mfd. is about correct. Several plates should be connected in series to prevent the dielectric from breaking down. We have shown on page 46 that a plate of glass $14'' \times 14''$ covered with foil $12'' \times 12''$ has a capacitance of approximately 0.0016 mfd.

Hence, four such plates in series will have $\dfrac{0.0016}{4} = 0.0004$ mfd. capacity

—about the correct value for a series condenser.

The short wave condenser should be connected in the earth lead and the wave length of the antenna circuit determined by trial with a wavemeter. If it exceeds 200 meters turns should be cut out at the secondary until 200 meters is obtained.

If the flat top portion of the aerial is more than 120 ft. in length, a short wave condenser of very small capacitance will be required to reduce the wave length to 200 meters. This will decrease the range of the set and it is therefore preferable to shorten the aerial. The author's experience proves the desirability of designing the antenna with dimensions not requiring a short wave condenser.

THE EARTH CONNECTION.—For amateur stations located in isolated districts where there are no water mains or steam pipes for connection to the earth a satisfactory earth plate can be made from 250 square feet of *galvanized sheet iron* or *copper* laid in moist earth to, let us say, a depth of from $5'$ to $8'$. A piece of copper strip $2''$ in width should be firmly riveted and soldered to the plates and led directly to the oscillation transformer of the transmitter. Additional contact with the moist earth can be made by driving several lengths of galvanized iron pipe with a sledge hammer to a depth of $8'$ or $10'$ under the copper plates. *Surface grounds or counterpoises* may be employed in localities having rocky soil. A good surface ground may be made of several copper wires laid on the surface of the earth directly underneath the aerial. Galvanized wire netting like that used for fencing is also satisfactory.

Connection *should not be made to the gas mains,* as in many cities, the pipes from the meter to the house chandeliers have an *insulating bushing* near the meter which insulates the pipes from the earth. At every station where it is possible to do so, a piece of *copper ribbon*, let us say 2″ in width, or a piece of heavily insulated No. 2 or No. 4 d.b.r.c. wire, should be attached to the water main on the street side of the house meter. In many homes and apartment houses it is not possible to do this, and, therefore, connection must be made to the *steam pipes*, to the *water mains* inside the house, or to the *steel frame* of the building if such exists.

In some localities the underwriters require that the earth lead from the apparatus be thoroughly insulated down to the point where actual connection to the earth capacity is made; under such restrictions special insulators should be constructed to hold the conductor away from the structure to which it is attached.

SAFETY GAP FOR THE HIGH POTENTIAL TRANSFORMER.— Suppose, for example, that the spark gap is widened out so that the transformer voltage cannot jump the gap. Both the transformer and the high voltage condenser are subjected to an abnormal potential difference, which may puncture the dielectric or break down the insulation of the secondary coil. The liability to breakdown under these conditions is far removed by placing a so-called *safety gap* in shunt to the transformer secondary.

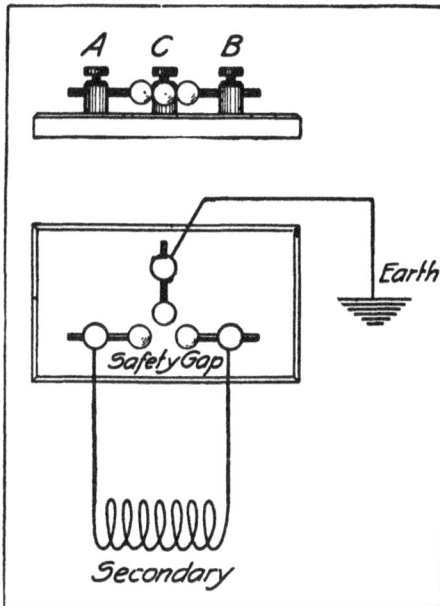

FIG 93. Safety gap for protecting the condenser and the secondary coils of the high voltage transformer from rupture.

Fig. 93 shows the construction of a safety gap. Two discharge balls ½″ in diameter are mounted on the ends of two brass rods ¼″ in diameter. A third ball of similar dimensions is centered between the other two and connected to earth.   The spacing between A and the center ball, and B and the center ball depends upon the potential of the transformer, but ordinarily it is not more than ½".   If now the main spark gap is widened out abnormally, the condenser discharges from A to C, and from B to C, protecting the transformer secondary and the condenser.

The space between electrodes A and B must, of course, exceed the effective spacing of the main discharge gap in normal operation or otherwise the safety gap will discharge continually.

PROTECTIVE DEVICES FOR POWER CIRCUITS.—The *electrostatic field* around an aerial charged to high voltages is very powerful. It is often of sufficient strength to induce disastrous potentials in nearby circuits which may rupture the insulation of the *power meter*, the *motor generator*. the *primary coil* of the *high voltage transformer* or the *lighting circuits* in the building.   These induced potential differences may be neutralized by protective condensers connected as shown in Fig. 94.

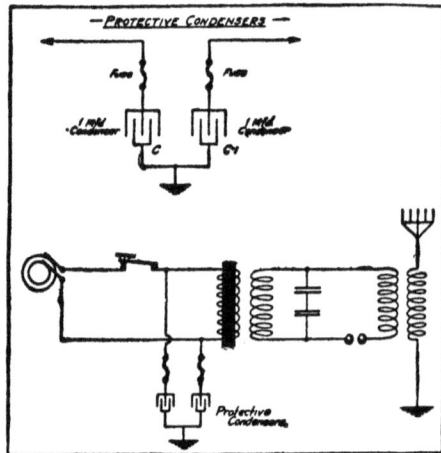

Fɪɢ. 94.   Protective condenser for the power circuits of a radio-telegraph transmitter.

Two condensers having a capacitance of *one* or *two microfarads each* are connected in series and earthed at the center connection.   The remaining terminals are connected through fuses of ½ ampere capacity to the leads from the alternating current source or across the power leads where they come out of the meter.   The fuses are necessary in case of an accidental short-circuit in one of the condensers.

A set of these condensers should be connected across any lighting leads that suffer from this source.   Graphite rods of 1000 ohms each are often used for the same purpose.   The ends of the rod are connected to the power leads and the center of the rod connected to earth.

THE LIGHTNING SWITCH.—The underwriters' rules in many cities require that the aerial of a radio station, when not in use, be connected to earth through a *100-ampere single blade switch*.   Without the proper set of tools the experimenter will hardly be able to construct a switch which will meet the requirements. Single pole, double throw switches of this current carrying capacity can be purchased at any electrical store. The switch should have a base of good insulating material and, if mounted outside the room, should be placed in a metal or asbestos lined box. *It is important that the base of the switch possess the very best insulating qualities* as otherwise the high voltage currents of the transmitter will leak to earth.

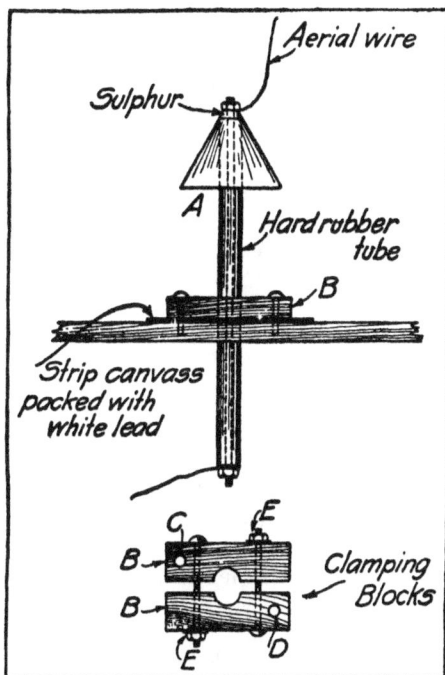

Fig. 95. Lead-in insulator for transmitting stations.

LEAD-IN INSULATOR.—This insulator is employed to bring the aerial leads through the roof of a building or the side of the house. It must be made of the very best insulation so that it will not allow the high voltage currents to leak through it. The construction should be such that it will be water-tight. The more elaborate insulators used by commercial companies are beyond the means of many amateurs and therefore the simpler type shown in the drawing, Fig. 95, may be adopted. It consists of a hard rubber tube, at least 16″ in length with a hole ⅜″ to ½″ in diameter, which is clamped to the roof by means of the wooden blocks, *B B* drawn together by the bolts, *E E*. The tube should fit snugly when the nuts on the bolts are drawn home. A second set of bolts is inserted at *C* and *D* through which the wooden blocks are drawn to the base board.

To make the joint watertight, use a strip of canvas slightly larger than the blocks, with a hole the size of the tube cut in the center. Then smear the canvas thoroughly with white lead. Insert the bolts at C and D and draw the blocks to the roof. If allowed to dry, a watertight joint will result, provided the strain of the lead-ins is removed by appropriate guys.

The wire extending through the insulator (or a brass rod if used) should be at least the diameter of a No. 6 d b.r.c. stranded conductor. It may be made watertight by filling the surrounding space with *melted sulphur*. A certain quantity of sulphur should be heated in a pan until it runs freely; the bottom of the insulator is then stuffed up with a quantity of waste and the sulphur poured in from the top until the insulator is completely filled. When dry, the sulphur hardens and possesses the requisite insulating qualities for high voltage currents. A *metal cone A* may be fastened to the top of the rubber tube to keep it dry in wet weather.

A very good lead-in insulator may be made by boring a hole in a thick *pane of glass* through which a brass bolt is placed. The strain of the lead-in wires must be taken off the bolt by a guy wire attached to the lead-ins near the point where they enter the station.

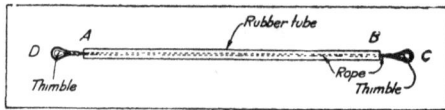

Fig 96　Home-made insulator for the antenna wires　This insulator is suitable for transmitting sets.

HOME-MADE INSULATORS FOR ANTENNA WIRES.—It is the usual practice to insulate the antenna wires from the "spreaders" by rods of hard rubber, bakelite, etc., fitted with an *eye bolt* at each end. They should be from 12″ to 24″ in length.

An inexpensive antenna insulator may be constructed as in Fig. 96. A piece of *marlin rope* about 16″ in length, with eyes at both ends made by serving the rope around *heart-shaped thimbles*, is covered with a *hard rubber tube* about 12″ long.

The tube should be large enough to afford a small air space between the rope and the wall of the tube. A quantity of melted sulphur is poured in the tube until the rope is thoroughly impregnated with it and the tube filled to the top. If this is properly done a watertight encasement will result. The matter of insulation with receiving aerials is not so important. Small porcelain knobs inserted at the ends of the antenna wires give all the insulation necessary.

TRANSMITTING KEYS.—The sending key should have contact points that will carry 15 to 18 amperes without fusing. *Platinum key points* are preferred but they are costly. *Silver contacts* are satisfactory and are widely used. Platinum contacts $\frac{3}{16}''$ in diameter will handle 15 amperes without difficulty. Silver contacts $\frac{5}{16}''$ in diameter will handle the same current

The constructional details of a transmitting key suitable for amateur sets are shown in Fig. 97. The lever and the structure for the lever bearings are of brass and the knob of hard rubber. The base is made of black fibre.

EXAMPLES OF 200 METER SETS.—Whether the panel or the isolated instrument type of transmitter is the most desirable depends upon the service to which the set is to be put. For economy of space and for a fixed installation with which no further experimenting is to be done, the panel set is to be preferred. The isolated instrument type appeals to the amateur who desires to experiment from time to time with various circuits and apparatus. The argument that the panel set is more efficient than the isolated instrument type carries no weight, for by proper design and by observing certain precautions in the installation the sets are equally efficient.

The panel set, in fact, introduces the difficulty of electrostatic induction from the radio frequency circuits, which may set up high voltages in the power circuits and puncture the insulation. Without due precautions in this respect the panel set may prove a rather expensive experiment.

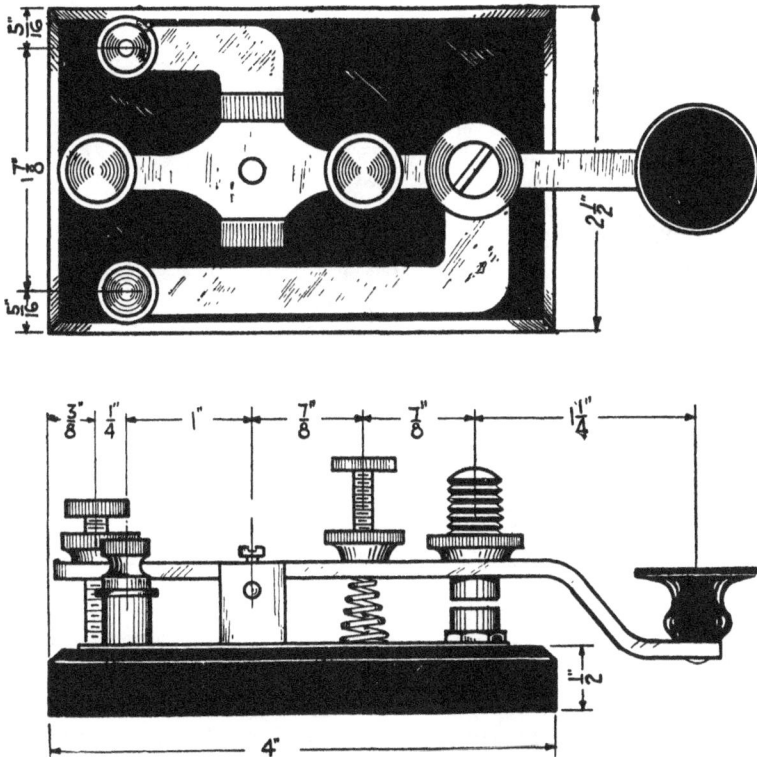

FIG 97   Constructional details of a transmitting key suitable for powers up to 1 kw.

In panel sets the leads to the measuring instruments and, in fact, those throughout the primary power circuits must be covered with a *metallic coating grounded to earth*.   Even the terminals of the meters must be covered with metallic caps connected to earth.

Protective condensers should be connected across the leads to the rotary gap motor.   All power leads should, if possible, be run at *right angles* to conductors carrying radio frequency currents, the more so, if high voltages around 20,000 or more are employed.

There is one distinct advantage to be credited to the panel set; if a *wave length changing switch* is to be employed, it can be fitted to the panel set with fewer mechanical difficulties than to the isolated instrument type.

Installation sketches of two sets using *isolated instruments* are shown in Figs. 98 and 99.   The complete wiring between the component instruments is indicated.   The set in Fig. 98 has a *glass-plate high voltage condenser*, while Fig. 99 has 4 *standard commercial type Leyden jars*. Open core transformers are shown but closed core types may be used as well.

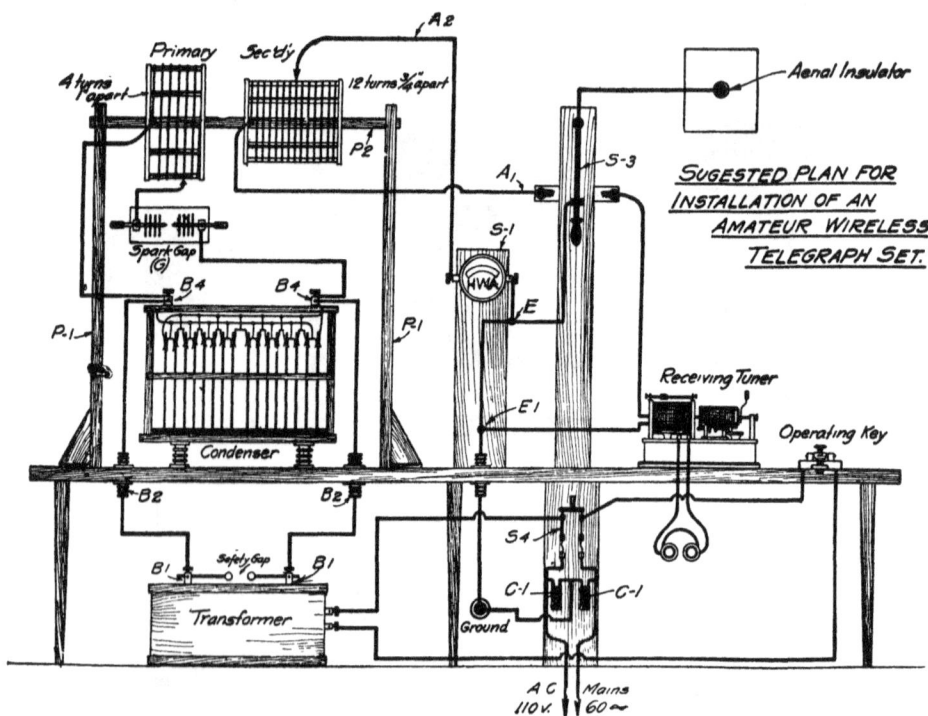

FIG 98   Showing the installation of an isolated instrument type of radio transmitter suitable for amateur work.

The operating table for the set of Fig. 98 is placed about 32 inches from the floor and is approximately 30″ in width.   It may be from 6 to 8 feet in length, according to the space available, and is, of course, supported at intervals by 2″ x 4″ uprights.

The high potential transformer is placed under the table and to the left in the drawing. The secondary terminals of the transformer are insulated by corrugated hard rubber bushings *B*-1, capable of withstanding potentials up to 20,000 volts without leakage. The safety gap connected across the terminals consists of 2 discharge electrodes connected to the terminals of the transformer and separated sufficiently to prevent discharge when the main gap is in action.

The condenser rack is insulated from the operating table by corrugated porcelain or electrose legs. For operation at the wave length of 200 meters, the condenser may consist of 20 plates of glass 14″ x 14″ with an average thickness of $\frac{3}{32}$″. They are covered with tinfoil 12″ x 12″. There are two 10-plate banks connected in series. Other dimensions may of course be used as long as the proper condenser capacity is provided.

The leads extending from the transformer secondary are led up through the table through the insulating bushings *B*-2, and are attached to the binding posts *B*-4, which are mounted on hard rubber slabs on the edge of the condenser rack. No. 18 d.b.r.c. wire will easily carry the secondary current, but a neater job will be done if $\frac{3}{16}$″ copper tubing is employed, as it is self-supporting when attached to the binding posts, *B*-1.

The tabs of tinfoil from the condenser plates are attached to the binding posts *B*-4. If the brush discharge is excessive the plates may be immersed in a metal or porcelain tank, filled with a good grade of insulating oil. However, oil immersed plates often do not prove practical, as it is difficult, even with the best grades of glue, to keep the tinfoil on the plates when they are oil-immersed, unless they are stacked up and bound together closely with insulating tape.

The *oscillation transformer* is supported by the pillars, *P*-1, and the cross rod *P*-2. The primary is permanently fixed to the rod, but the secondary may be slid backward and forward for variation of the coupling.

The primary coil has about four turns of $\frac{3}{8}$″ copper tubing, with the turns spaced $\frac{3}{4}$″ apart. It is from 8″ to 10″ in diameter. The secondary coil contains from 8 to 10 turns, spaced $\frac{3}{4}$″ apart, and may be the same size of tubing. To secure a pure wave according to the U. S. law it is usually undesirable to have the secondary coil inside of the primary coil; generally there is an air space intervening of from 2″ to 3″. Both the primary and secondary turns must be well insulated and, in consequence, it is necessary that the supports of each be made up of some good insulating material, such as a high grade of hard rubber, porcelain, bakelite or micarta or any of the well known insulating materials.

A *spark gap G*, is placed between the condenser and oscillation transformer and it may take one of several designs. Amateurs are accustomed to use the *non-synchronous* gap, which usually consists of a disc of insulating material, 6″ to 8″ in diameter with a series of equally spaced discharge electrodes mounted on the shaft of a motor, rotating from 1800 to 2400 r.p.m.

The terminals of the gap motor are connected to the top terminals of a 10-ampere power switch by means of which it can be started and stopped. Various methods can be devised for starting the motor; one is to fit the antenna switch with an extra set of contacts so that when it is thrown into transmitting position the circuit to the motor is closed. It is much simpler, however, for the beginner to employ the method shown in the drawing.

The antenna switch S-3, is a single blade, double throw switch, which, when thrown to the left, connects one terminal of the secondary winding of the sending oscillation transformer to the antenna and, when thrown to the right makes contact with a line leading to the primary of the receiving tuner. In the center position the blade connects the antenna to the *earth lead* at the point *E*, making an efficient *lightning switch*. To comply with the underwriters' regulations, this switch must be capable of carrying a current of 100 amperes and, to accord with the requirements of wireless installations, all contact studs should be spaced at least 6″ to prevent the direct discharge of antenna potentials over the base. It is, of course, evident that the lever and studs of the switch must be mounted on a good insulating material, which in *no case should be of slate*. A good grade of hard rubber, bakelite, or micarta will fulfil the requirements.

A connection is extended from the aerial lead-in wires to the lever of the change-over switch through an insulator, which is a hard rubber tube with walls ⅜″ in thickness, with a ¼″ hole. A brass rod threaded at both ends and fitted with nuts and connecting lugs, is passed through the hole and firmly fastened in place. As an alternative a hole can be

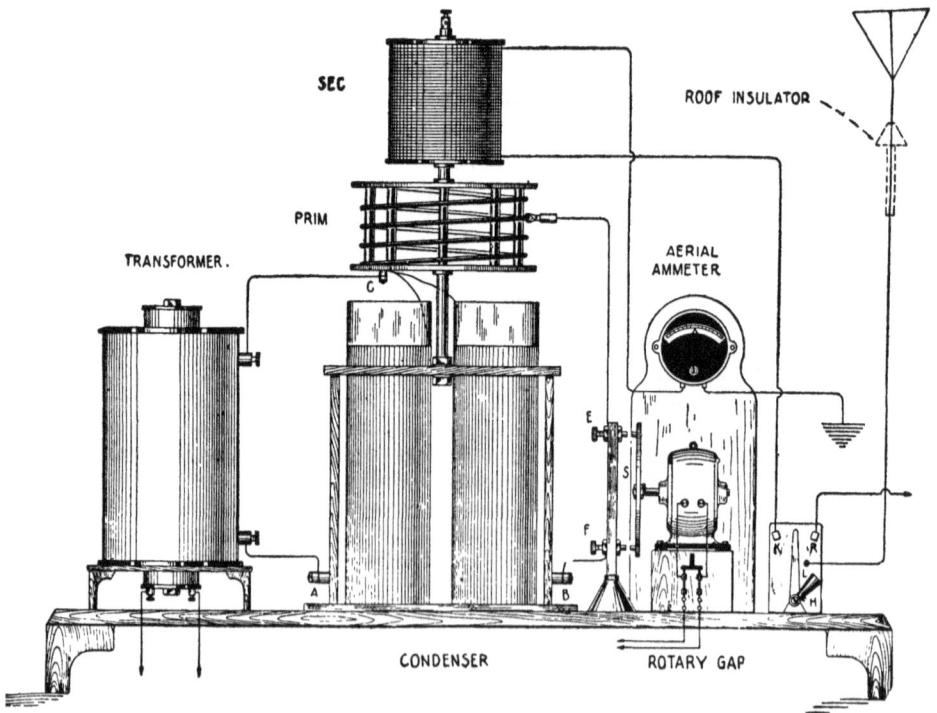

FIG 99   Modified design for an amateur transmitter

drilled through a window pane and a brass rod passed through it, but the incoming leads of the aerial must then be well backstayed to remove the strain from the glass.

The right hand stud of the antenna change-over switch is connected to the antenna post of the receiving tuner and the earth connection from the latter makes contact at the point *E*-1, to the wire to which the transmitting ground lead is attached. With this type of switch in use the receiving tuner should be fitted with a shunt switch *to protect the detector* from the local oscillations of the transmitter during the period of transmission.

The *transmitting key* is mounted immediately to the right of the tuner, and is placed far enough from the edge of the table to allow the sending operator's elbow to rest on the ledge.

The hot-wire ammeter should be mounted on an insulating stand made of a sheet of hard rubber or other suitable material of the correct dimensions to support the meter. The base should be insulated from the wall by hard rubber legs. It is not necessary for this meter to be mounted on an insulating support when connected in series with the earth lead, but the Electrical Inspectors Code in many cities requires that this be done, regardless of the point at which it is connected.

A ½ kw. *transmitting set* of slightly different design is shown in the sketch of Fig. 99, which includes a ½ kw. open core transformer, four standard copper plated Leyden jars of the type manufactured by the Marconi Company (capacity 0.002 mfd. each), an oscillation transformer mounted directly above the condenser, a non-synchronous rotary spark gap, an aerial change-over switch and a hot wire aerial ammeter.

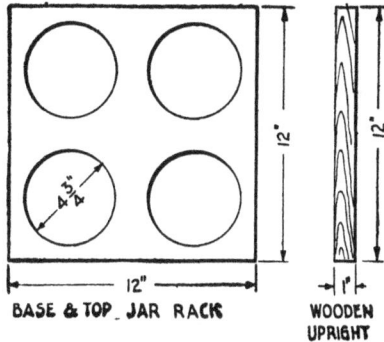

Fig 100. Details of the Jar rack for the installation of Fig 99.

Note that the condenser jars are mounted in a wooden rack, the base of which is covered with a sheet of copper. Small upright posts 1″ x 12″ support the top piece shown in Fig. 100. The holes to take the jars are 4¾″ in diameter.

Fastened to the top of the rack is an upright rod of metal, hard rubber or of wood, which supports the primary and secondary coils of the oscillation transformer. It will be seen that with this arrangement of apparatus the connections in the closed circuit are very short. The complete *closed oscillation circuit* can be traced out in the following manner: The four leads from the inside coatings of the four Leyden jars are connected to the binding post *C*, which is mounted on the base of the primary winding of the oscillation transformer. The circuit continues through the turns of the *primary coil*, out the top to the terminal *E*, of the rotary spark gap. The circuit then goes through the disc of the gap, out the electrode *F*, and finally to the binding post *B*, which in turn is connected to the copper strip at the base of the *jar rack*. With the apparatus mounted in this way no more than two or three turns of the primary winding are required for the wave length of 200 meters.

The primary winding is made of copper tubing $\frac{3}{8}''$ in diameter, spaced $\frac{3}{4}''$ from center to center of the turns. This tubing is fastened to hard rubber supporting posts by means of brass machine screws, the copper tubing and the hard rubber post being drilled accordingly.

The secondary winding of the oscillation transformer comprises from 6 to 12 turns of No. 6 d.b.r.c. wire which are closely wound. The top terminal is connected to one binding post of the aerial meter while the opposite terminal of the meter is connected to earth. The other terminal of the secondary winding leads to point *K* of the aerial change-over

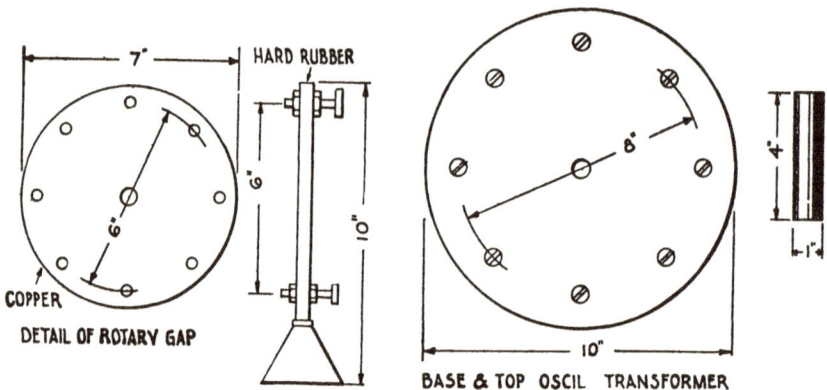

FIG 101. Details of the rotary gap.

FIG. 102 Details of the oscillation transformer.

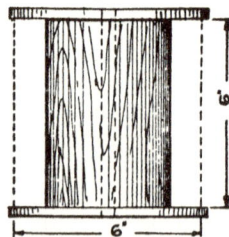

FIG. 103. Details of the support for the secondary coil.

switch, extend to the contact blade L, and thus on to the aerial. When the switch is thrown to the contact point R, the aerial is connected to the receiving equipment.

Drilling and over-all dimensions for the rotary gap disc appear in Fig. 101. A motor should be selected that revolves 1800 r.p.m. The disc, mounted on the shaft of the motor, should be about $7''$ in diameter and should have eight or ten sparking points equally spaced about the circumference. These may be made of machine screws $\frac{1}{8}''$ or $\frac{3}{16}''$ in diameter. The sparking points should be placed on a radius $3''$ from the center and the stationary electrodes E and F should accordingly be separated about $6''$. With this design 240 sparks per second are obtained.

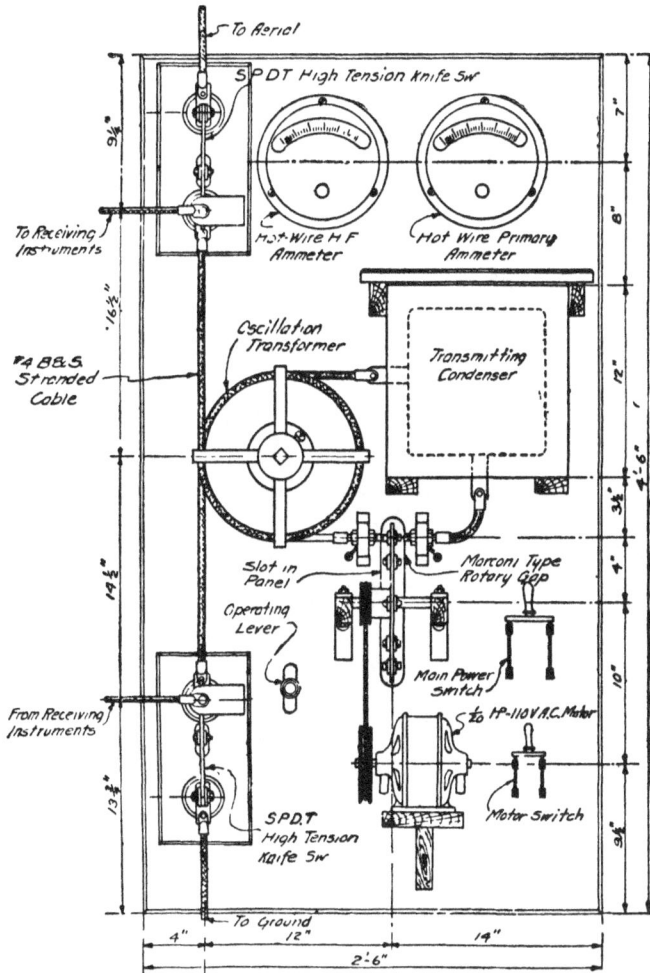

FIG. 104   Panel transmitter designed by a U S. experimenter. The good workmanship indicated in this design deserves special attention.

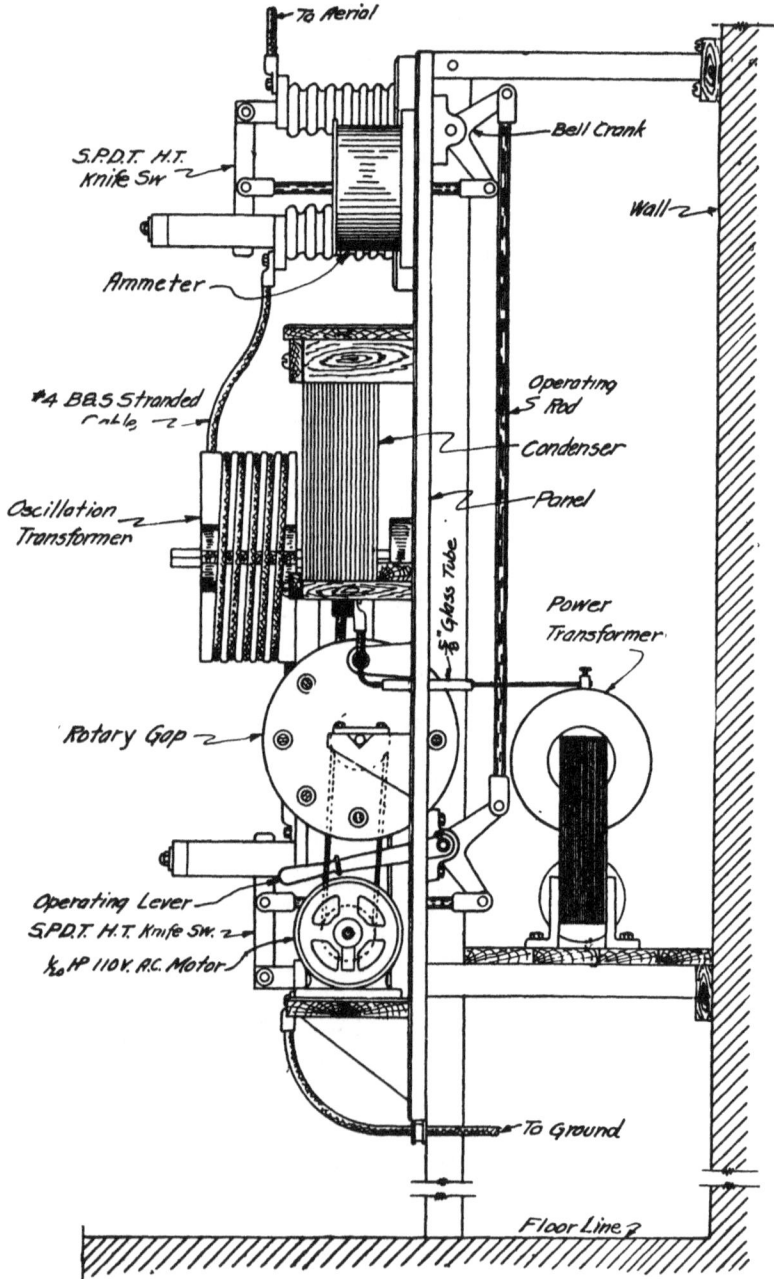

To Aerial

S.P.D.T. H.T.
Knife Sw

Bell Crank

Wall

Ammeter

#4 B.&.S Stranded
Cable

Operating
Rod

Condenser

Oscillation
Transformer

Panel

8" Glass Tube

Power
Transformer

Rotary Gap

Operating Lever
S.P.D.T. H.T. Knife Sw.
⅒ HP 110V. A.C. Motor

To Ground

Floor Line

FIG 105.   Right hand side elevation of the panel of Fig. 104

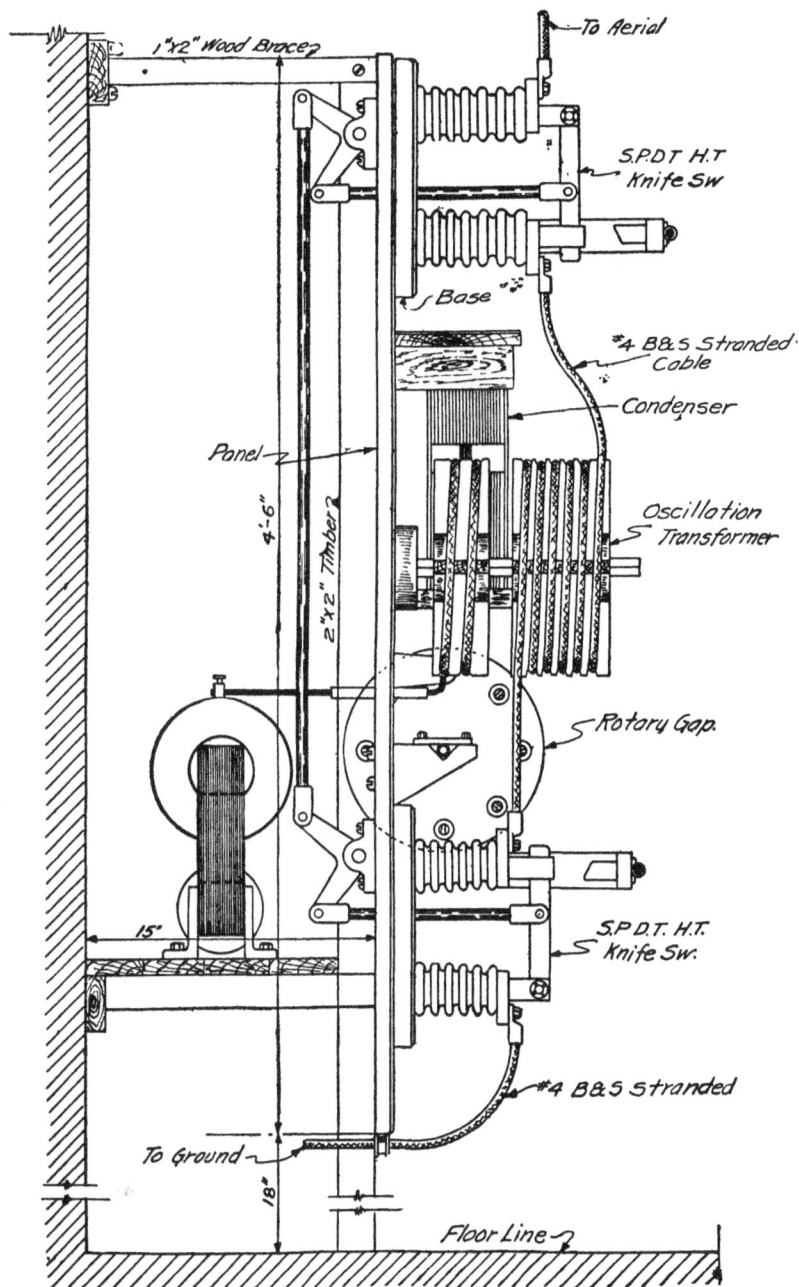

To Aerial

1"x2" Wood Brace

S.P.D.T H.T
Knife Sw

Base

#4 B&S Stranded
Cable

Condenser

Panel

4'-6"

2"x2" Timber

Oscillation
Transformer

Rotary Gap.

15"

S.P.D.T. H.T.
Knife Sw.

#4 B&S Stranded

To Ground

18"

Floor Line

FIG. 106. Left hand side elevation of Fig. 104.

The top and base for the primary winding of the oscillation transformer are constructed as shown in Fig. 102.   Fig. 103 gives the dimensions of the drum for the secondary coil, which may be of wood.

The aerial ammeter is mounted on a board immediately behind the rotary spark and at such a height that the reading of the scale is directly visible.   The aerial change-over switch is similar to that shown in Fig. 98.

A representative *panel transmitter* designed by A. R. Zahorsky of New York City is shown in Figs. 104, 105, and 106.   The panel indicates first-class workmanship throughout and shows what may be done when mechanical as well as electrical details are given careful attention.   The over-all dimensions are given in Fig. 104.   They may have to be altered somewhat to suit the instruments and material which the amateur has at hand.   The sketches, however, suggest other designs that the builder may work out at his convenience.

The panel shown in the accompanying figures, may be made of *transite asbestos wood* from $\frac{1}{2}''$ to $\frac{3}{4}''$ thick.   This is the best kind of material to use, because it is cheaper than slate or marble, is more easily worked and is *fireproof*.   However, a neat panel may be built of 1″ oak or pine boards and given two coats of floor varnish.   The panel is mounted by means of wood screws on two, 2″ x 2″ wooden uprights, or on iron brackets braced 15″ from the wall.

For switching from a transmitting to a receiving position two single pole double throw high tension knife switches are employed.   The blades as shown in the side elevations, Figs. 105 and 106, are interlocked by means of *bell-cranks* and *levers*.   They are operated by a lever approximately 2′ long—7″ from the floor.   This places it in a convenient position for a man sitting in a chair.

The conductor for the oscillation transformer shown in Fig. 106 should be no smaller than No. 4 B&S stranded bare or insulated cable.   If none is available, the amateur may build up an equivalent cable by twisting together 27, No. 18 B&S bare wires, or 40, No. 20 B&S wires.   The secondary coil has two turns, the primary 5 turns.   The conductor for both primary and secondary of the oscillation transformer are wound on crosses made of pine boards impregnated with paraffine.   Both coils are supported so that they may slide back and forth on a square brass rod set into the front of the panel.   The coils are 10″ in diameter.

The condenser in the primary circuits is built up of thirteen $\frac{1}{4}''$ glass plates 12″ square.   Half the plates are coated with extra heavy tinfoil 9″ square applied to both sides.   Terminal lugs are placed on the case, one on the lower edge, the other on the side.   They are preferably made of thin sheet copper   In assembling the condenser a plain glass plate is placed between two coated plates to vary the capacity.   This method avoids taking taps from the helix, the more common method.

The *rotary gap* is one of the Marconi type of dischargers   The disc is made from a piece of red fibre $\frac{1}{4}''$ thick, 10″ in diameter and impregnated with paraffine.   It has eight $\frac{3}{8}''$ brass studs spaced equally around the circumference and fastened by means of a thin nut on each side.   These nuts are made thin by cutting a $\frac{3}{8}''$ brass nut in two with a hack saw.   The driving pulley should be about 3″ in diameter.   The disc and pulley are mounted on a piece of $\frac{3}{8}''$ drill-rod 6″ long which serves as a shaft.   The rotary gap should be driven by a small induction motor and run about 1800 r.p.m. giving 240 discharges per second. The size of the pulley on the motor depends upon the motor speed. A leather belt serves as a driving medium.

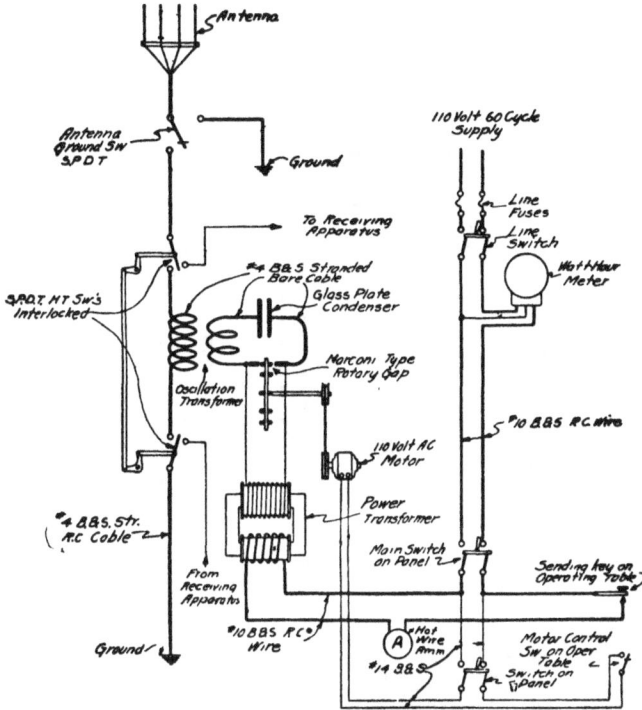

Fig. 107. Wiring diagram of the apparatus in Fig. 104.

The *hot-wire ammeter* on the left hand side of the panel should have flexible leads with spring clips so that it may be inserted in the aerial circuit when required. The other ammeter is placed permanently in the 110 volt supply line as shown in the wiring diagram Fig. 107.

The special features of this transmitter are the short leads in the oscillating circuits and the lack of sharp bends or kinks which would cause leakage at high voltages. All sharp corners on both conductors and insulators should be rounded off with a file. The insulators shown on the high tension knife switches may be of porcelain, electrose, or hard wood baked dry and then boiled in paraffine. High voltage insulation is necessary.

The diagram Fig. 107 does not show protective condensers. A protective unit consists of two one-microfarad condensers connected in series. The central wire is connected to earth and the two outside wires across the power line close to the meter. A protective spark gap might also be connected across the secondary of the high voltage transformer

A panel set of less expensive construction has been shown by H. R Hick of Connecticut of which the front and side views are given in Fig.

108. No. 1, on the front view, is an aerial hot wire ammeter; No. 2 is a shunt switch for cutting it out of the circuit when not in use. No. 3 is a control knob which, when pulled in and out, varies the coupling of the oscillation transformer.

A single blade antenna change-over switch is indicated at No. 4. The blade should clear the knob of No. 3. No. 5 and No. 6 are large size snap switches; the former closes the power circuit to the transformer, the latter starts and stops the rotary gap motor.

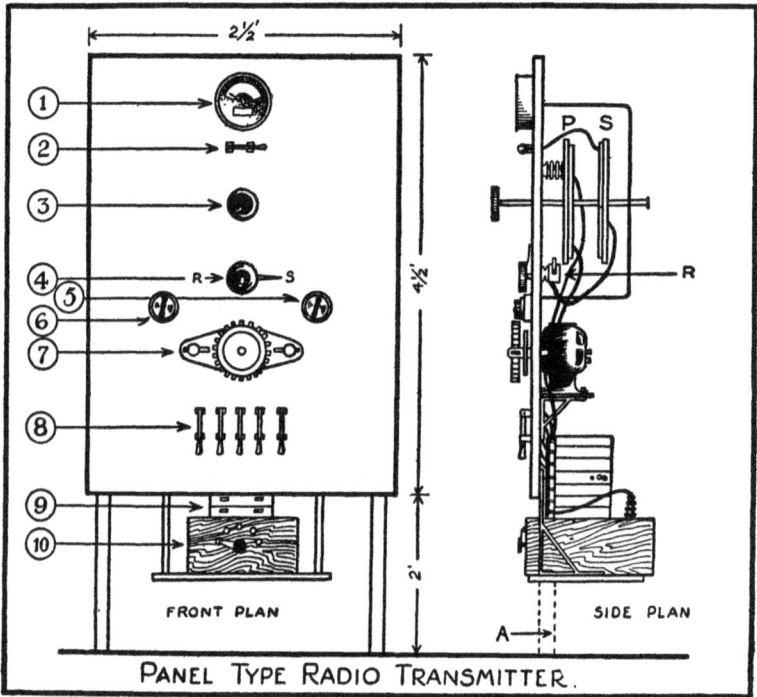

Fig. 108. Simple panel set designed by an experimenter

The driving motor for the rotary gap No. 7 is mounted behind the panel with the shaft extending through the panel. The switches No. 8 vary the capacity of the high voltage condenser which comprises several units of the moulded type. The high voltage transformer is indicated at No. 10. A variable input is obtained by the multi-point switch mounted on the transformer box.

Outlining the construction of a simple panel transmitter suitable for the amateurs financial resources and mechanical ability, C. S. Ballantine suggested the arrangement of Figs. 109 and 110. A quenched gap is employed. The oscillation transformer is one of the helical type (shown at *A* Fig. 110) and immediately above it is an aerial tuning coil *B*. The condenser at the bottom of the panel has two sections of the moulded type, the capacitance per section being 0.0017 mfd.

The source of high voltage is an induction coil fitted with a magnetic interrupter. The plan presented is equally applicable to an a.c. transformer set.

In Fig. 109, the condenser busbars are indicated at 3. The changeover switch and coupling adjustment handle are shown at 4, while 5 and 6 are three-point wave length changing switches; one for the primary turns and the other for the secondary turns. Taps from the two coils for *three suitable wave lengths* are brought to the contact points on the switch.

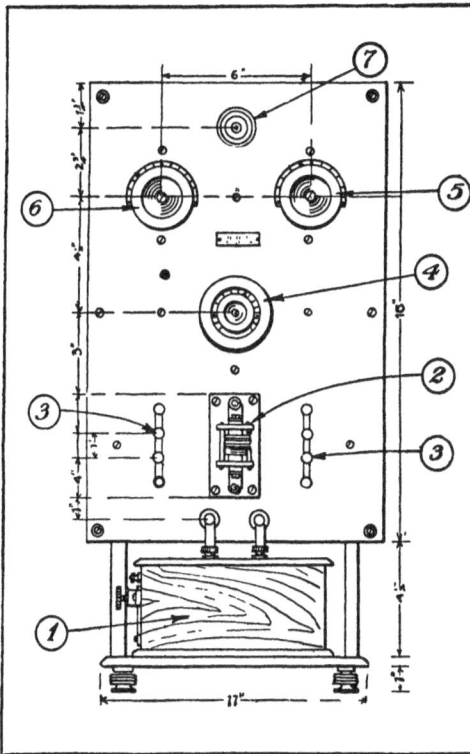

FIG 109   Panel transmitter using a spark coil and a small quenched gap

The notation 7 is a binding post for connection to the aerial lead-in wire. A small quenched gap of 4 to 6 plates suitable for spark coils is indicated at 2.

The secondary coil of the oscillation transformer is wound with ordinary flexible lamp cord, the equivalent of a No. 14 wire. The primary turns are of No. 8 wire. The details of the wave length changing switches shown in the side view of Fig. 110 are not very clear, but the fundamental idea of construction is treated in the following paragraph.

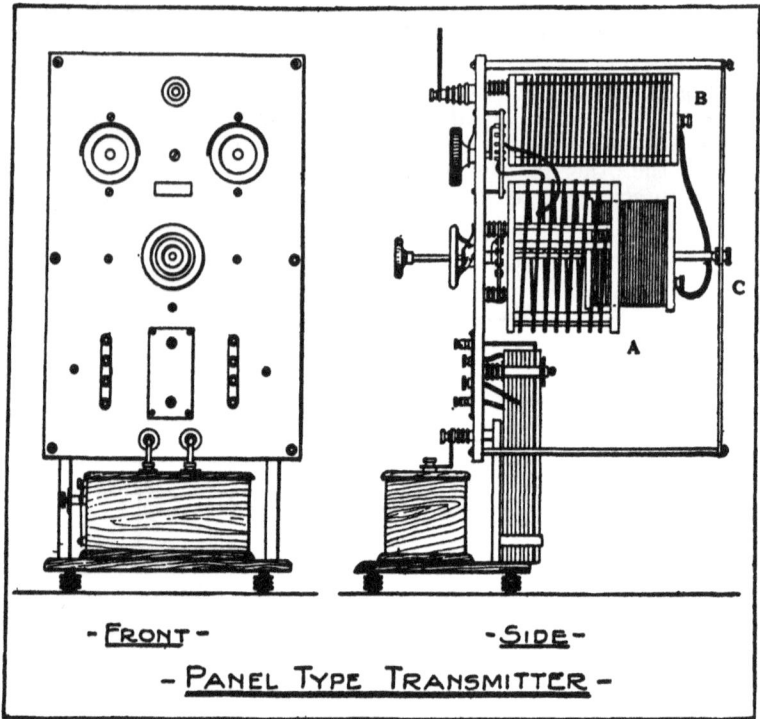

Fig. 110.—Showing the position of the wave length changing switch and other instruments in the set of Fig. 109.

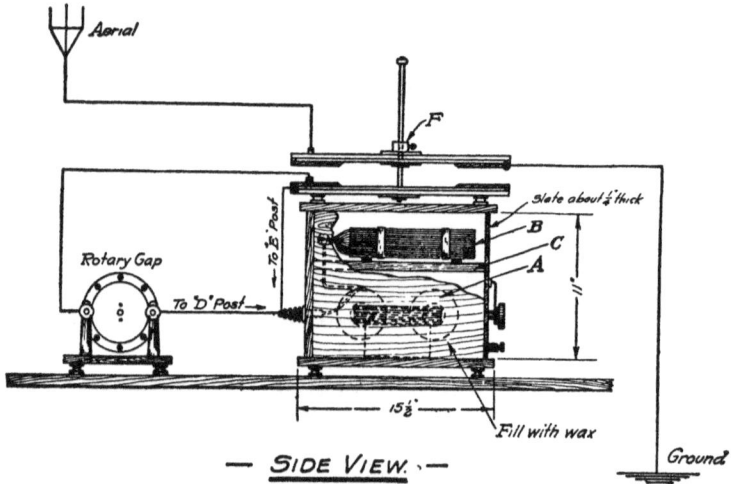

FIG. 111. A spark transmitting set compactly mounted.

Fig. 111 shows a compact spark transmitting set not of the panel type, designed by Roger G. Wolcott of Virginia. The cabinet has a slate front for the power control switches. A pancake oscillation transformer is mounted on the top. The position of the secondary is fixed by the set screw *F*.

A is a closed core transformer, *B* a high voltage condenser, and *C* a partition between. The rotary gap is mounted external to the cabinet. It should be placed nearby to keep the leads in the closed oscillation circuit at a minimum length.

WAVE LENGTH SWITCH FOR TRANSMITTERS.—It is a distinct advantage to fit a panel transmitter with a wave length changing switch. These switches are now commonly used in government and commercial wireless telegraph transmitters.

A mechanical switching arrangement of any type which will change the inductance or will instantly change the inductance and capacity of the open and closed circuits for several wave lengths, will permit the operator to use one of these wave lengths to "feel" his way through "jamming" in a way which the ordinary design does not afford.

It must be understood that for *each wave length* not only a *new value of primary and secondary inductance* must be found, but the *transformer coupling* must be changed as well. It might seem at first hand that two simple multipoint switches, one in the primary and the other in the secondary, would fulfil the requirements, if the taps on the oscillation transformer were correctly located for each wave length. But in addition

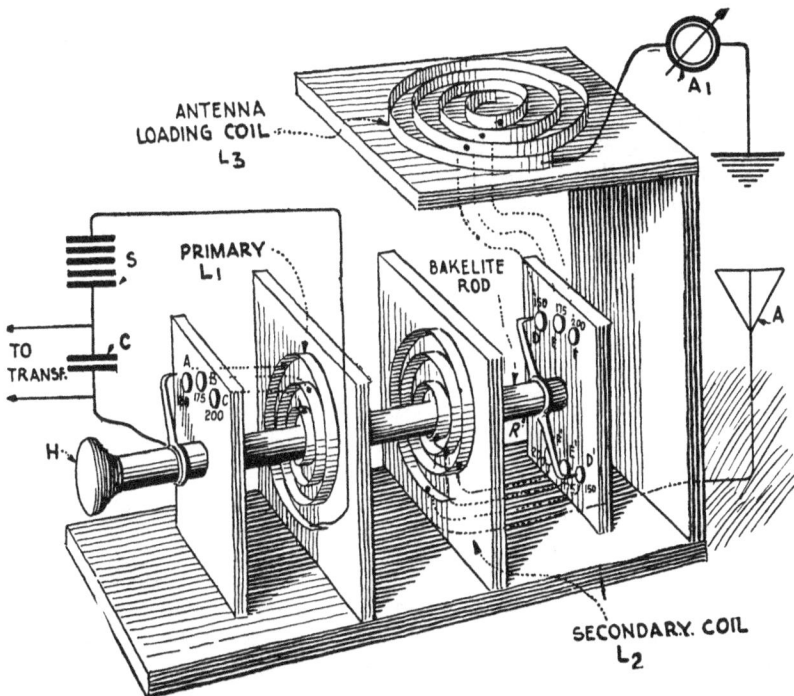

Fig. 112. Details and circuits of a wave length changing switch suitable for the amateur transmitting set.

to this the *transformer coupling* would have to be altered for each wave length and this would necessitate an added operation. However, if the scheme of switches and connections shown in Fig. 112 is used, the primary coil L-1 and the secondary coil L-2 may remain a fixed distance apart for all wave lengths and the coupling changed by varying the number of turns in the secondary L-2 and the loading coil L-3.

No mechanical details are given in Fig. 112. These are left to the builder. The object of the sketch is to show the basic principles of a wave length changing switch. L-1 and L-2 are spiral coils mounted parallel. The primary wave length switch is mounted in front of the primary L-1, and the secondary switch immediately behind L-2. The former has three contact studs while the latter has six.

From the switch points A B C of the primary switch, leads are extended to the primary inductance tapped on the coil at whatever points will give the desired wave length. The positions for the taps are, of course, found by a wave meter. The wave lengths for a 200-meter set might, for example, be 150, 175 and 200 meters and with the switching arrangement here provided the operator may use any of these at his discretion.

Note that for the 200-meter wave the secondary switch closes the circuit between coils L-2 and L-3, at the contact points F and F'; for the 175-meter wave at E and E', and for the 150-meter wave at D and D'. The switch points D', E', and F' are connected successively to an increasing number of turns in L-2; as are D, E, and F on L-3.

This affords a progressive increase of inductance as the wave length is raised although these cannot be said to be the exact relative positions of the contact clips on the oscillation transformer for maximum antenna current at the several wave lengths when in actual use. This will be explained presently.

The blades of the primary and secondary switches can be attached to a rod R controlled by the handle H which may extend through the front of the transmitter panel. This rod may pass through the center of the supports of the primary and secondary coils. It should be made of some good insulating material. The coil L-1 may be fastened rigidly to the panel, but L-2 should slide on the rod R. It is not necessary to provide a control handle to vary the spacing between L-2 and L-1 as this may be fixed when the correct position for L-2 is found. For quenched gaps the spacing is about 2″, but for rotary gaps it may have to be 4″ to 6″.

The contact studs of the wave length changing switches must be mounted on good insulating material. Sheets of *dilecto* or *bakelite* erected parallel to the coils will be suitable.

For the 200-meter set operated at the three wave lengths which we have already mentioned, the capacity of the condenser C cannot exceed 0.004 to 0.006 mfd. unless the connections in the spark gap oscillation circuit are extremely short. For it can be readily seen that if the capacitance of the condenser were 0.008 mfd. for example, at the shorter wave, (150 meters) substantially no turns could be cut in at the primary.

In regard to the use of the loading coil L-3 the experimenter should know that there are two ways of obtaining some desired degree of coupling between L-1 and L-2. *First, the spacing between L-1 and L-2 may be varied until maximum antenna current and a pure wave is secured; or, second, L-1 and L-2 may be placed fairly close to one another and the coupling varied by cutting out turns at L-2 and adding a similar amount of inductance at L-3 so as not to change the wave length of the complete antenna circuit.*

To summarize, if L-1 and L-2 are separated one or two inches and the coupling is too close for maximum antenna current cut out turns at L-2 and add an equivalent number at L-3. If the coupling is too loose, add turns at L-2, and take out an equivalent number at L-3.

It is now clear that the use of the loading coil L-3 permits variation of the transformer coupling without disturbing the spacing between L-1 and L-2. It is equally clear that the wave length switches must be mounted on the panel as near as possible to the radio frequency circuits in order that all connections may be of a minimum length.

The only way to find the correct number of turns to be cut in at L-2 and L-3 for each wave length is to try them out, in the meanwhile noting the reading of the aerial ammeter for maximum current The primary circuit should be calibrated by a wavemeter. This insures the correctness of the wave length of the antenna circuit when it has been tuned to the closed circuit. Just how far L-2 is to be placed from L-1 for its final position must also be found by experiment If the set is tuned according to the foregoing instructions the operator need only throw the switch in order to change the wave length. The advantages of such a switch are obvious.

BUZZER TRANSMITTING SETS.—The simple vibrating buzzer serves many useful purposes around a wireless station. A so-called "*test buzzer*" is a part of every receiving set.

For many years buzzers have been employed *to energize oscillation circuits.* Small buzzers have been used to transmit over distances of several miles. In fact, the author recalls distinctly a buzzer used in 1904 to transmit radio signals between two stations four miles apart. A *high power buzzer transmitting* set permitting transmission over 100 miles was developed during the war. Installed in an airplane the signals were distinctly readable 25 miles distant, using a simple crystal rectifier as a detector.

Fig. 113 shows the basic idea of the circuits. By giving heed to the following details, any experimenter with the necessary tools will be able to construct a vibrating spark transmitter that will give satisfactory communication up, to let us say, 50 miles.

The principle feature of Fig. 113 is the *buzzer* itself, consisting of a large *open core magnet* in front of which is mounted the vibrator spring S-1 carrying a silver contact C-1 about 1″ in diameter. The stationary contact C-2 is of the same material and of the same diameter. The spring S-1 should be rather stiff and if it is not an adjusting screw may be attached near the bottom to increase its rigidity. It is well to have adjustable spacing between the magnet core and the iron armature I.

For use with a 500-volt d.c. source the magnet should be wound with 2800 feet of No. 32 d.s.c. wire on a core about $5\frac{1}{2}$″ long and $1\frac{1}{4}$″ in diameter. The core is made of a bundle of No. 22 iron wires.

The condenser C should be of rather large capacity while L, the prim-

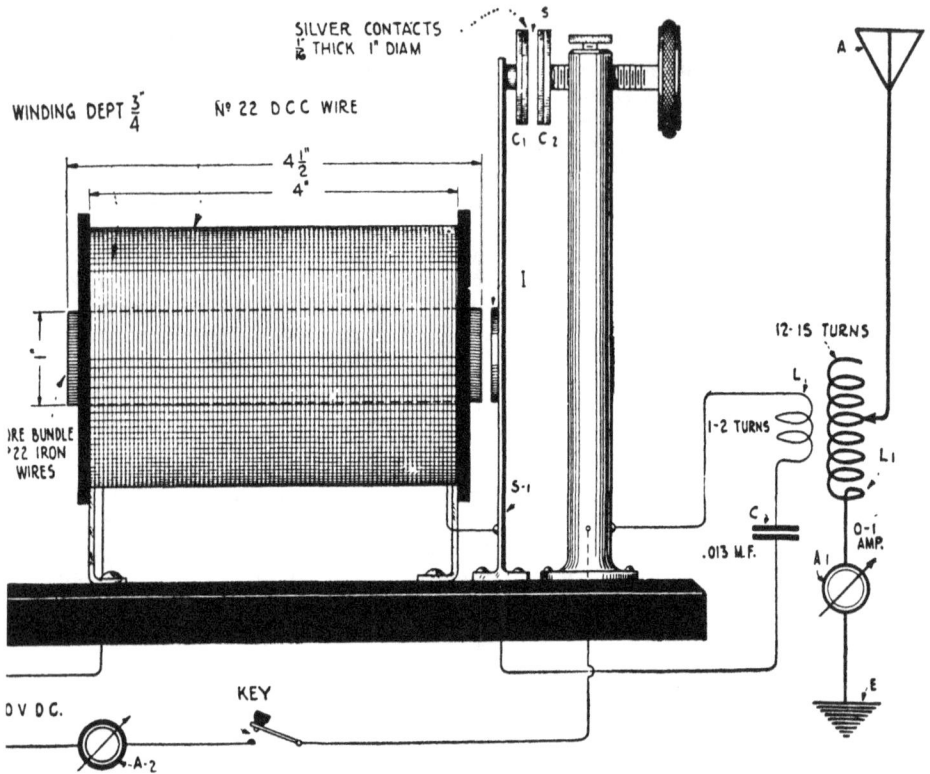

FIG. 113. Fundamental circuit and constructional details of a high power buzzer transmitter suitable for 20 to 100 miles communication. The source of power is either 110 v. or 500 v. d. c.

ary of the oscillation transformer need only have a single turn. It is claimed that the oscillations in the spark circuit are highly damped, so that *impact excitation* is obtained. It is claimed further that the primary and secondary circuits need not be in exact resonance, although better results are obtained if they are in resonance. The capacity of C should be changed for different wave lengths and the exact value for maximum antenna current found by trial. A capacitance of 0.013 mfd. is suitable for 200 meters.

When the vibrator is in operation an audio frequency discharge occurs between the electrodes C-1, C-2. Although the pitch of the note is not high it is fairly uniform. As the separation between the electrodes C-1 and C-2 increases (during vibration of the gap) the oscillations in the first part of the wave train have undamped characteristics but as the separation becomes greater their amplitudes fall off, giving the oscillation damped characteristics. The undamped characteristic becomes more prominent, at the shorter wave lengths. Successive trains do not fall to zero and hence the transmitting range, using receiving detectors suitable for damped oscillations, is materially reduced. For maximum signals some form of the *heterodyne* or *"beat"* receiver may be used.

Increased range may be secured with damped wave detectors by inserting a mechanically operated *"chopper"* in the antenna circuit of the transmitter. This should be constructed to interrupt the antenna currents about 200 times per second. Very good transmitting ranges may then be obtained by using crystal rectifiers or any type of oscillation detector suitable to *damped oscillations.* Some form of a commutator type of interrupter may be used as a "chopper."

When a 500-volt d.c. source is not available, good results may be obtained from 110 volts d.c. For operation at the latter voltage, the magnet in Fig. 113 is made of a soft iron core 1″ in diameter, 4½″ long, wound with No. 22 d.c.c. wire. The winding space is about 4″ long and ¾″ deep making the diameter of the magnet about 2½″. The vibrator is made of a piece of cold-rolled steel about 1″ wide, 5½″ long, and 1⁄16″ thick. The soft iron armature mounted on the spring is about 1″ in diameter and ¼″ thick. The silver contacts are 1⁄16″ thick and about 1″ in diameter.

The most expensive items in the buzzer transmitter are the silver contacts, which are sweated on to copper bases. This is a difficult job and is best done by a jeweler. The oscillating current in the closed circuit is approximately 40 amperes which explains the necessity for the large contacts.

The oscillation transformer secondary may consist of 15 turns of bare No. 12 B&S wire wound on a form 12″ in diameter. The primary may have *one* or *two* turns placed inside the secondary.

The 110-volt set takes about 0.75 ampere d.c. and the 500-volt set about 1 ampere.

This type of transmitter should prove useful for inter-city radio communication, though it is also practical for greater distances. The 75-watt buzzer set will give antenna current of about 1 ampere. The range is about 20 miles with crystal rectifiers at the receiving station. With a chopper in the antenna circuit of a 500-watt buzzer set, to gain a more effective modulation of the antenna currents, 200 miles have been covered.

CHAPTER V

CONSTRUCTION OF AERIALS AND MASTS

TYPES OF AERIALS.—The *inverted "L" flat top and the "T" type* are the most common of the four principal types of aerials in use to-day. The *"fan"* and *"umbrella"* types are rarely seen at amateur stations.

Phosphor bronze aerial wire is employed at commercial stations, because of its tensile strength, combined with a fair degree of conductivity. A single wire usually consists of 7 *strands of No.* 19 *wire* **or** 7 *strands of No.* 21. Hard drawn *copper* wire or *aluminum* wire may be used. Galvanized iron or steel wire has been employed but the resistance losses are rather high.

For *receiving work* two wires spaced 4 to 10 feet are sufficient, but for transmitting use 4 to 8 wires. The *fundamental or natural wave length* of transmitter aerials obviously must be *less than* 200 *meters* for some inductance must be inserted at the base to absorb energy from the spark gap circuit. As shown in Chapter III, pages 52 and 54, the flat top portion of an inverted "L", for operation at 200 meters, cannot exceed 100 feet in length; while for a "T" aerial, the length cannot exceed 140 feet. In either case the flat top portion with the lengths given, cannot be more than 40 feet above the earth. In any case it is better *to measure the natural wave length of an aerial with a wave meter.* If it is too long, the flat top can be cut off until the desired fundamental is obtained; or a *short wave condenser* may be employed to reduce the wave length. But this last is not recommended.

The same applies to the receiving aerial; for efficient reception at 200 meters its fundamental wave length must be less than 200 meters. Although very long receiving aerials cannot be used for the reception of short wave lengths a small 200 meter aerial may be employed to receive the longest waves; for because of the amplification obtainable from the vacuum tube very weak incoming signals, which would be obtained from such small aerials, may be brought up to audibility.

Fig. 114 shows the general design of an *inverted "L" aerial* supported between two masts on separate buildings. Fig. 115 shows a *"T" aerial* with the *lead-in wires* attached to the center of the *flat top.* The wires are strung on *spruce spreaders* 8' or 10' in length. Each wire has one insulator 12" to 24" in length at each end of the flat top. The ends of the bridle for each spreader are attached to eye-bolts at the ends of the spreader, and the middle of the bridle fastened to the lifting *halyard* which runs through a *pulley block* fastened to the top of the mast. The lead-in wires are connected to a *lead-in insulator* like that described on page 131.

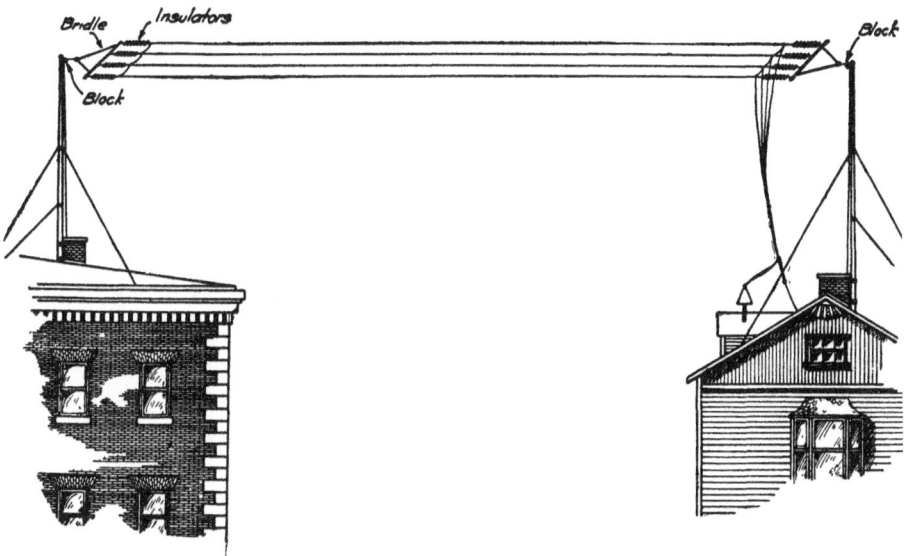

Fig 114.   An inverted "L" aerial suitable for the amateur station

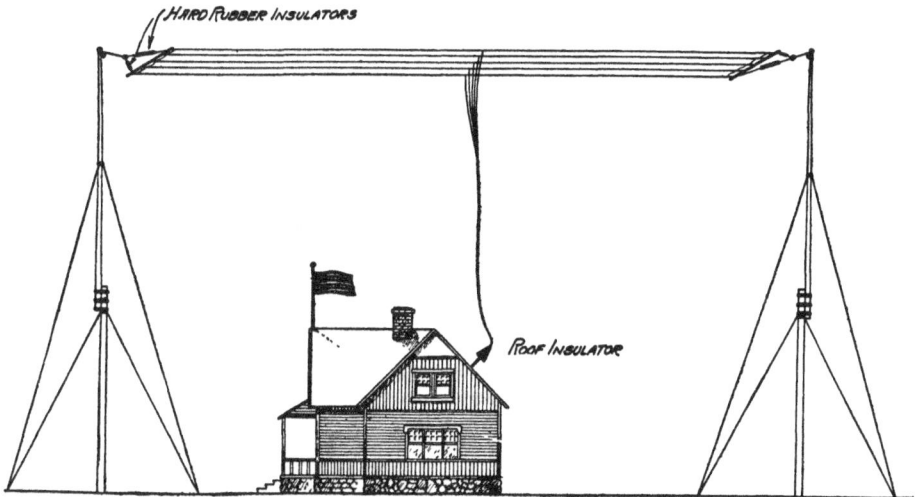

FIG. 115.   A "T" aerial which is generally used where it is more convenient to take the lead-in wires off the center of the flat top rather than off the end.

The lead-in wires should be "bunched"; that is, twisted into a cable. There is no advantage in using spreaders longer than 10'.

PRECAUTIONS.—If the transmitter operates at high voltages the flat top *should not be erected over power, light or telephone wires*, as high voltage radio frequency currents may induce disastrous potentials in such circuits.   The flat top should be swung just as far as possible from such electrical circuits.   If it is impossible to erect the antenna at a considerable distance from local power circuits try and place the antenna at right angles to them.

Only the *best grades of insulators* should be used in transmitting aerials.   Insulators made of cheap materials will leak the antenna currents in wet weather and thereby destroy the efficiency of the set.

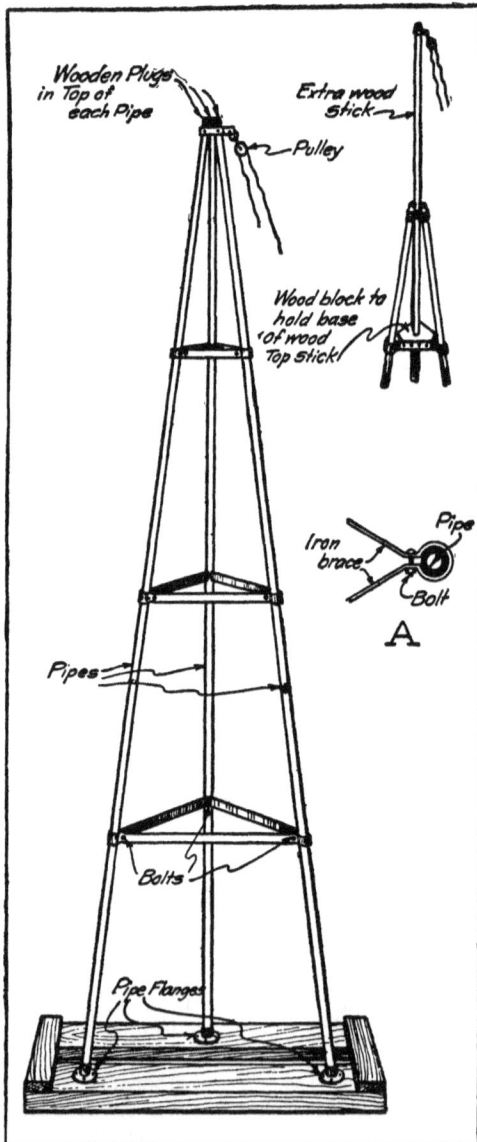

FIG 116.   Self-supporting triangular tower designed by an amateur experimenter.

MAST CONSTRUCTION.—The amateur is accustomed to attach his aerials to any high structure available. Often the antenna wires are seen stretched between a water tower and a tree, between the gable of a barn and the roof of a wireless cabin, or between two high trees.

In cities the roofs of apartment houses are dotted with flat top aerials, supported on *gas pipe masts* 20′ to 30′ in height. Any one can easily install, single handed, a couple of lengths of *gas pipe* joined by a *coupling* with a *pipe stand* fitted to the base. The pipe stand is bolted to heavy pieces of 4″ x 4″ timber, and further support given to the mast by iron wire guys attached to convenient fastenings on the roof.

The *triangular self supporting tower* in Fig. 116 is neat in appearance and very desirable if there is insufficient room for guy wires. Robert Kennedy of New Jersey suggested the design. The essentials of the design will be evident at a glance. Three lengths of 1″ gas pipe are separated at their base by one-fifth their length. They are bound together at equi-distant spaces, by an iron strap rigidly attached to each pipe as is shown in the detail *A*. The strap passes part of the way around the pipe. A bolt is inserted and screwed up tightly as shown. In order to increase the height, a *wooden top-mast* may be fastened to the top. This is supported by a *piece of heavy wood* placed at the first brace from the top as is shown in the upper right hand part of the drawing.

As an alternative, fit a *pipe stand* to the base of the top-mast and fasten it to the wood base by lag bolts. Attach an iron band with an eye to take the pulley block to the extreme top, as shown in the illustration.

Allan Lawson shows a simple method of supporting an aerial between a wooden mast on a building and a tree. His designs are presented in Figs. 117 and 118. The mast is supported solely by guys and this construction is recommended where it is not permissible to cut holes in the roof. The mast is built of 2″ x 2″ timber. For a height of 30′, 6 pieces, 10′ in length are required. The legs are separated 8′ at the base. They rest on 1″ boards and are fastened to them by *cleats*.

Wood cross pieces made of one inch board are placed at regular intervals to brace the structure. Further bracing is secured by the cross wires which may be pulled taut by turn-

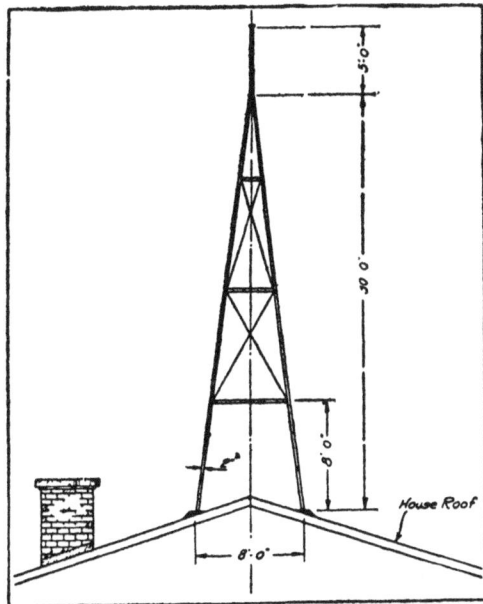

FIG 117. Showing the construction of a mast which can be erected on the roof of a building without boring holes for support of the base.

buckles. Two guy wires are fastened to the top and two more at the middle. These are then fastened to eye-bolts inserted in the roof.

Fig. 119 shows the amateur how he may erect a wooden mast with the aid of his neighbors. H. E. Lange of Missouri suggested this method. The particular mast to which this process of erection was applied was 54' high and built up in three sections. The bottom section was made of 3" x 3" stock, the second section of 3" x 2" stock and the top section

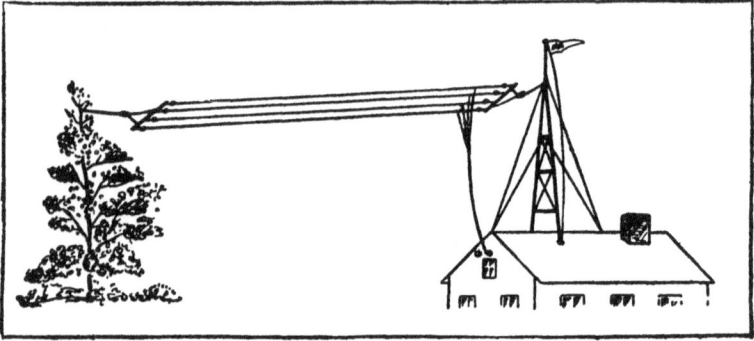

Fig 118. Showing the completed mast and antenna.

2" x 2¼". Although the sections are bolted together the strain is taken off the bolts by cutting away the overlapping part of the lower section so that the weight of the section above rests on the section below it. Beginning at the bottom, three sets of guys are placed 17, 18, and 19 feet apart.

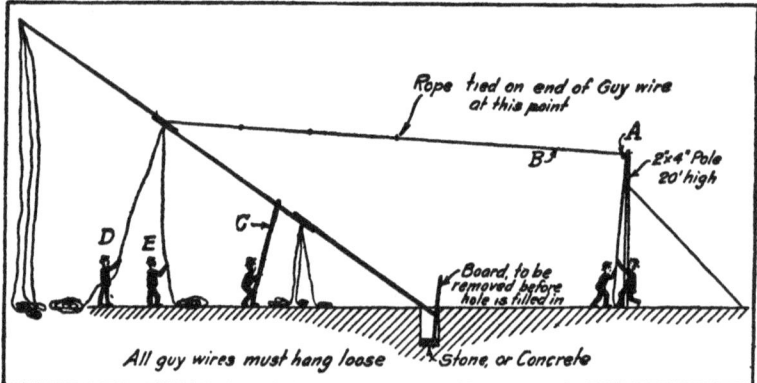

Fig 119   One experimenter's method of erecting a "wireless" mast

B shows the halyard for pulling up the mast. It is attached to a guy wire and the rope goes through the pulley *A*. In order to prevent side sway the men stationed at *D* and *E* hang on to the mast guys. While the men at *A* are pulling on the halyard, the man at *C* is inserting a prop to relieve them between pulls. The men holding the guys at *D* and *E* must stand back at a distance so that the mast will not topple in the opposite direction when it drops into the hole.

Fig. 120 shows a 70 foot *gas pipe mast* of unique construction designed by Geo. H. P. Gannon of Massachusetts. The complete specifications appear in the sketch. The bottom section is made of 2″ gas pipe, successive sections are of lesser diameter until for the top section the diameter is ¾″. The mast is braced by steel telegraph wire and turnbuckles

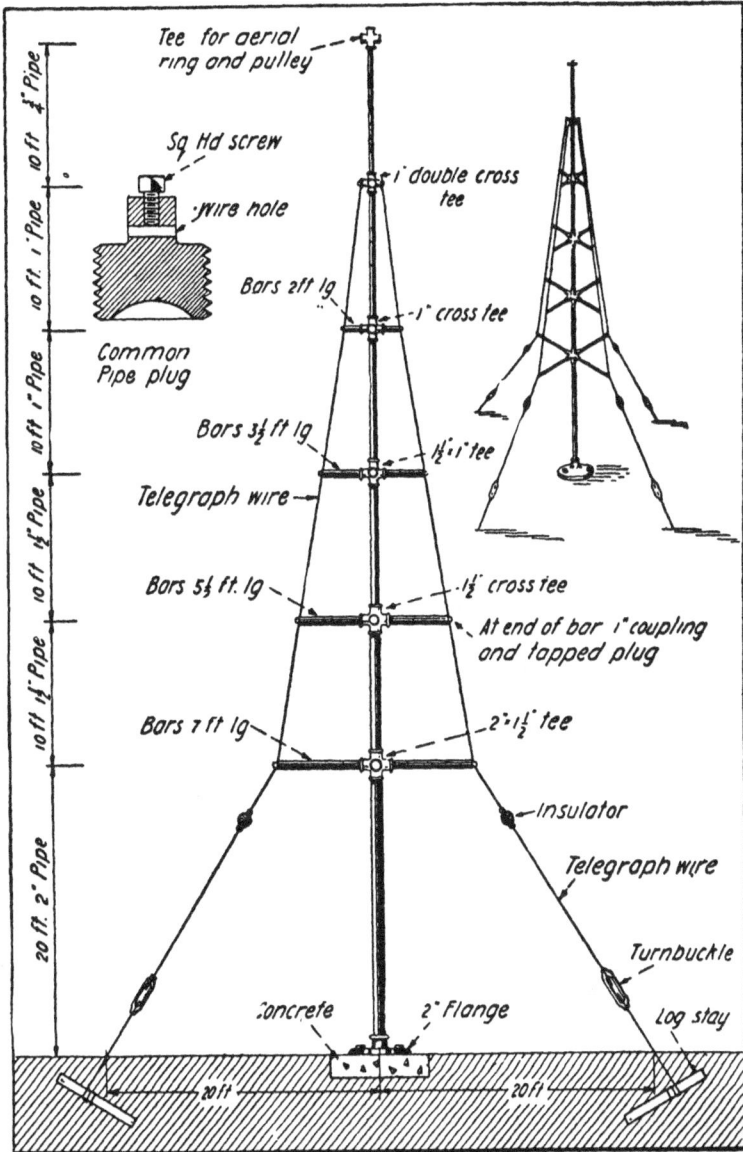

Fɪɢ 120   A 70 foot mast constructed of gas pipe

inserted in the guys at the base as shown.   The drawing makes clear
the remainder of the construction.

Fred Jameson of Kansas erected a 110-foot mast of the type shown
in Fig. 121.   It is built up of discarded *steel boiler tubes* which may be
secured from junk yards at small cost.   The tubing weighs about 70 lbs.
per 14-foot section.   In Fig. 121, No. 1 is a detail of the sectional joints;
No. 2, a cut away portion of the topmast; No. 3, a detail of the base, and
No. 4, the completed mast.   Fourteen foot lengths of $3\frac{1}{2}''$ tubing are
most desirable.   The connecting sleeves are made from a discarded piece
of water pipe 20 feet long, cut into 7 sections.   If $3\frac{1}{2}''$ boiler tubes are
used the sleeve should be $3\frac{1}{2}''$ water pipe, as boiler tubes are measured
on the outside and water pipe on the inside.

The guy wires are made of No. 8 or No. 10 galvanized wire.   The
completed mast weighs about 700 lbs.   Precautions must be taken
during the erection.   First bolt the two top sections together and then

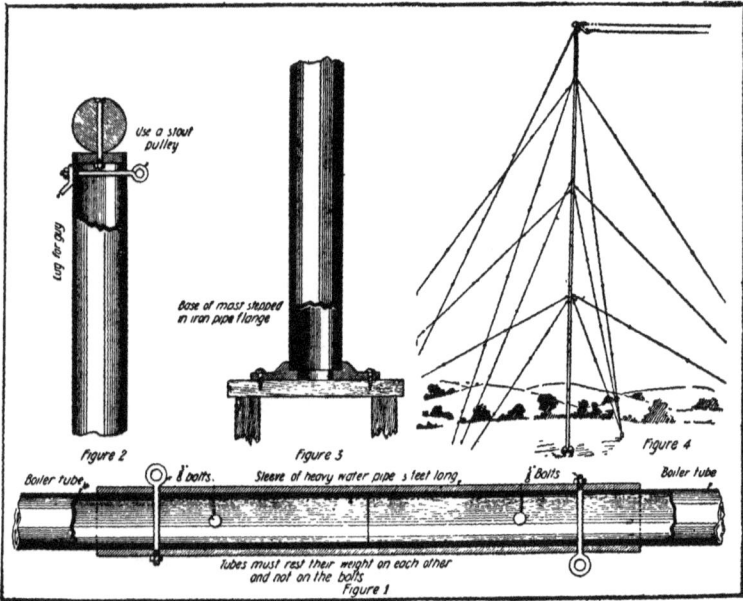

Fig. 121.   An aerial mast 110 feet in height designed and erected by an ambitious experimenter.   It was
made of discarded steel boiler tubes and water pipes.

pull them up in a perpendicular position by means of a 30-foot gin pole
erected alongside the base.   Then raise the two top sections high enough
so that the next lower section can be bolted to the bottom and so on.

The guy wires are attached before the erection and are "payed out"
through eye-bolts fastened to "dead men" buried, about 40' from the
base.   A helper must be stationed at each "dead man" to pay out the
guy ropes slowly as the mast goes up so that it may be held in a strictly
vertical position.   This is a "ticklish" job and should not be undertaken
by any one without previous experience in this line of work.   Each guy
wire is insulated from the earth by ordinary porcelain knobs inserted in

the guys near the "dead men." To raise the mast a one-ton hoist is necessary.

Fig. 122 shows another way of raising a mast with a *gin pole* set alongside the hole for the mast. The method is clear from the sketch. All guys are attached previous to the erection. To prevent it from snapping off, the mast can be strengthened, by the brace *F*, over which is passed a cable fastened to both ends of the mast.

ANTENNA INSULATORS.—A cheap and simple *antenna insulator* with sufficient insulating qualities for transmitting aerials is shown in Fig. 123. It is made up of porcelain cleats, porcelain knobs and iron bolts as shown. The drawing shows two parallel cleats of the ordinary type separated by two porcelain knobs. The aerial wires are fastened to the bolt on the end, farthest from the spreader. All the necessary material can be purchased at any electrical supply store. The drawing explains the assembly of the insulator and the method of attaching the antenna wires. If the strain on the wire is excessive, on account of a lengthy span, the insulators may be strengthened by using two cleats on each side instead of one.

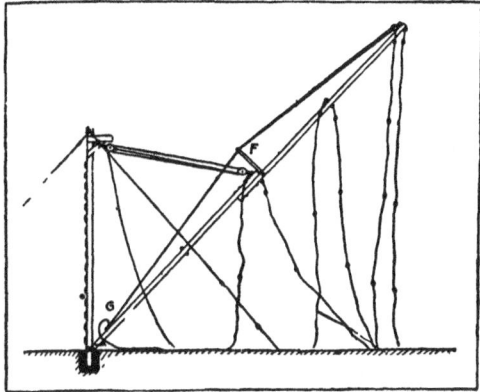

FIG. 122. Showing the use of a "gin" pole in raising a mast.

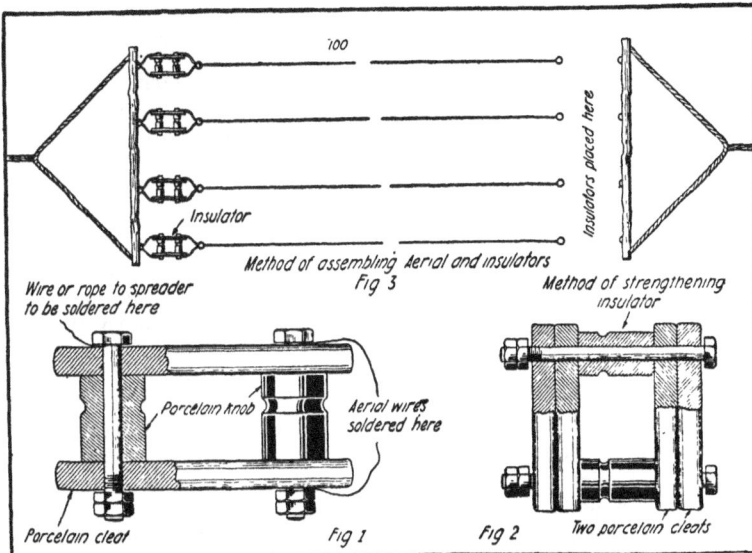

FIG 123   Constructional details of a home-made antenna insulator made up of common porcelain cleats and knobs

CHAPTER VI

RECEIVING TUNERS AND OSCILLATION DETEC-TORS—GENERAL THEORY OF OPERATION—PREFERRED CIRCUITS—GENERAL DE-TAILS OF CONSTRUCTION

GENERAL CONSIDERATIONS.—There is a certain similarity between receiving and transmitting circuits in wireless telegraphy. Take for example, an *inductively coupled transmitter:* The closed circuit generates radio frequency currents which are passed on to the open or aerial circuit from which a wave motion is propagated. In an *inductively coupled receiving set* the antenna or open circuit absorbs a certain amount of energy from the passing wave, and passes the resulting currents to the closed circuit where they are detected and, in some cases, amplified. If at the transmitter an oscillation detector is substituted for the spark gap, and the secondary of the high voltage transformer is replaced by a telephone receiver the transmitting radio frequency circuits will function as a receiver. With this change of apparatus, the same circuit may be used for both purposes. This, however, is not done in practice.

The necessity for resonance between the closed and open circuits of the transmitter has already been pointed out. Similarly, if any useful results are to be expected, the closed and open circuits of the receiver must be *tuned* to each other and in electrical resonance with the frequency of the wave motion radiated by the transmitter. The process of obtaining resonance is called *tuning* and in receiving circuits maximum signals are obtained only when the receiver circuits are accurately *tuned* to the transmitter.

The ordinary telephone receiver is used to translate radio frequency currents into audible signals. Certain facts about the operation of the telephone will be reviewed. Though it might be thought that since the telephone receiver will indicate the passage of very feeble currents it may be connected in series with the receiving aerial to detect the flow of radio frequency currents, the fact remains that the telephone diaphragm cannot follow the extremely rapid variations of radio frequency currents of the higher frequencies. Even if this were possible no sounds would be detected by the ear, since the average ear is unresponsive to vibrations above 16,000 per second. The ability of some ears to detect sound vibrations fails at frequencies considerably below that value—at, say 8,000 vibrations per second.

160

When a receiving antenna is tuned to a 200-meter transmitter the frequency of the antenna currents is 1,500,000 cycles. The lowest frequency employed for transmission and reception up to date is about 15,000 cycles; the highest about 3,000,000 cycles. The first value is about on the boundary line between an audible and inaudible current. Since the ear will not respond to such high frequencies and since the average telephone responds best to lower frequencies—between 500 and 1000 per second—it is necessary to modify the wave form of radio frequency currents into a form suitable for maximum response in the telephone receiver.

One way of making damped oscillations of radio frequency audible in the head telephone, is to rectify the currents into d.c. pulses. There are several *minerals* and *compounded crystalline substances* which have the ability to rectify high frequency currents. When used in this way they are called *oscillation detectors*. It cannot be said that they detect anything for what they really do is convert radio frequency currents into uni-directional currents to which the telephone will respond. To obtain response in a telephone the amplitudes of the rectified radio frequency currents must be modulated at audible frequencies. In spark transmitters the antenna currents are generated in audio frequency groups determined by the number of spark discharges per second. The telephone is then impulsed once for each spark discharge.

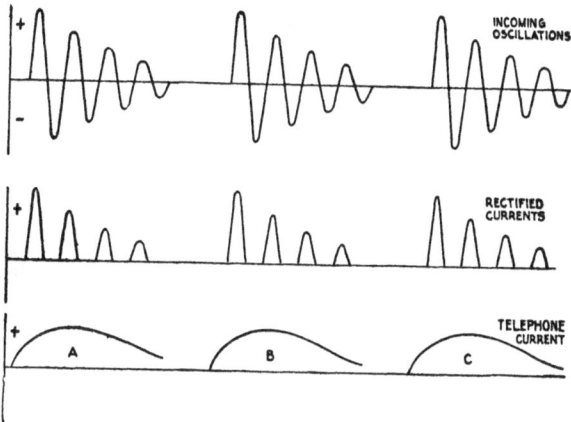

Fig 124   Graphs outlining the phenomena involved in the detection of damped oscillations by rectifiers

The curves of Fig. 124 show the phenomena of rectification and reception. The upper curve shows three groups of incoming radio frequency oscillations. The second curve shows the uni-directional currents resulting from rectification, and the lower curve shows the telephone current. The *rectified pulses* of each group in the second line are, of course, *radio frequency pulses* to which the diaphragm cannot respond. The pull on the diaphragm is the result of a sort of an average effect of each group of pulses as shown by the lower graphs.

Then if the amateur's transmitter has, let us say, a non-synchronous gap giving 240 sparks per second, 240 groups of radio frequency currents

flow in the antenna circuit of the transmitter; a similar number are induced in the receiver circuits where they are rectified and the diaphragm
of the telephone will be impulsed 240 times per second.   The listener
will hear in the telephone a note of the same pitch that he would hear
were he standing beside the transmitting spark.

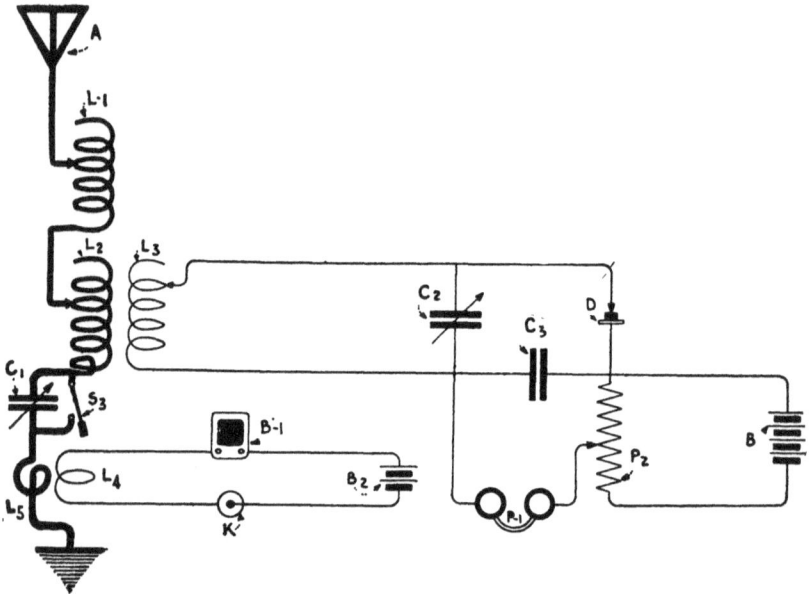

FIG 125   Fundamental wiring diagram of the inductively coupled receiving set, often called the 'two
circuit'' receiver. The diagram shows the circuits of a buzzer tester for pre-adjustment of the oscilla
tion detector.

THE INDUCTIVELY COUPLED RECEIVER.—We must next con-
sider the fundamental circuits of an inductively coupled receiver utilizing
a *simple rectifier* as an oscillation detector.   The fundamental diagram
is shown in Fig 125 L-1 is the aerial tuning inductance, L-2 the primary coil of the tuning transformer, C-1 the short wave condenser, A the
antenna which is connected to earth at the lower end.   These are the
elements of the *open circuit.*

The *closed circuit* comprises the secondary of the tuning transformer
L-3 and the shunt variable condenser C-2. D is a carborundum rectifier
and C-3 a telephone condenser. B is a battery of 2 to 4 volts, P-2 is a
400 ohm potentiometer, and P-1 are head telephones of 2000 ohms resistance

The tuning elements for establishing resonance with the transmitter
are L-1, L-2, L-3, C-1, C-2   The circuit of Fig. 125 operates practically
as follows: If the distant transmitter radiates at 200 meters the
frequency of the radiated wave motion is 1,500,000 cycles per second.
As the antenna A is in the path of the wave it is acted upon by the static
and the magnetic components of the electric wave which tend to induce
in the antenna system oscillations of the frequency radiated by the

transmitter. It is hardly possible that the antenna circuit of the receiving system will have the correct values of inductance and capacitance for resonance with the radiated wave; therefore, resonance must be established by the tuning elements shown in the diagram.

We have seen in connection with the transmitter that the tuning of a radio frequency circuit to a given impressed frequency resolves itself into the problem of making the *inductance reactance and capacitance reactance equal to a particular frequency.* And that is just what the operator at the receiving station does in order to establish resonance with the transmitter. By varying L-1, L-2, or C-1, he makes the total inductance reactance of the antenna equal to the total capacitance reactance for a particular frequency, say of 1,500,000 cycles. This has been accomplished when maximum signals are obtained in the head telephone.

Assume now that the antenna circuit is in resonance with the transmitter; the antenna current has reached a maximum, and it now remains to be "detected." The coil L-3, the secondary, is brought into inductive relation with the primary L-2, and consequently an e.m.f. of radio frequency is impressed across L-3. By varying the capacitance of C-2, a value can be found that will just neutralize the inductive reactance of L-3, and the current in the closed circuit will reach a maximum. In other words both the closed and open circuits are in resonance with the transmitter.

Referring now to the detector and associated apparatus, note that the current from a local battery B flows through the rectifier D, through the secondary L-3, through the telephones P-1, through the contact on the potentiometer P-2 back to the negative side of the battery. The object of the battery is to increase the efficiency of the detector. The detector is a much better conductor of electricity flowing in one direction than in the opposite direction. Assume then that the current from battery B flows through the crystal in the direction in which it is the better conductor and the detector is subjected to potential differences at radio frequencies. Due to a peculiar property of the *carborundum rectifier* it can be demonstrated that when the e.m.f. due to the incoming signal is in the same direction as the battery e.m.f. there will be a large increase of current in the circuit which includes the head telephone. When the e.m.f. of the incoming signal (for the next half oscillation) opposes the battery e.m.f., there will be a small decrease in the normal current through the head telephone. Since the current increases to a greater degree than it decreases, the *preponderance of current*, during a group of incoming oscillations, *flows in one direction.* This in effect amounts to a direct current to which the head telephone readily responds.

Each spark at the transmitter induces a group of radio frequency currents in the receiver; and each group, according to the actions outlined above, sends a uni-directional current through the telephone. Hence if the transmitter produces 240 sparks per second, the diaphragm in the telephone will be pulled 240 times.

The *aerial tuning inductance* L-1, in Fig. 125 is required only when very *long waves* are to be received on short aerials. For wave lengths up to two or three times the fundamental of the receiving aerial the necessary antenna inductance may be included in the primary L-2.

The *short wave condenser C*-1 is used to establish resonance with wave lengths below the fundamental wave length of the antenna. The variation of this condenser necessarily gives a small range of wave lengths, for it is not possible to reduce the fundamental wave length of antenna by quite one-half with a series condenser. As an illustration if the fundamental wave length of an aerial is 600 meters about the lowest wave length that can be obtained for practical working is about 350 meters.

On the other hand an antenna may be loaded by inductance to many times its natural wave length. An aerial whose fundamental is 300 meters may be loaded to 20,000 meters and will give good signals from high power stations if a multi-stage vacuum tube detector set is employed.

THE USE OF A BUZZER TESTER.—The adjustment of an oscillation detector of the "contact" type depends largely on skill gained through practice. As shown in the following designs these crystals are usually mounted in a container which is one terminal of the detector. The opposite terminal is a *sharp pointed contact* which presses against the crystal and is designed so that contact can be effected with practically any spot on the surface of the crystal.

The spot on the crystal which gives the loudest signals in the head telephone from a given transmitter is called the "most sensitive" point of contact. This may be found by trial while receiving from a distant station; but the operation is facilitated by a *"test buzzer"* the circuits for which are shown in Fig. 125.

A common electric buzzer B-1, energized by a battery B-2, includes in its circuit a push button K and a coil L-4. The latter is placed in inductive relation to the antenna coil L-5. Each coil consists of a few turns of bell wire wound on a spool 2″ in diameter.

When the buzzer is active an e.m.f. of audio frequency is impressed across L-5, setting the antenna circuit into oscillation at whatever frequency it happens to be adjusted to. By coupling L-3 closely with L-2 radio frequency currents are impressed upon the detector D. The operator then varies the position of the point on the crystal, trying varying pressures on the contact point and simultaneously adjusting the position of the contact on the potentiometer P-2, until the loudest possible signals are obtained in the telephone from the buzzer. Care must be exercised to send the battery current through the crystal in the right direction. The proper polarity is readily determined by experiment.

TUNING THE RECEIVER.—If the receiver has been calibrated by a wavemeter it is an easy matter to tune a receiving set to any transmitter. If a table of wave lengths is not provided the distant station must be "found" on the tuning box by trial. The closed circuit can be calibrated in the factory but since the primary coil will be used with aerials of different values of inductance and capacitance the antenna circuit cannot be calibrated until the set is installed.

If the set is calibrated by a wavemeter a table of wave lengths corresponding to the dials on all the tuning appliances can be prepared to which the operator may refer from time to time.

If a receiving tuner is not pre-calibrated resonance with a distant transmitter may be established by taking progressively the following steps. Referring to Fig. 125:

(1) Set condenser C-2 at zero.

(2) Set coil L-1 (if used) at zero.

(3) Close the switch S-3 around the short wave condenser.

(4) Couple the coil L-3 closely to L-2 using half of the turns of L-3.

(5) Set the buzzer in operation by closing the push button K.

(6) Adjust the pressure of the contact point on the detector and vary the position of the slider on the potentiometer until loud signals are obtained.

(7) If it is definitely known that the transmitter is in operation vary the inductance of L-2 until signals are heard.

(8) Then reduce the inductance of L-3, add a little capacity at C-2, and try various couplings between L-2 and L-3 for louder signals.

(9) If the signals are weak move the potentiometer slider to ascertain if still better signals can be obtained.

(10) If the fundamental wave length of the receiving aerial exceeds the wave length of the transmitter, cut in the short wave condenser C-1. Vary its capacitance until the desired station is heard.

(11) If the wave length of the distant transmitter exceeds the fundamental wave length of the receiving antenna with all the turns of coil L-2 cut in, add turns at L-1 until the desired signal is heard.

To reduce interference proceed as follows:

(1) Reduce the coupling of the tuning transformer by drawing L-3 away from L-2.

(2) Then increase the capacity of C-2.

(3) If the interfering signal still remains cut condenser C-1 in the circuit, and add turns at L-2 or L-1 to maintain resonance.

If the interfering signal cannot be eliminated in this way it indicates that:

(1) the interfering signal is a highly damped wave or,

(2) that the interfering station operates on the same wave length as the station it is desired to receive.

Crystal rectifiers will not hold a "sensitive" adjustment indefinitely, hence frequent use of the test buzzer is necessary to maintain communication.

OTHER METHODS OF COUPLING.—While the inductively coupled tuner is generally preferred in radio work there are other methods of coupling the detector to the antenna circuit.

Fig. 126 shows the *plain aerial connection*. The detector D is connected directly in series with the antenna circuit. This is an inefficient method for the resistance of the detector introduces losses of energy and prevents the establishment of sharp resonance with the transmitter.

Electrostatic or *capacitive coupling* between the closed and open circuits is shown in Fig. 127. Coils P and S are coupled through the variable condensers C-1 and C-2. Variation of their capacity is said to vary the transfer of energy from P and S. Condensers C-1 and C-2 are operated simultaneously from a single control handle. They may

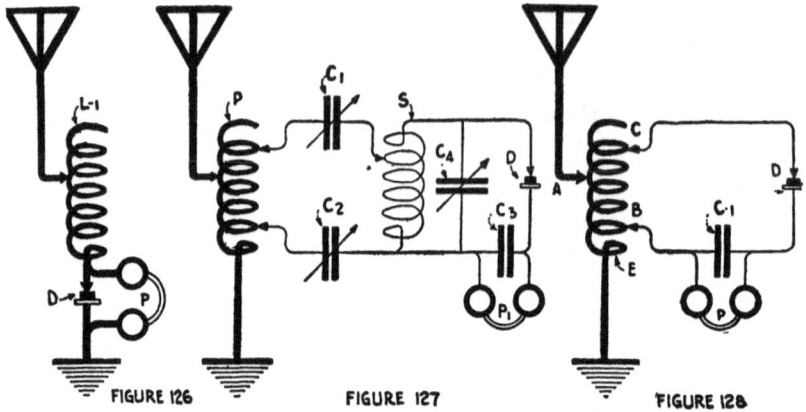

FIGURE 126   FIGURE 127   FIGURE 128

Fɪɢ 126. "Plain aerial" receiving system—perhaps the simplest detection circuit that is operative.
Fɪɢ. 127. Two circuit receiver with "electrostatic" coupling used extensively by the U. S. Navy
Fɪɢ. 128   Receiving circuits utilizing direct or conductive coupling. An auto-transformer is used as the coupling element.

be of fixed capacitance for a limited range of wave lengths. In fact one of the condensers is sometimes eliminated and the other used to vary the coupling. The chief advantage of this circuit is that it permits a more rapid change of coupling than the usual tuner mechanism where the secondary is drawn in and out the primary.

Direct or conductive coupling is shown in Fig. 128. A single coil is used as an auto-transformer. Tap *A* varies the antenna wave length, while taps *B* and *C* vary the frequency of the detector circuit and permit the coupling between the antenna and detector circuits to be varied. For instance if the turns between taps *B* and *C* are cut in at a distance from those between *A* and *E*, loose couplings are secured just as in the inductively coupled receiver. Just as good signals will be obtained with this method of coupling as with the inductive coupling, but obviously, it is less difficult to change the coupling of the inductively coupled set than the conductively coupled set.

The beginner is advised to practice with the circuit of Fig. 128.

RECTIFIERS OF RADIO FREQUENCY CURRENTS.—There are many minerals and compounded elements which will rectify radio frequency currents. Crystals of *galena, silicon, iron pyrites and carborundum* are most common.

Zincite in contact with bornite is a good combination and *molybdenite* is often used while many recommend a piece of *graphite* in contact with *galena. Carborundum* is generally preferred. These crystals are rugged and tend to hold their adjustment over long periods. They are sufficiently sensitive for several hundred miles working in commercial practice.

It is well to remember that carborundum is not a native mineral, but a product of the electric furnace. It is a combination of sand, salt, sawdust and coke. The finished crystal is known to chemists as carbide of silicon. Galena (lead sulphite Pb S), iron pyrite (Fe S_2), bornite (3 Cu_2 Fe$_2$ S_3) and molybdenite (Mo S) are natural sulphides. Zincite (Zn O) is a natural oxide.

The majority of these crystalline elements differ from carborundum in that the opposing contact must bear on the crystal with very light pressure. Hence we have the familiar "cat whisker" detector. With carborundum, on the other hand, it is possible to apply very heavy pressure.

Some crystals of silicon, when ground down to a polished surface on an emery wheel, will withstand a fairly heavy pressure at the opposing contact and remain sensitive to incoming signals. Generally, however, this does not hold true, particularly in the case of galena which on account of the lightness of contact is difficult to adjust to a sensitive condition and to maintain in a sensitive condition for an indefinite period.

Suitable holders or "detector stands" for various crystals will presently be described.

CHARACTERISTIC CURVE OF THE CARBORUNDUM CRYSTAL.

—It is interesting to study the phenomena exhibited by carborundum crystals under various d.c e.m.f.'s. The characteristic curve shows that the resistance of the crystal changes with the current flowing through it; i.e., a circuit with such a crystal in eries, does not obey Ohm's law. This can be demonstrated by the experiment outlined in Fig. 129.

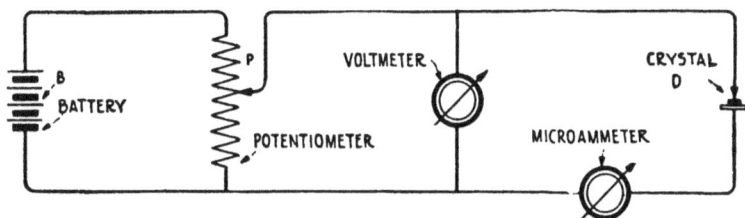

FIG. 129. Circuit and apparatus for obtaining the characteristic curves of crystal rectifiers. The data is obtained by noting the readings of the microammeter at several voltages

A battery B of 4 to 6 volts is shunted by a 400-ohm potentiometer P. D is a carborundum crystal in the circuit of which is connected a low reading voltmeter and a microammeter. The impressed e.m.f. is varied progressively from 1, to say, 4 volts, and the readings of the microammeter noted at successive steps. When the data obtained in this way are plotted in the form of a curve they may have the appearance of Fig. 130. It is seen that as the voltage is increased from 1 to 2.5 volts, the current rises slowly, but for potentials in excess of that value, the current rises rapidly, showing that the resistance of the crystal has dropped rather suddenly. The bend in the curve is approximately where the change takes place and it is here that the crystal usually functions best as an oscillation detector.

Referring now to Fig. 125, assume that contact on P-2 is adjusted to correspond to the lower bend on the curve*: Assume further, that the e.m.f. of radio frequency impressed by the incoming signal has a potential of 0.5 volt. If the steady e.m.f. is 2.5, the potential across the crystal

---

*The large values of current indicated by these curves were purposely selected to make the phenomena more appreciable

will vary from $(2.5+0.5)=3$ to $(2.5—0.5)=2$. The current through the telephone will change from the normal of 25 microamperes to 150 microamperes when the potential difference is 3 volts, and will drop to 20 microamperes when the potential difference is 2 volts.

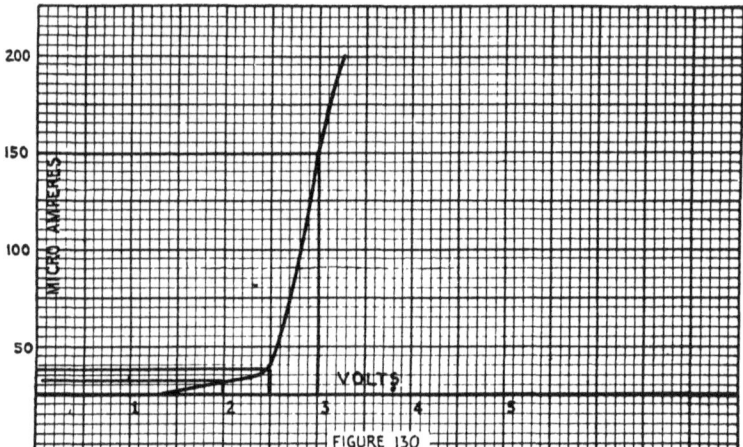

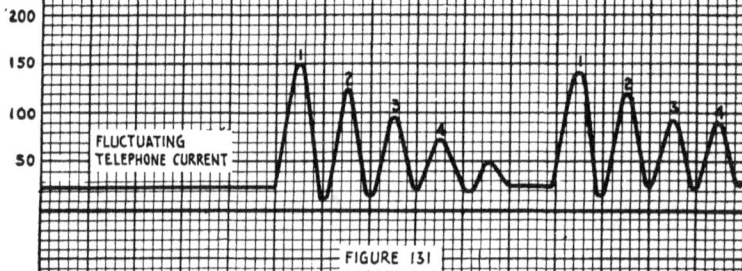

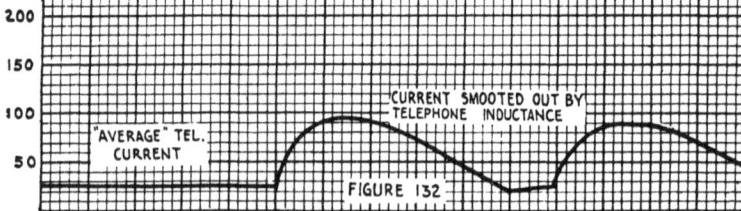

FIG 130. A characteristic curve of the carborundum rectifier indicating its non-uniform conductivity.
FIG. 131   Showing the unilateral currents produced by the carborundum rectifier and local battery during the reception of radio frequency oscillations.
FIG 132   Graphs of the resulting telephone currents in radio reception.

It is easily seen that the preponderance of current is in one direction so that a group of oscillations sends through the telephone what amount to rectified currents. The resulting uni-directional currents in the telephone are shown graphically in Fig. 131. Successive maxima decrease in amplitude according to the damping of the incoming signal. The telephone responds to an average of the maxima 1, 2, 3, 4.   The resulting telephone currents for two groups of incoming oscillations are shown in Fig. 132.

The student with the proper instruments may plot for himself characteristic curves of various carborundum specimens.   Obviously the crystal with the steepest curve, will give the best response.   The bend in the curve of Fig. 130 has purposely been magnified to illustrate more effectively the rapid drop of resistance at that point.

While the foregoing curve explains the action of a carborundum crystal when functioning as a detector of radio frequency currents, it perhaps does not explain the facts fully enough.

TUNER DESIGN.—The dimensions of a receiving tuner to cover a definite range of wave lengths is a matter of first importance.   Putting too many turns on the primary and secondary coils for the shorter waves is an error made by some experimenters.   The *unused turns* of a coil introduce "*end turn losses*," which tend to decrease the efficiency of a set.

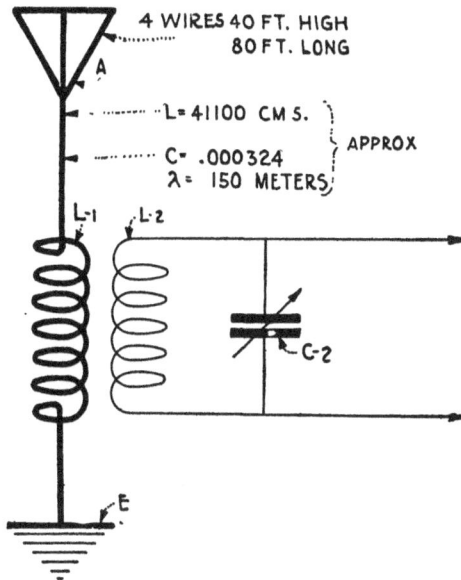

Fig 133   Diagram showing the circuits of radio frequency in an inductively coupled receiving system. The dimensions of the tuning elements for any range of wave lengths may be calculated by simple formulae

Assume for example that the aerial A of Fig. 133 is made of four wires spaced 2', the flat top being 80' long and 40' high.   From the table Fig. 20, its inductance $L_0 = 41,110$ centimeters, and capacitance $C_0 = 0.000324$ mfd.   The fundamental wave length is approximately 150 meters.   We now wish to calculate the inductance of L-1 to raise the wave length to 200 meters.   We also want the inductance of L-2 and the capacity of C-2 to give the closed circuit a frequency equivalent to the wave length of 200 meters.

When very large values of inductance are inserted at L-1, sufficient, let us say, to increase the wave length many times the fundamental wave length, there is no great error in using the wave length formula for lumped circuits, viz.:

$$\lambda = 59.6 \sqrt{LC}$$

Here L may represent the inductance of coil L-1 in Fig. 133, the distributed inductance of the antenna being ignored. C is the capacity of the antenna in microfarads. Some experimenters let L represent the sum of the distributed and the concentrated inductance of the antenna circuit

When small amounts of localized inductance are required, as called for in the problem under consideration, formula (30) is more accurate for it takes into account the inductance of the antenna.

$$\lambda = 59.6 \sqrt{(L_1 + L_a) C_a}$$

Where

$L_a = \dfrac{L_0}{3} =$ low frequency inductance of the antenna.

$L_1 =$ inductance of the loading coil at the base.
$C_a =$ low frequency capacity of the antenna system.

If, then it is desired to determine the value of L-1 for any given wave length we may transpose (30)

$$L_1 = \frac{\lambda^2}{3552 \, C_a} - L_a$$

which is formula (39) on page 49.

Substituting the values given in connection with Fig. 133:

$$L_1 = \frac{200^2}{3552 \times 0.000324} - 13{,}703 = 21{,}227 \text{ cms.}$$

$$\left(L_a = \frac{L_0}{3} = \frac{41{,}110}{3} = 13{,}703 \right)$$

In other words it requires but 20 microhenries to raise the wave length of the antenna in Fig. 133 to $\lambda = 200$ meters.

Before we can determine the magnitude of L-2 in Fig. 133 some value must be assigned to the condenser C-2. Because of the short wave length and high frequency of the circuit C-2 can be dispensed with, as the distributed capacitance of the coil, i.e., the capacitance between turns, is sufficient to establish resonance. Such a circuit will function particularly well when a vacuum tube detector is used as it will provide a maximum potential difference across the input terminals of the detector.

However a small variable condenser across the secondary is of some advantage in eliminating interference. Hence in this problem we will assign to C-2, a maximum value of 0.0001 mfd. and calculate the value of L-2 for 200 meters. This is a circuit with lumped values, therefore

$$\lambda = 59.6 \ \sqrt{L\,C} \qquad \text{(formula (23) page 40)}$$

and

$$L = \frac{\lambda^2}{3552\ C} \qquad \text{(formula (33) page 45)}$$

Hence

$$L = \frac{200^2}{3552 \times 0.0001} = 112,612 \text{ cms.} = 112.6 \text{ microhenries.}$$

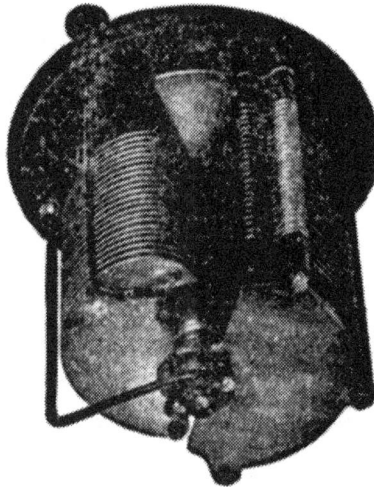

FIG. 133a.   DeForest variable air condenser suitable for tuning purposes in receiving apparatus.

CALCULATION OF INDUCTANCE.—We now desire the dimensions of a primary and secondary coil to have inductance of 21.1 and 112.6 microhenries respectively.

Nagaoka's formula (**40**) on page 54 is applicable. It is here repeated.

$$L = 4\pi^2 \ \frac{a^2\,n^2}{b}\ K$$

L = inductance in cms.
a = mean radius of the coil in cms.
b = equivalent length of coil in cms.
n = total number of turns

K = a factor varying as $\dfrac{2a}{b}$

As shown on pages 62 and 63, this formula may be transposed to give the value of n if the length of the coil and its diameter are first decided upon. (Note formula (42) page 62). But since tuning coils are wound closely, the problem may work out so that the pitch of the winding is less, or perhaps greater, than the overall diameter of the wire. A series of tedious trial computations would be necessary to effect a close winding.

It is about as practical for the amateur experimenter to assume a coil of a certain mean diameter, of a certain length and a definite number of turns, and then to make trial computations with formula (40) and see how near the result comes to the desired value.*

After working out a few problems of this kind, the designer will be able to approximate the dimensions of a coil for a given wave length. If the coil is too long it can be reduced to the desired number of turns.

The possible wave length range may be checked up accurately by a wavemeter as explained in chapter XI.

In the particular problem under consideration, we will make the primary coil $L\text{-}1$, $3\frac{1}{2}''$ in diameter. The secondary $L\text{-}2$ will be $3''$ in diameter wound with No. 30 d.s.c. The primary will be wound with 12 turns of No 26 d.s.c. The following data obtains:

From the table Fig. 37, the diameter of No. 26 d.s.c. $= 0.022''$. Hence, length of the coil $= 0.022 \times 12 = 0.264'' \times 2.54 = 0.6705$ cms.

$$a = 1.75 \times 2.54 = 4.445 \text{ cms.}$$
$$b = 0.6705 \text{ cms.}$$
$$2a = 8.89 \text{ cms.}$$
$$\frac{2a}{b} = \frac{8\ 89}{0\ 6705} = 13.26$$

From the curve Fig. 27, $K = 0.186$ (approx.).

Hence,

$$L = 39.47 \times \frac{4.445^2 \times 12^2}{0.6705} \times 0.167 = 27,970 \text{ cms.}$$

From the above trial computation we see that less than 12 turns of No. 28 d.s.c. are required to raise the fundamental wave length of the antenna in Fig. 133 to 200 meters. It is well to have a few extra turns in order to locate the exact point of resonance. The correction term of formula (41) page 57 has not been applied to the problem above. The method outlined in connection with the transmitter problem may be followed if this correction factor is to be applied. The results obtained by formula (40) are sufficiently accurate for amateur working.

We must next determine a secondary coil to have inductance of 112.6 microhenries. We will place 30 turns of No. 30 d.s.c. wire on a tube $3''$ in diameter.

From the table Fig. 37, the diameter of No. 30 d.s.c. $= 0.015''$. Hence, length of the coil $= 30 \times 0.015 = 0.45'' \times 2.54 = 1.14$ cms.

---

*A fairly rapid method of calculating the dimensions of a coil for a given inductance is described on pages 286-292 in "Radio Instruments and Measurements "

Hence

$$a = 3.81 \text{ cms.}$$
$$b = 1.14 \text{ cms.}$$
$$2a = 7.62 \text{ cms.}$$
$$\frac{2a}{b} = \frac{7\ 62}{1\ 14} = 6.6$$

From the curve, Fig. 27, $K = 0.2685$

Hence,

$$L = 39.47 \times \frac{3.81^2 \times 30^2}{1.14} \times 0.2685 = 121,401 \text{ cms.} = 121 \text{ microhenries.}$$

While this exceeds the value desired—112.6 microhenries; a few extra turns are desirable.

The experimenter will now observe that a tuning transformer for operation at 200 meters has very small coils. In fact, the primary and secondary windings are less than ½″ in length in the problem just presented. The experimenter who uses tuning coils 5″ to 6″ in length for reception at the wave length of 200 meters, should contrast his tuner with the one just designed.

The question whether all of the antenna inductance should be included in the transformer primary is mainly a mechanical one. For very long wave lengths the transformer coils become unwieldy if made of single layer windings. If all the turns are in the primary an end-turn switch should be provided to cut off the unused turns.

It is also more feasible for long wave lengths to use a secondary loading coil mounted separate from the transformer secondary.

RECEIVING TUNERS FOR VARIOUS WAVE LENGTHS.—The table in Fig. 134 has been prepared to aid the busy experimenter in finding the dimensions of tuning coils suitable for definite ranges of wave lengths. The data in this of most interest to the amateur are the dimensions of the secondary coils and the possible wave length with a definite capacity in shunt.

It is difficult to give the dimensions for the primary coils unless the inductance and capacitance of the antenna with which they are to be used are definitely known. The table gives the wave lengths of *primary* circuits, with coils of different dimensions connected in series with antennae of 0.0004 mfd. and of 0.001 mfd. capacitance. The capacitance first mentioned is an average value for the amateur's aerial. The value of 0.001 mfd. is a fair average for large aerials, like those aboard ship or at land stations where the flat top varies in length from 200′ to 400′. In the computations for the table of Fig. 134, the *distributed inductance of the aerial was ignored*, and the *wave length formula for lumped circuits used*, i.e., the antenna was treated as a simple closed circuit.

The wave length of the secondary has been computed for two capacities in shunt, one of 0.0001 mfd. and the other, 0.0005 mfd. With vacuum tube detectors it is rarely permissible for maximum signals to use more than 0.0005 mfd. in shunt to the secondary. In fact better

TABLE XII

	Diameter primary coil	Diameter secondary coil	Size primary wire	Size secondary wire	Length of primary coil	Turns primary coil	Length of secondary coil	Turns secondary coil	Primary loading coil and dimensions	Turns and size of wire of primary loading coil	Secondary loading coil dimensions	Turns and size of wire for secondary loading coil	Wave length of secondary with 0.0001 mfd. in shunt	Wave length of secondary with 0.0005 mfd. shunt	Wave length of primary with antenna of 0.0004 mfd. capacity	Wave length of primary with antenna of 0.001 mfd. capacity	Total inductance primary in cms.	Total inductance secondary in cms.
1	5″	4″	No. 24 D. S. C.	No. 28 D. S. C.	4″	152	4″	212	None	……	……	……	1050 meters	2350 meters	1810	2862	2,307,470	3,102,550
2	5″	4″	No. 24 D. S. C.	No. 28 D. S. C.	4″	152	6″	318	None	……	……	……	1360 meters	3040 meters	1810	2862	2,307,470	5,200,000
3	5″	4″	No. 24 D. S. C.	No. 28 D. S. C.	4″	152	8″	424	5″ dia. 4″ long	152 No. 24 D. S. C.	……	……	1605 meters	3620 meters	2557	4050	4,615,000	7,372,600
4	7″	6″	No. 24 D. S. C.	No. 28 D. S. C.	14″	532	14″	742	None	……	……	……	3250 meters	7270 meters	5330	8495	20,316,600	29,782,500
5	7″	6″	No. 24 D. S. C.	No. 28 D. S. C.	14″	532	14″	742	7″ dia. 24″ long	923 No. 24 D. S. C.	6″ dia. 24″ long	1856 No. 32D. S. C.	7200 meters	16100 meters	9150	14470	58,936,200	114,536,000
6	4″	3½″	No. 26 D. S. C.	No. 32 D. S. C.	5½″	247	5½″	380	……	{Time Signal λ=2500 meters}	……	{Tuner}	1546 meters	3460 meters	2182	3450	3,350,000	6,737,530
7	3½″	3″	No. 24 D. S. C.	No. 28 D. S. C.	3″	114	3″	159	……	……	……	……	682 meters	1525 meters	1090	1755	867,700	1,308,950
8	3½″	3″	No. 26 D. S. C.	No. 30 D. S. C.	0.26″	12	0.45″	30	……	{Suitable for 200 meter reception}	……	……	200+ meters	……	200+	……	31,149	121,401

Fig. 134. Table showing the dimensions of tuning coils for definite ranges of wave lengths.

signals are generally obtained with smaller capacities. With crystal detectors, secondary capacities up to 0.002 mfd. may be used.

The thirteenth and fourteenth columns of Fig. 134 show the possible wave length range of the secondary coils at two capacities which really represent the lower value and the upper value of a shunt secondary condenser of about 0.0005 mfd. maximum capacitance. It is assumed, of course, that for ranges of wave lengths shorter than those obtainable with the maximum number of turns, the primary and secondary coils will be tapped at intervals.

The secondary coils may be conveniently *tapped every half-inch* while the primary should be fitted with a *"units"* and *"tens"* switch described in a paragraph following.

Design No. 3 in Fig. 134 is suitable for the 2500 meter time signals from Arlington. It will be noted from the data that the primary is 5″ in diameter, 4″ long, wound with 152 turns of No. 24 d.s.c. An antenna loading coil, of exactly the same dimensions, is connected in series with the antenna circuit. The secondary coil is 4″ in diameter, 8″ long, and wound with 424 turns of No. 28 d.s.c. With the average amateur's antenna (0.0004 mfd.) the primary coils will raise the wave length to 2500 meters, and with larger antennae to 4000 meters. With 0.0005 mfd. in shunt, the secondary will respond up to 3600 meters and at some lesser capacity to 2500 meters.

Design No. 6 in Fig. 134 is also suitable for the reception of the *time signals.* Design No. 1 is suitable for wave lengths between 600 and 1600 meters, although, as the data shows, longer wave lengths can be tuned in. In design No. 5, observe that both *primary and secondary loading coils* are employed. With primary inductances of the dimensions given, the primary circuit will respond to 14,000 meters with an antenna of 0.001 mfd. Should the amateur wish to tune to a wave length in excess of 14,000 meters, he may place a *small variable condenser* in shunt to the primary inductance, or across both the primary inductance and the loading coil.

Design No. 7 is particularly applicable to the reception of 600 meter waves, although, as the table shows, longer wave lengths can be tuned in with a coil of those dimensions.

LORENZ INDUCTANCE FORMULA.—The inductance of the coils in Fig. 134 was calculated by Lorenz's formula, but the Nagaoka formula (40) page 54 may be used. A *correction term* should be applied to the Lorenz formula, but it was not used in the data obtained in the table of Fig. 134. The *correction formula* (41) page 57 originally given in connection with the Nagaoka formula is correct for the Lorenz formula.

Lorenz's formula for the calculation of inductance is as follows:

$$L = a\,n^2\,K \qquad\qquad (57)$$

Where

L = inductance in cms.
a = mean radius of coils in cms.
n = total number of turns
K = a factor varying as $\dfrac{2a}{b}$,

where b is the length of the coil.   Values of K for $\frac{2a}{b}$ appear in Fig. 135.

To illustrate the use of Lorenz's formula, assume that we have a tuning coil 5″ mean diameter, 4″ long, wound with 152 turns of No. 24 d.s.c. wire.   What is the inductance?

Remembering that

$$1'' = 2\ 54 \text{ cms.}$$

Then,

$$a = 6.35 \text{ cms.}$$
$$n = 152$$
$$\frac{2a}{b} = \frac{5}{4} = 1.25$$

and from the table, Fig. 135,

$$K = 15.7$$

Hence,

$$L = 6.35 \times 152^2 \times 15.7 = 2{,}303{,}347 \text{ cms.}$$

We may then calculate the wave length of the primary circuit with an aerial of 0.0004 mfd.

TABLE XIII

$\frac{2a}{b}$	K
.2	3.63240
.3	5.23368
.4	6.71017
.5	8.07470
6	9.33892
.7	10.51349
.8	11.60790
.9	12.63059
1.0	13,58892
1.2	15.33799
1.4	16.89840
1.6	18.30354
1.8	19.57938
2.0	20.74361
2.2	21.82049
2.4	22.81496
2.6	23.74013
2.8	24.60482
3.0	25.41613
3.2	26.18009
3.4	26.90177
3.6	27.58548
3.8	28.23494
4.0	28.85335

FIG. 135.   Values of K in Lorenz's inductance formula for the ratio $\frac{2a}{b}$.

We will use the formula for lumped circuits and ignore the distributed inductance of the antenna.

$$\lambda = 59.6 \sqrt{L\,C}$$

Then

$$\lambda = 59.6 \times \sqrt{2,303,347 \times 0.0004} = 1800 \text{ meters (approx.)}$$

For a capacity of 0.001 mfd.,

$$\lambda = 59.6 \times \sqrt{2,303,347 \times 0.001} = 2850 \text{ meters (approx.)}$$

The wave length of the closed circuit may be computed in the same way.

When he learns to use the above formulae, the amateur may first measure the *inductance and capacitance of an antenna* by the process described on pages 297 to 299, and afterward determine the amount of inductance required in the primary and secondary circuits to obtain any desired wave length. The problems of finding the correct dimensions for the primary inductance of the tuning transformer, and the proportion of turns assigned to the primary and to the loading coil resolve themselves for solution simply to matters of space and convenience. Ordinarily the primary coil is made about the same length as the secondary coil, although this is not absolutely necessary.

The same remarks apply to the secondary circuit. A secondary tuning coil more than 10″ or 14″ in length becomes unwieldy and if one of such dimensions does not possess sufficient inductance to give a desired wave length, a loading coil must be included in the circuit

Here is a good design for a receiving set to operate at wave lengths from 4,000 to 18,000 meters. First construct a *coupler* of the correct dimensions for response up to 3,000 meters. Then, construct primary and secondary loading coils, fitted with multi-point switches and appropriate cut-off switches, of the proper dimensions to permit whatever upper wave length is desired.

Although it is possible to design a single receiving set that will respond to all wave lengths from 200 to 20,000 meters, such a design necessitates the use of end turn switches for cutting off the unused turns from the circuit (as otherwise appreciable losses will occur) and therefore complicates the construction.

The author prefers two sets, one for the range of the shorter waves and the other for the longer waves. The first set should respond to waves between 200 and 3,000 meters and the second should respond to waves between 4,000 and 18,000 meters. The first set covers the wave lengths of all spark stations from the 200-meter amateur transmitter to the 2,500-meter time signals from Arlington. The second set covers all wave lengths from undamped wave high power stations, using arc transmitters, high frequency alternators, etc.

CAPACITANCE OF A VARIABLE CONDENSER.—The following formula applies for a variable condenser of semi-circular plates.

$$C = 0.1390 \, K \, \frac{(n-1) \, (r^2{}_1 - r^2{}_2)}{r} \tag{58}$$

Where

n = total number of parallel plates
r_1 = outside radius of plates
r_2 = inner radius of plates
r = thickness of dielectric
K = dielectric constant
C = capacitance in microfarads for the position of maximum capacitance.

The capacity of such a condenser for other positions is proportional to the angle through which the plates are rotated.

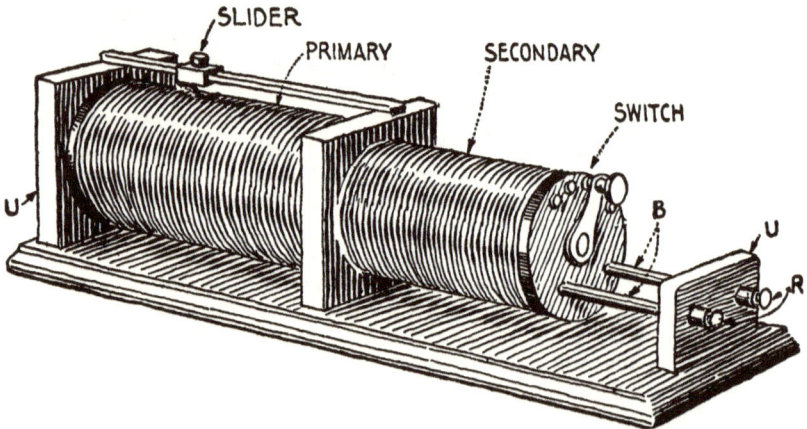

FIG. 136.   Design of a simple inductively coupled receiving transformer.

COUPLING.—It makes no difference whether the primary slides into the secondary or vice versa. Ordinarily the primary is set in a stationary position and is given a larger diameter than the secondary. The secondary coil is mounted on rods extending through its center which are supported by uprights on the base of the coupler. The rods rest on bushings mounted in the ends of the coupler which act as bearing surfaces and as means of connection to the movable secondary coil. A general idea may be obtained from Fig. 136. At B are two bushings inserted in wooden or hard rubber discs set in the ends of the coil. The rods R are supported by the uprights UU, the rods terminating in binding posts for external connection to the secondary. The bushings mounted in the secondary head are connected to the terminals of the secondary coil.

The tuner should be designed so that the *used* turns of the primary and secondary are in *direct inductive relation*. The necessity for this is apparent from Fig. 137. When the primary and secondary switches are set as shown in that figure, the *used* turns of the primary are those between A and B; and of the secondary those between C and D. If the used turns between A and B for the shorter range of wave lengths, are on the opposite end of the primary, then when the necessary coupling for an effective transfer of energy is obtained, the *unused* turns of both windings will overlap providing undesirable electrostatic coupling between them. This may prevent sharp resonance at some wave lengths

Following Fig. 137 in detail: The secondary inductance is varied by a simple multi-point switch. Some amateurs take off taps every 10 or 15 turns. Resonance between taps is established by the shunt secondary condenser *C*-1.

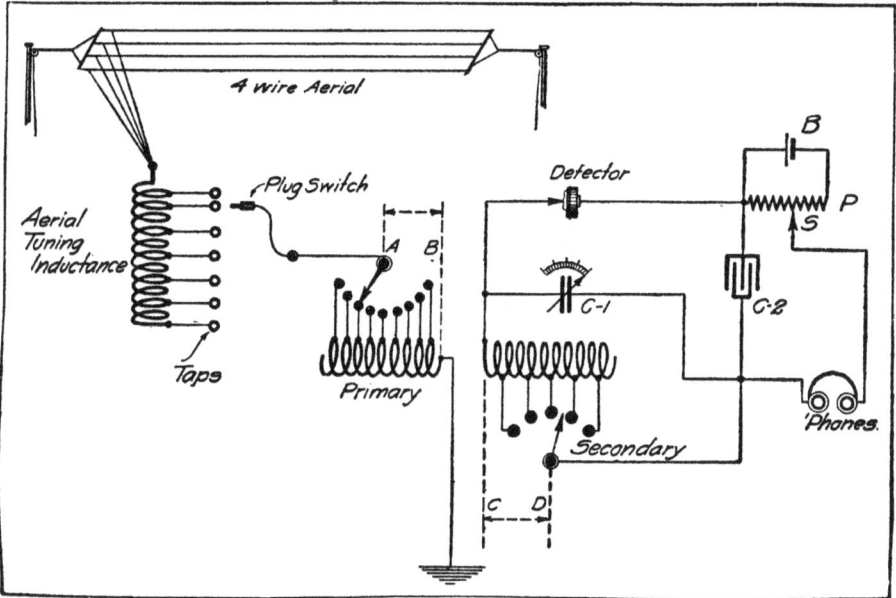

FIG. 137. Fundamental circuit of the inductively coupled receiving set using a crystal rectifier with a local battery. The diagram shows the relative positions of the "used" turns of the primary and secondary coils.

The plan of Fig. 137 appeals to the amateur who does not care to go to the expense of a "tens" and "units" switch for the primary coils. The number of turns in the aerial tuning inductance is varied by a simple plug contact switch, while the primary has a simple multi-point switch. The primary coil is tapped, at about every 10 turns. Resonance to any wave length occurring between the primary taps may be secured either by shunting the primary with a *variable condenser*, or by placing a *variometer* in the antenna circuit. *Considerable tuning between taps can be effected by change of coupling between the primary and secondary coils alone.*

If the detector in Fig. 137, is a carborundum rectifier, the battery *B* should have an e.m.f. of 4 volts. The potentiometer *P* is one of 400 ohms. Condenser *C*-2 may be one of fixed capacity—some value between 0.002 and 0.005 mfd. The telephones should be of 3000 ohms resistance.

The beginner should study the practical application of the foregoing inductance and wave length formulae and make several calculations of the possible wave lengths with various coils. This will instill in his mind a relative sense of the dimensions of inductances for various wave lengths.

He will then be able to account for some of the ludicrous designs found in the experimental station, many of which were worked out on a "guess work" basis.

It is true that the experimenter may build up primary and secondary coils of almost any dimensions and then determine the correct values for different wave lengths by "tuning in" transmitting stations whose wave lengths are definitely known. Turns may then be added to the coil or taken off to meet the requirements. The better way, however, is to predetermine the dimensions as this saves both labor and material. Such calculations have a certain educational value that cannot be ignored.

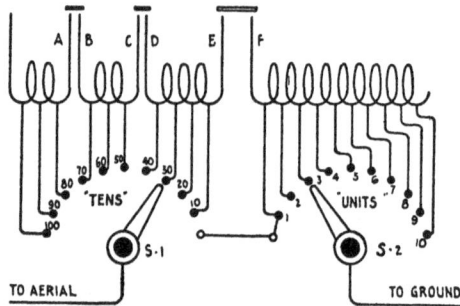

Fig. 138. Showing how the inductance of a coil may be varied a turn at a time by a "tens" and a "units" switch.

"TENS" AND "UNITS" SWITCH.—The use of a sliding contact on tuning coils for inductance variation is not recommended for, with continual use, the turns will be cut through, adjacent wires pushed into direct contact, and the result will be a *short circuit*. Unless a variable condenser is placed in shunt with the primary coil or in series with the antenna a switch must be provided that will permit the *antenna inductance to be varied by a turn at a time*. Such variation may be secured by the combination of a *"units and a tens"* switch. The use of *"units"* and *"tens"* switches is shown in Fig. 138, where switch S-2 cuts in *ten single turns* on one end of the primary coil. The switch S-1 cuts in *ten turns in a group*. Thus to cut in 37 turns in the primary, S-2 is placed on the contact stud marked 7 and S-1 on the contact stud marked 30. If three turns are required for resonance with a distant station, S-1 is set at zero and S-2 on the tap marked 3. If a coil of 410 turns, for example, be employed, the first 10 turns should be connected to the switch S-2 and the remaining 400 divided between 40 taps on S-1.

END TURN SWITCH.—As we have said, the unused turns of primary and secondary coils (when the maximum number of turns are not used) should be cut off as otherwise considerable energy losses are introduced into the circuit. This loss is occasioned by the *self-capacity* of the coil; that is its capacitance between turns, which may give it a defined period of oscillation so that the unused turns will absorb energy in a way that may be compared to coupled primary and secondary circuits. If the unused turns are disconnected from the used turns the losses from this source are reduced and a more efficient set is the result.

The position of the end turn switches in the circuit of a tuning coil is shown in Fig. 138. The breaks marked *A B*, *C D*, *E F*, connect to brushes on the end turn switch of Fig. 139. These breaks are located at points convenient to the range of wave lengths over which it is desired to work.

Assume that the primary coil in Fig. 138 applies to a tuner to be operated at wave lengths up to 3,000 meters. The turns to which the taps on the "units" switch are connected should be in direct inductive relation to the used secondary turns. The break at *E F* should be placed so that the first group of turns will resonate at some wave length just above 200 meters, say 225 meters. The next break *C D* should be located to include a sufficient number of turns between the end of the coil and the point *D* so that the circuit will resonate to some wave length slightly above 600 meters, let us say about 700 meters.

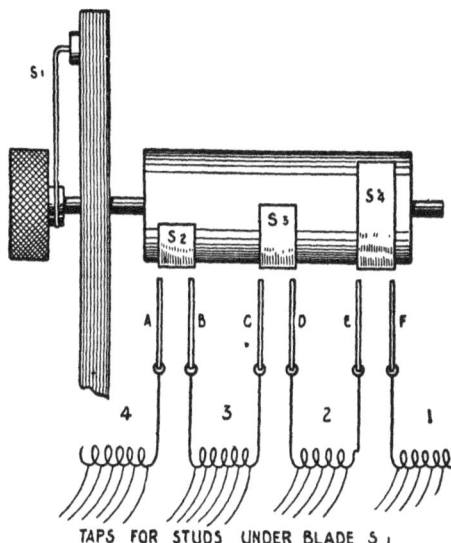

Fig. 139. Showing the position of the blades of an "end-turn" switch on a tuning coil.

A sufficient number of turns should then be included between break *C D* and break *A B* so that the antenna circuit will resonate at wave lengths up to 1,100 meters. The last group of turns should then add enough inductance to the complete primary circuit for it to resonate at the wave length of 3,000 meters. It is now evident that the first group of turns covers *amateur wave lengths*, the second group, *commercial wave lengths* up to 600 meters, the third group, *Naval wave lengths* up to 1,000 meters, and the last group enables the tuner to respond to the *time signals* at 2,500 meters.

This switching arrangement can be applied to tuners having greater numbers of turns. The signals are sometimes increased 20% by cutting off the unused turns in the circuit.

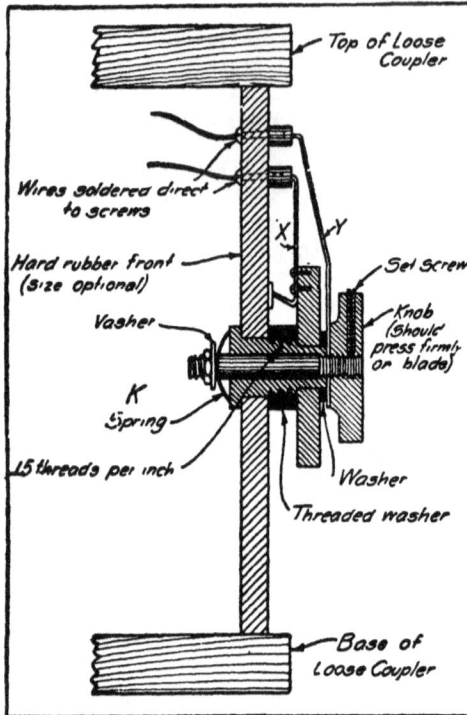

FIG. 140. A "tens" and "units" switch designed by an amateur experimente

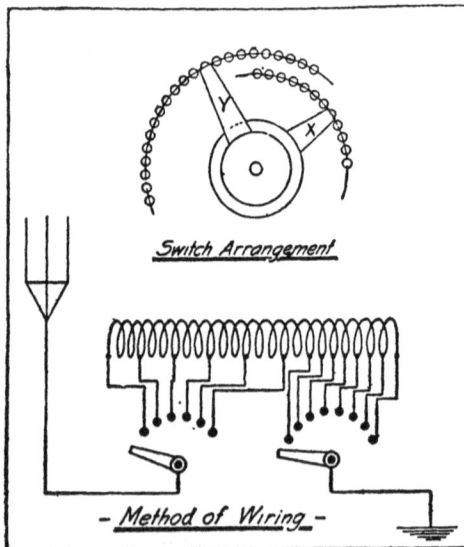

FIG. 141. Front view of the "end-turn" switch and a wiring diagram.

Explaining the end turn switch in Fig. 139, brushes A B, C D, E F, are connected to the interruptions in the winding of Fig. 138. S-2, S-3 and S-4 are copper segments mounted on a hard rubber drum which, in turn, is mounted on a shaft rotated by the knob shown at the left of the drawing. This same shaft carries the blade S-1 of the "tens" switch, which makes contact with the studs leading to the taps brought from the primary winding. The switch S-2 in Fig. 138 is mounted separately from S-1 and usually beside it. When the segments S-2, S-3, S-4 are placed as in Fig. 139, the switch S-1 must be turned counter-clockwise to increase the inductance. The taps at the bottom of Fig. 139 (extending from coils 1, 2, 3, 4) lead to the studs on the "units" and "tens" switches.

CONSTRUCTION OF "TENS" AND "UNITS" SWITCH.— R. Neupert of Pennsylvania has shown the design for a "tens" and "units" switch given in the drawings, Figs. 140, 141, and 142. The general idea of the design is more readily understood from the side view in Fig. 140. Switch X is the "units" switch, switch Y is the "tens" switch

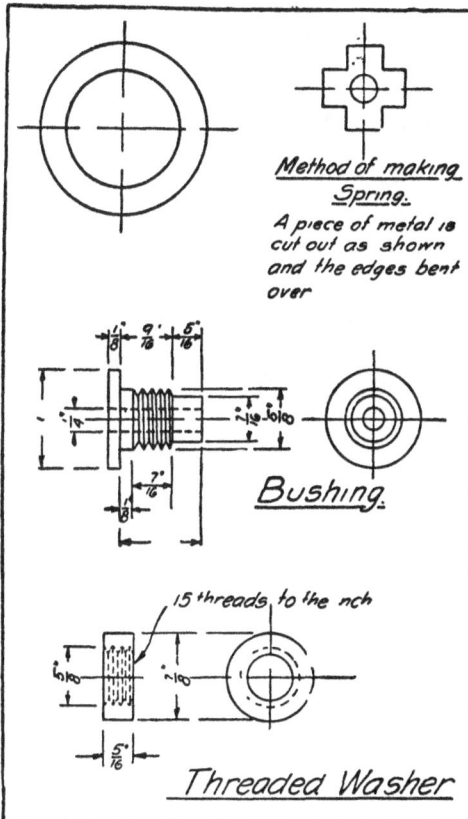

Fɪɢ. 142. Details and dimensions of the "end-turn" switch of Figs 140 and 141.

and both are mounted integrally. The front view appears in Fig. 141.
Fig. 142 gives the dimensions and details for construction. The dimensions must, of course, be altered if the coil used necessitates a greater number of taps.

Note in the side view of Fig. 140 how the blade of switch X presses against a brass ring held in place by two or three machine screws. The end of the blade is bent as shown to make contact with the ring. A lead extends from this ring to a binding post on the outside of the tuner. Connection to the switch Y is made through a brass lever upon which it is mounted. The switch blades are preferably made of *spring brass* or *phosphor bronze*. Good contact for the switch blade Y is assured by the spring K mounted on the supporting rod. This is made of a piece of metal, cut as shown, with the edges bent over. If a variable condenser is placed in shunt to the primary coil, it may be found practical to include as many as two turns in each contact of the switch X; otherwise adhere to the connections outlined above.

The wiring diagram of Fig. 141 lacks a "zero" stud for the "tens" switch. One should be connected in the circuit according to the wiring diagram of Fig. 138.

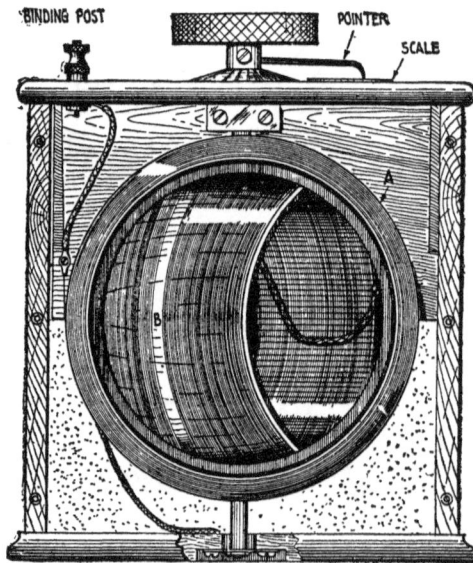

Fɪɢ 143.   View of an assembled variometer

VARIOMETER.—This instrument affords a means of obtaining a continuous variation of inductance and is therefore useful in radio frequency circuits. The principal advantage that the variometer has over the ordinary variable inductance is its absence of sliding contacts or complicated switch mechanisms.

The construction of a variometer for a small range of inductance variation is shown in Fig. 143 where a stationary coil A mounted inside a wooden frame is in inductive relation to a second coil B rotatable on an axis. The two coils are mounted concentrically and connected in series. When the inner coil is turned so that its axis coincides with the axis of the outer coil and the current circulates through both windings in the same direction, their magnetic fields add and the *self induction of the variometer is maximum.* If the inner coil is then turned 180°, the current circulates through the two coils in opposite directions, their magnetic fields oppose and the *self induction* of the unit is a minimum or nearly zero. When the inner coil bears any other angle to the outer coil the self induction is a function of the angle of rotation.

It is not customary to design variometers of the type shown in Fig. 143 to give a large range of inductance variation. In this case multi-layered coils or coils of large diameter are called for. Single layer coils are generally used, the variometer as such being employed to give small ranges of inductance variation.

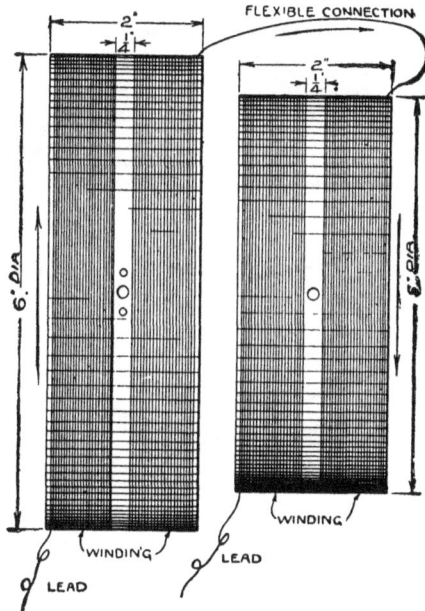

Fig. 144. Construction of a simple variometer made up of two concentric cardboard rings. This instrument will be found suitable for experimental work.

The variometer, for instance, is useful in the antenna circuit of Fig. 137 where only a multi-point switch and a plug contact switch are used to vary the antenna inductance. In this case it affords *close tuning* between the taps on the tuning coils.

The design of Fig. 143 is rather elaborate but a less expensive type of variometer as the one shown in Figs. 144 and 145, is almost equally suitable. The inner and outer coils are wound on concentric cardboard tubes one of which may be rotated on an axis as shown. The details of construction are as follows: Two cardboard tubes, one 6″ in diameter, the other 5″ in diameter and each 2″ in length, are wound with a single layer of No. 24 s.s.c. wire. A space of ¼″ is left in the middle of the windings to pass the shaft.   The same amount of wire should be placed on each tube and the turns should be thoroughly shellaced to prevent them from loosening up.   The inner and outer coils are connected as in Fig. 144.

A hole is drilled in both sides of each tube to take a piece of $\frac{1}{4}''$ round brass rod, $8''$ in length.   The inner tube should be fastened to the brass rod so that it will revolve with the knob.   A pointer may be fastened to the knob to work over a scale as shown in Fig. 145.

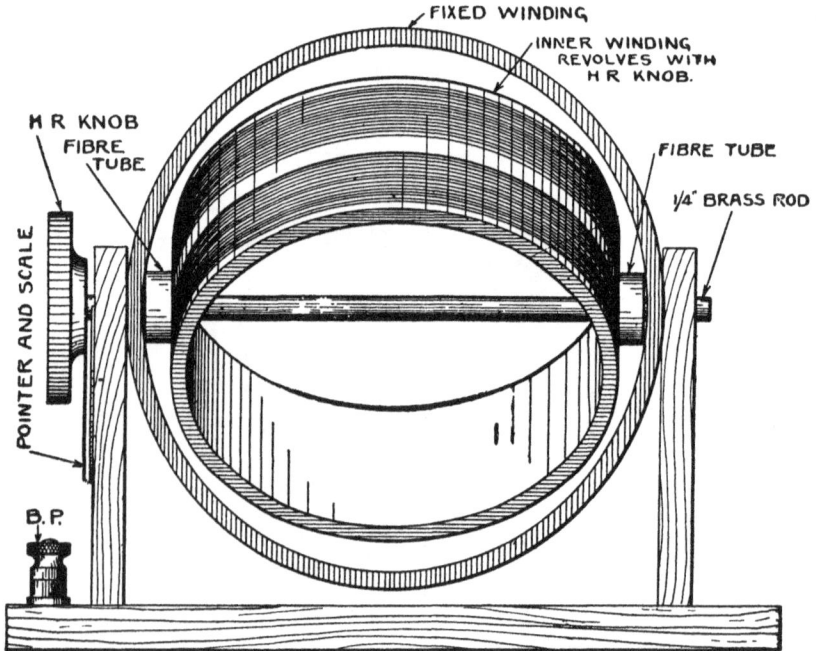

FIXED WINDING

INNER WINDING
REVOLVES WITH
H R KNOB.

H R KNOB
FIBRE
TUBE

FIBRE TUBE

1/4" BRASS ROD

POINTER AND SCALE

B. P.

Fig. 145.   View of the assembled variometer of Fig. 144.

If a greater range of inductance variation is required with the same size coils they may be wound with No 30 s.s.c. wire, instead of No. 24 s.s.c.

Two methods of using the variometer as a tuning element in simple receiver circuits are shown in Figs. 146 and 147.   The variometer $V M$ in Fig. 146 is connected in series with the antenna circuit to provide close tuning.   In Fig. 147 it is used both as a *tuning* and *coupling* element. The circuit is suitable for 200-meter reception.   A large range of wave length variation cannot be obtained from a variometer when used as the sole tuning element unless it be made very large and wound with a great number of turns.   The circuit of Fig. 147 has been used for long wave reception using a large variometer and a vacuum tube detector.   The simplicity of manipulation is obvious.

The dimensions and assembly of a ball type variometer like that in Fig. 143 is shown in the drawings Fig. 148, 149 and 150.

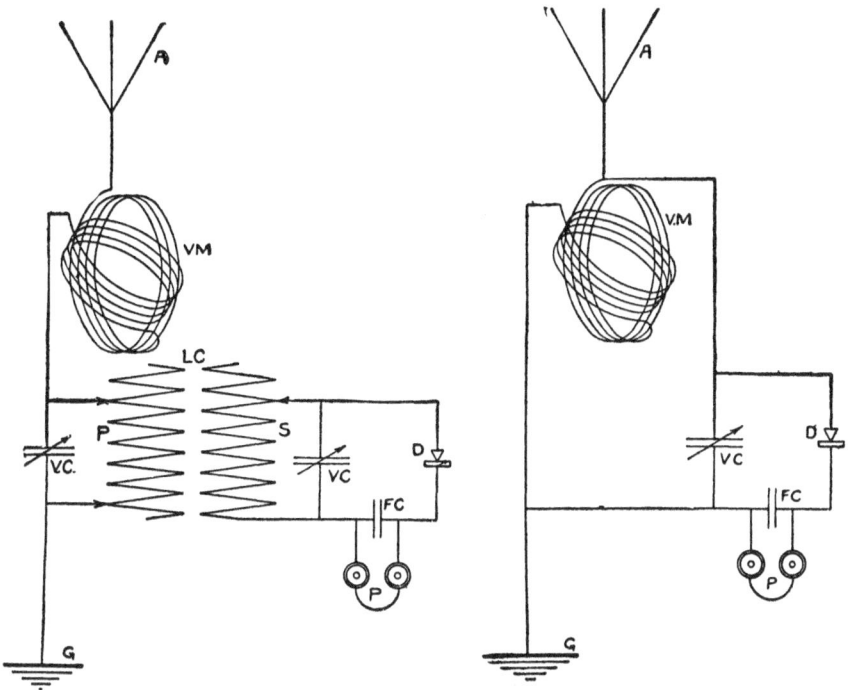

FIG. 146. The use of the variometer as a variable antenna inductance.
FIG. 147. The use of the variometer as a direct-coupled tuning transformer. This circuit is particularly suited to short wave reception

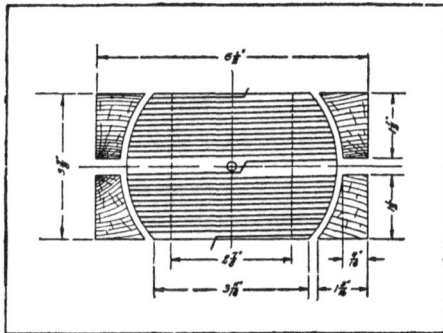

FIG 148. Plan view of ball type of variometer giving a larger range of inductance variation than that shown in Figs. 144 and 145.

Fig. 148 is a plan view looking down from the top giving the dimensions of the ball and the support for the stationary coil. Both the support and the ball are made of hard wood and are of course, turned out on a wood-turning lathe. The ball for the inside winding measures 5″ at its greatest diameter, gradually tapering to $3\frac{11}{16}$″ at both edges. Although the ball represented is hollow, it may also be made of solid wood.

The winding on the ball is left vacant at the center to allow the brass rod to pass through. If the variometer is to be used in the antenna circuit for short wave reception, each coil is wound full to the edge with No. 22 s.s.c. wire. The supports for the stationary winding are turned from a solid, square piece of wood and are of course, gouged out on a wood-turning lathe The two halves of the stationary frame may be held in position by screws set in from the top and base of the box or they may be strapped together by a brass strip. The support for the stationary coil and the ball should have a slight flange (not shown) to assist in holding the turns in place.

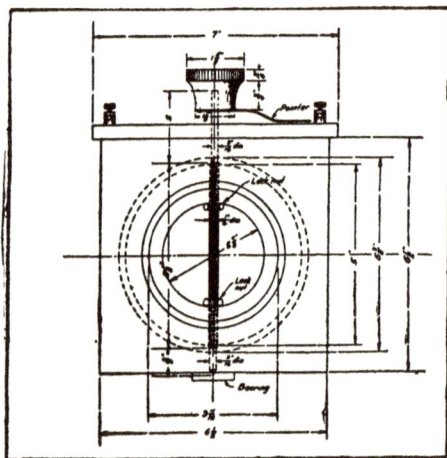

FIG. 149. Side view of the variometer in Fig. 148.

The inside coil is mounted on a brass rod $8''$ in length, tapered at the bottom, where it rests on the brass bearing. This bearing may be made to suit the builder. The rod has a diameter of $\frac{3}{16}''$ from the end inserted in the hard rubber knob, for a distance of $2''$; for the next $5\frac{1}{8}''$, it has a diameter of $\frac{1}{4}''$ and is threaded as shown; for the remaining $\frac{3}{4}''$ or $\frac{7}{8}''$ it has a diameter of $\frac{3}{16}''$, and is tapered at the extreme end. Lock nuts hold the ball for the inside winding in place. A $180°$ scale may be mounted on the top underneath a pointer. Pins should be placed at either end of the scale to limit the movement of the rotating coil. The inside coil is connected to the outside coil by flexible conductors.

It is not difficult to place the turns on the inside of the stationary coil if the following instructions are heeded. Construct a wooden ball, split in the center which just fits inside the stationary coil. The object of the split is to permit a wedge to be driven into it. Wind the turns on the split ball and afterward cover the inside of the stationary support with two or three coats of shellac. Before the shellac dries, force the turns on the split ball into place and press them firmly against the sides of the stationary support by driving the wedge into the split. The wedges are kept in place until the shellac has thoroughly dried when they may be removed. It will then be found that the turns of the stationary coil are

firmly placed.   They should be given a further coat of shellac on the
outside so there is no possibility of the wire coming loose.

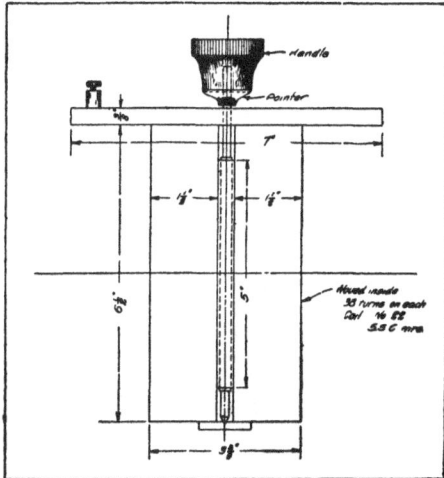

FIG. 150.   Side elevation of the variometer of Fig. 148.

One terminal of the stationary coil is connected to a binding post
mounted on the top of the box.   The other terminal of the stationary
coil is connected to a brush which is in contact with the rod carrying
the hard rubber knob.   This rod also connects to one terminal of the
ball winding, the second terminal of the ball winding being connected

FIG. 150a   Clapp-Eastham type S S Variometer for receiving circuits.

to the second binding post by means of a flexible piece of cord. The stationary and rotary coils can also be connected in parallel with each other when small inductance values are required.

CRYSTAL HOLDERS.—Experimenters have shown scores of designs, many of them ingenuous and providing very accurate adjustments for sensitiveness. The chief consideration in the design of a crystal holder is to provide means of obtaining a contact with any spot on the crystal. A contact with a variable tension is desirable with most crystals.

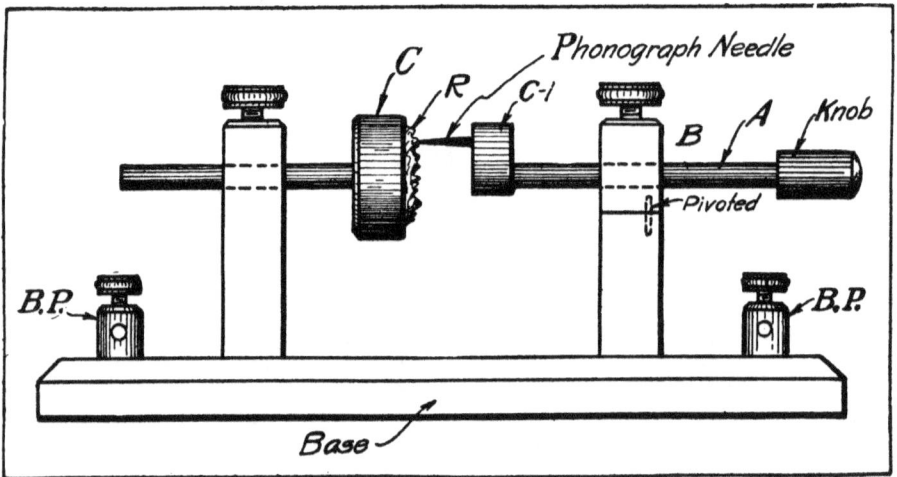

FIG. 151. Detector stand for carborundum crystals.

A detector stand suitable for carborundum crystals is shown in Fig. 151. The crystal *R* is imbedded in a retaining cup *C* by means of *Woods metal* or some other readily fusable compound. The opposing contact is a steel phonograph needle mounted in a cup *C*-1 which is set off center on the arm *A*. The latter has a universal movement, as a sidewise motion is provided by the pivot and a "round and round" movement at the cup *C*-1. A spring tension may be provided for the arm *A* but the details for this are not shown. The sketch of Fig. 151 is drawn approximately to scale.

B. B. Alcorn of New Jersey has shown an excellent design for the use of several crystals of which the details are given in Fig. 152 and are self-explanatory. The dimensions may be varied to suit the builder. The cups for the crystals are mounted on a brass or copper disc $\frac{1}{8}''$ thick, fitted with a knurled hard rubber ring for turning the various crystals into position.

The supporting rod and overhanging arm are made of square brass rod. The method of supporting the phosphor bronze contact point deserves particular attention.

Daniel O'Connell of New York City has developed a detector stand of the "cat whisker" type suited particularly to *galena crystals*. The details of construction shown in Fig. 153 are self-explanatory. The crystal is placed in a small glass tube between two "cat whisker" contact points mounted on rods supported by vertical binding posts.

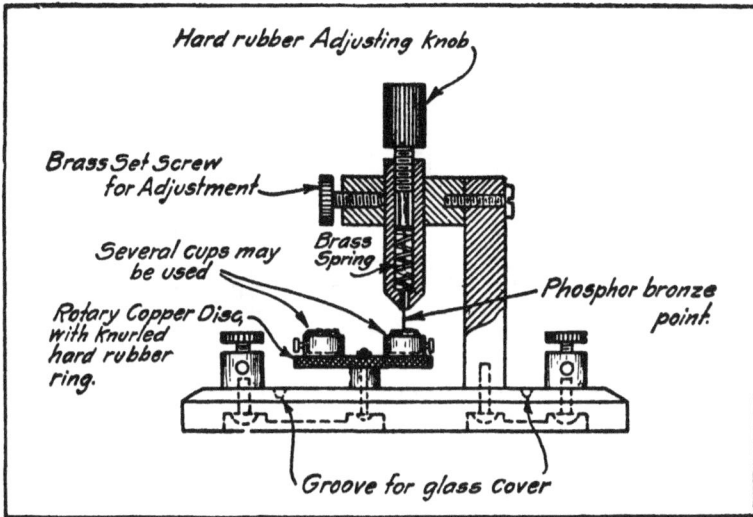

FIG. 152. Detector stand for the use of several crystals.

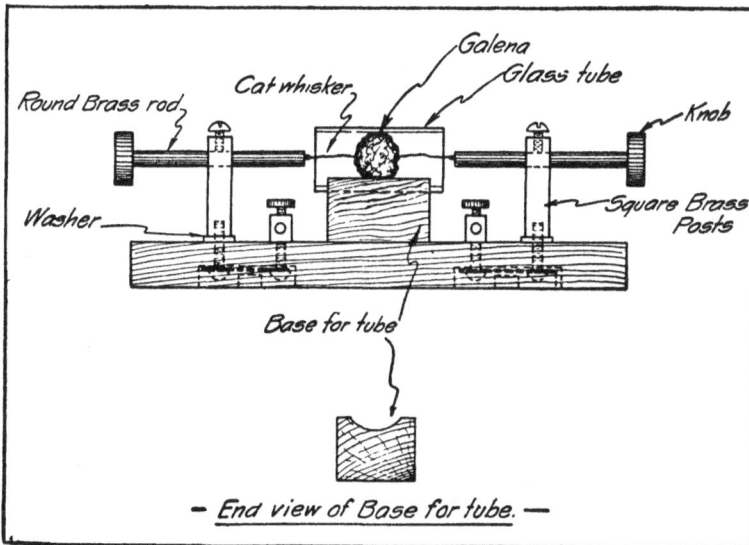

FIG. 153. "Catwhisker" detector stand.

There are several ways of providing an adjustment for this detector. The upright standards may have offset swivel joints on their tops to permit the "whiskers" being moved sidewise; the whiskers may be moved in or out; or the glass tube holding the mineral may be revolved until loud signals are obtained. If the mineral is so small that the tube permits too free a movement, it should be wound with tin foil until it fits the tube snugly.

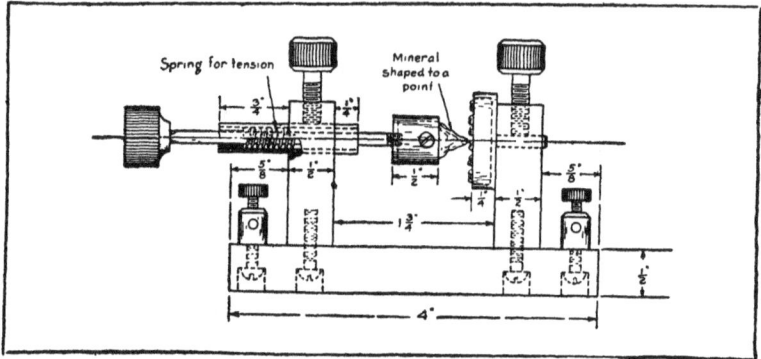

FIG. 154.   Detector stand for bornite and zincite crystals.

Fig. 154 shows a detector stand for *zincite and bornite crystals* designed by John E. Finn of New York City   The large cup in which the zincite crystals are imbedded with Woods metal is 1″ in diameter, ¼″ deep. The other cup which is ½″ in diameter and ½″ deep contains a single crystal of *bornite* shaved to a point.   The remaining dimensions are given in the drawing.

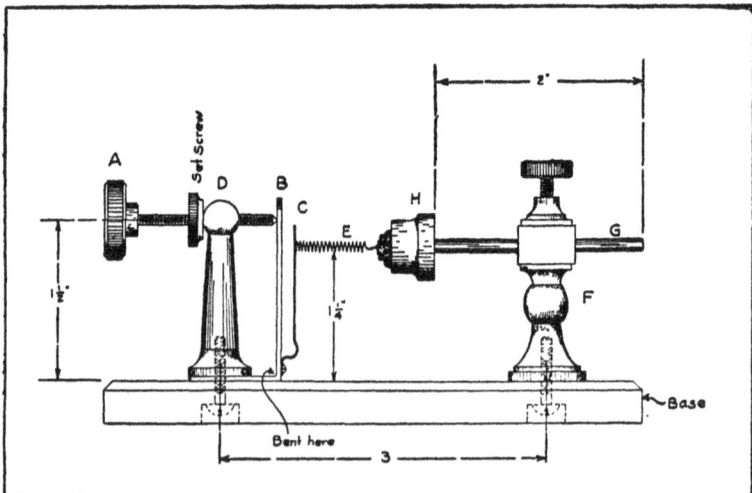

FIG. 155.   Detector holder for crystals requiring light contact pressure.

John Shaler of California regards the detector holder of Fig. 155 as particularly suited to crystals requiring light contact pressure The principal dimensions appear in the drawing.   The material for construction can generally be found around the amateur's workshop.

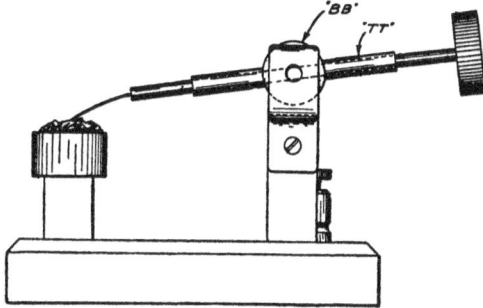

FIG 156.   Detector stand for universal adjustments.

Fig. 156 shows a detector stand the arm of which is fitted with a ball joint to permit contact at points over the entire surface of a galena crystal.   It was designed by J. F. Bernard of New York City. $B\,B$ is a brass ball $\frac{5}{8}''$ in diameter, with a $\frac{3}{16}''$ hole bored through the center. $T\,T$ is a brass tube of the correct diameter to slip inside the hole in the ball.   The whisker is a No. 30 copper wire.   The ball is clamped between two uprights which are of exactly the correct tension to hold it firmly.

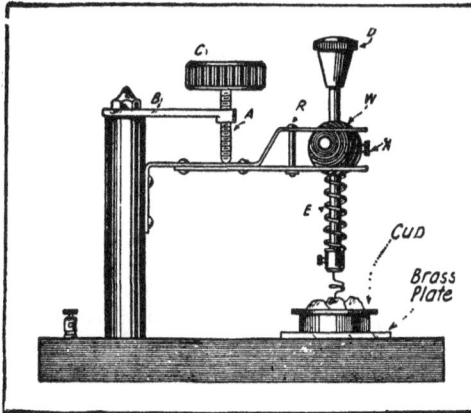

FIG 157   Detector holder for accurate adjustment of contact pressure.

Another detector holder with a universal ball adjustment, designed by J. G. Stelzer of Michigan, is presented in Fig. 157.   Tension is given to the contact point by the spiral spring E.   Additional tension is provided by the elastic arm supporting the ball joint and contact piece, through the adjusting screw A operated by the knob C.   The finished instrument should have dimensions about 1½ times those given in the drawing. A universal movement is provided by the ball W.

ELECTROLYTIC DETECTOR.—A receiving detector which has rapidly fallen into disuse since the discovery of the sensitive crystals and vacuum tube amplifiers is the *"whisker point" electrolytic responder*, though this played an important part during the early days of wireless telegraphy in the United States. The essential elements of this detector are indicated in Fig. 158.

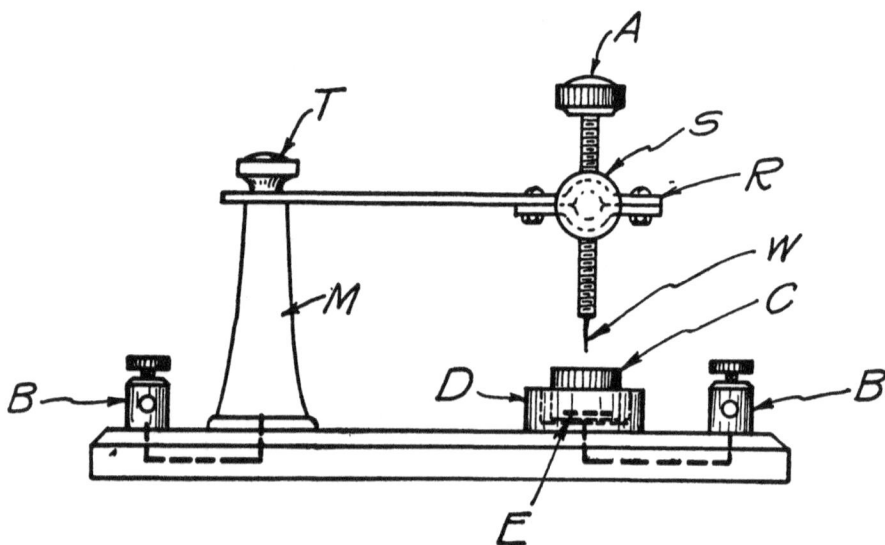

FIG. 158.   Fundamental design for the electrolytic detector.

Referring to Fig. 158: The upright binding post, M, has the extended arm R, through which is screwed rod A. This has a piece of platinum wire W, with a diameter varying from 0.0001″ to 0.000038″ soldered to the lower end. This wire is known as *"Wollaston wire"* and was formerly used for another purpose. It is coated with silver to permit easy handling.

The fine platinum point dips into the glass cup C, which need only be large enough to hold eight or ten drops of a 20% solution of *nitric acid*—the electrolyte. In the base of the glass cup is placed a small sheet of *platinum E*, about ³⁄₁₆″ square, which is connected to the right hand binding post B by means of a connecting wire placed under the hard rubber base.

The circuit for this detector is the same as that used with carborundum rectifiers. The local battery should be one of five or six volts, and should be shunted by a wire-wound potentiometer of 300 to 400 ohms

resistance. The telephones may vary from 75 to 2000 ohms, the important thing is that they be well constructed and sensitive. The positive
pole of the local battery should be connected to the fine wire point of the
detector.

Two rules must be observed in order to adjust this detector to a
sensitive condition. First, the fine wire platinum electrode must just
touch the acid, in fact barely make contact with it; second, the silver
must be removed from the platinum wire.

The silver coating may be removed by one of three methods: First,
an abnormal current from the local battery may be sent through the
electrolyte with the platinum point slightly submerged. A hissing,
grumbling sound will be heard in the head telephone which, after a
period of two or three minutes becomes of less intensity. The current
from the local battery should then be reduced until a loud response is
obtained in the head telephone from the buzzer tester. If the silver has
been thoroughly removed, good signals will be obtained. The second
method of removing the silver is to take the platinum wire and place it
for a moment in a very strong solution of chemically pure *hydrochloric
acid*, the point being removed at intervals and examined by the aid of a
magnifying glass to determine the degree to which it has been "trimmed."
When an extremely fine "whisker" is observed, the point is ready for use,
and it should be dipped in fresh water to prevent further action of the
hydrochloric acid. The third method is to heat a small amount of
mercury over a Bunsen burner. If the platinum point is dipped into the
hot mercury, the silver will immediately be removed.

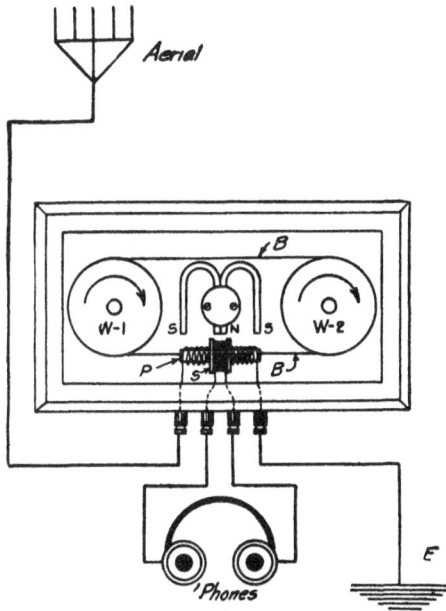

FIG 159   The Marconi magnetic detector

The amount of acid in the containing cup and the area of the surface of the lower electrode of the coil are unimportant, except that the upper point and the lower electrode should be separated no more than $\frac{1}{4}$".

MARCONI MAGNETIC DETECTOR.—This detector offers some interesting possibilities to experimenters of a scientific turn of mind. Contrary to the belief held by some amateurs, it is rather sensitive particularly at wave lengths around 2500 meters where it will give as good signals as a carborundum detector. The magnetic detector shown in Fig. 159 consists essentially of an iron wire band B (several strands of No. 36 insulated iron wire twisted in the form of a cable), with a total length of from 18" to 24", which is slowly drawn through the glass tube P by the ebonite or wooden pulleys, W-1 and W-2, which, in turn, are set in rotation by clock work or by a small direct or alternating current motor. The glass tube P is wound with from 6 to 10 feet of No. 36 s.s.c. The bobbin S is wound with a number of turns of No. 36 s.s.c. wire such as will give it resistance of about 150 ohms. The head telephones joined across the terminals of S should have a resistance close to 150 ohms. Immediately above the two bobbins of wire are placed two horseshoe magnets with like poles adjacent. These magnets generally have a spacing between their poles of about $1\frac{1}{4}$" and are placed at a distance of approximately $\frac{3}{4}$" to 1" from the band B.

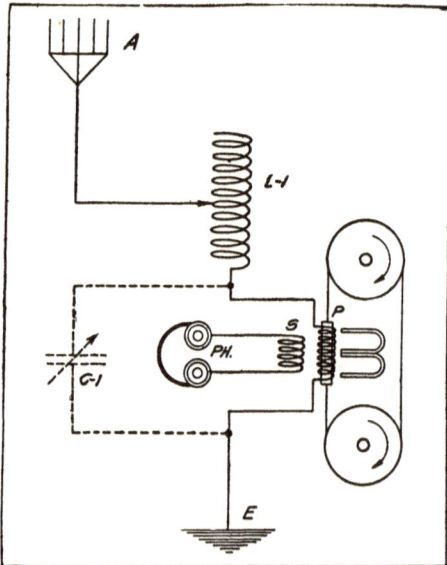

Fig. 160   "Standby" circuits for the Marconi magnetic detector.

The glass tube P is usually about 2" in length and $\frac{3}{16}$" in diameter. The bobbin of wire for the head telephone circuit has a diameter of $1\frac{1}{2}$" and an overall width of $\frac{1}{2}$".

When the band B is set into movement by the clock-work some parts of it are magnetized as they approach the south pole of the magnet, and accordingly are demagnetized when approaching the north pole of the magnet; however, the change of flux does not take place on any part of the band as it is directly under a magnetic pole, but is effected at a point beyond, and the final result is that the iron band exhibits marked hysteresis which cause it to be susceptible to changes of magnetic flux when excited by radio frequency currents. When high frequency electrical oscillations flow through the winding P for each group, a momentary change of the flux takes place, which in turn induces a current pulse in the coil S, creating a single sound for each wave train.

The "plain aerial" circuit for the magnetic detector is indicated in Fig. 160, in which an aerial tuning inductance L-1 is joined in series with the primary winding of the detector P. The circuit then continues to the earth connection E. As indicated by the dotted lines a variable condenser C-1 is sometimes connected in shunt to the antenna coil P. With this circuit it is only necessary to tune the antenna to the distant transmitting station, to secure an audible response.

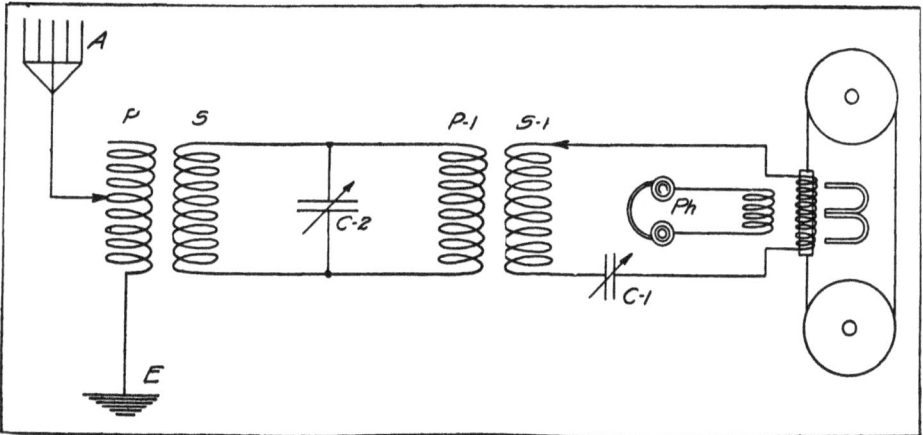

Fig. 161. Circuits of the multiple tuner for use with the Marconi magnetic detector

For sharper resonance effects the *multiple tuner circuits* shown in Fig. 161 are preferable. The *intermediate* circuit of this tuner comprises the winding S in inductive relation to the primary winding P, the variable condenser in shunt C-2, and the second coil P-1 which is in inductive relation to S-1. The secondary circuit then continues through the variable condenser C-1 to the primary winding of the magnetic detector. Since this detector is a *"current-operated"* device the secondary coil S-1, must have comparatively low values of inductance and resistance. For example, 18 turns of No. 18 bell wire wound on a form 4″ in diameter provide sufficient inductance for waves up to 2600 meters. Litzendraht wire is preferred In this circuit the variable condenser C-1 is an active part of the oscillatory circuit and the wave length can be changed over a considerable range by variation of its capacity. Usually a condenser of about 0.005 or 0.01 mfd. is employed at this point.

The primary and intermediate circuits are wound with the sizes of wire commonly employed in receiving tuners.

THE FILINGS DETECTOR.—During the past three or four years a new form of crystalline detector, modeled somewhat after the original Marconi coherer has appeared in the amateur market. It consists essentially of a cartridge C, Fig. 162, in which are placed two brass lugs, D and E, which fit snugly into the inside of the cartridge. The space between the two brass lugs is filled with a mixture of *scrapings or filings* taken from some of the well known crystal rectifiers. The manufacturers are generally reluctant to give the constituents but good results have been obtained with mixtures of *galena* and *silicon* filings to which has been added a slight amount of *brass* and *nickel* filings. In fact it is possible to use the scrapings of any of the crystalline detectors known to-day. The brass lugs are separated by a distance of about ¼″ and the intervening space half filled with a mixture of filings. The cartridge is then sealed up tightly and in some cases exhausted by a vacuum pump. The detector is then connected in the circuit of an ordinary receiving tuner in the standard manner and the tube revolved by hand until a "sensitive" adjustment is found.

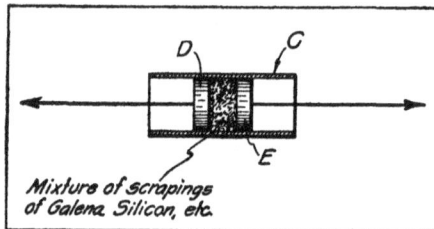

FIG. 162. Crystal rectifier made up of scrapings from well known crystals compressed between two electrodes.

In order to facilitate the finding of a sensitive point, a buzzer tester is employed, one terminal of which is connected directly to a terminal of the detector. When the buzzer is turned on certain of the filings will cohere and thus automatically locate a sensitive point of rectification. It is rather difficult to state off-hand just how these detectors operate, but it is probable that the action is no different from that in an ordinary rectifying mineral; however it may be that with so many filings in contact a sensitive point of rectification is more readily located than by the ordinary means.

Some detectors of this construction are found to have a rather low value of resistance and consequently, are used in a "*series*" tuning circuit like that used with the magnetic detector. It then becomes necessary to change the usual type of stopping condenser of fixed capacity to one of variable capacity for it is now an active element of the oscillation circuit.

It has been found that certain mixtures of filings require a rather high local e.m.f., much in excess of that called for by crystals of zincite, bornite. etc.

PRIMARY CELL DETECTOR.—Another form of the electrolytic detector which is sensitive and fairly reliable in operation is known as the primary cell detector or the *"Shoemaker Electrolytic."* It not only performs the functions of a detector, but has the added property of generating its own local e.m.f.; in fact it is really a small chemical battery constructed so it can be used as a radio frequency oscillation detector.

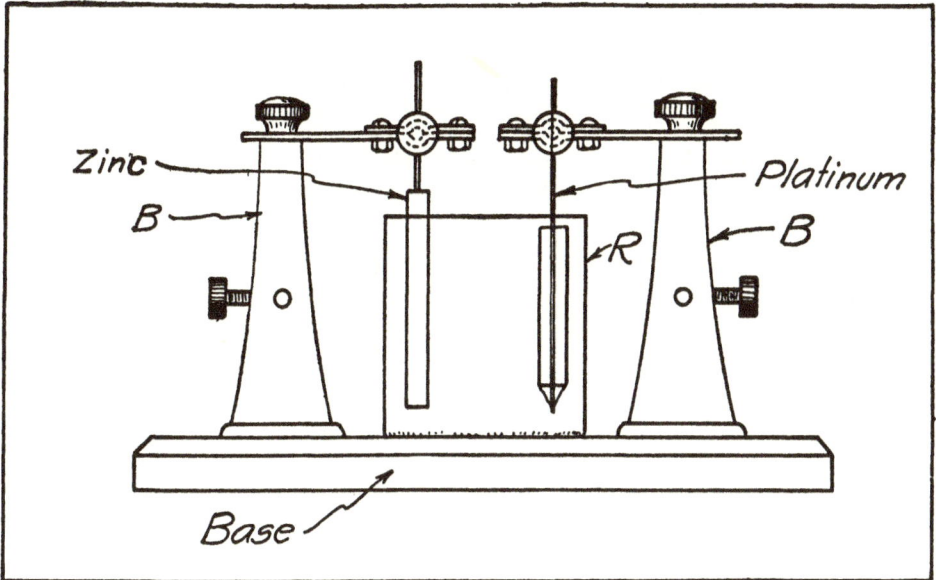

Fɪɢ. 163. Fundamental construction of the "primary cell" detector.

The details of the device appear in Fig. 163. To a hard rubber base are fitted two upright binding posts B, and the hard rubber or glass cup R, having a capacity of about four or six ounces of electrolyte. The left hand element of the cell is a small piece of *amalgamated* zinc about one-third the size of that usually employed in wet cells. The right hand element is a small platinum electrode having a diameter of $0.001''$ or $0.0001''$ sealed in glass, with only the tip of the wire exposed to the acid. The electrolyte is a 20% solution of sulphuric acid in which the two elements are immersed. A click is heard when a telephone is joined across the two terminals of the primary cell, indicating that an e.m.f. is being generated. In fact, such cells have often shown an electromotive force of 0.5 volt.

THE USE OF A LOCAL BATTERY.—The carborundum rectifier, requires a local battery for best response. The sensitiveness of the zincite-bornite detector is improved by a slight local e.m.f. of a small fraction of a volt. Some crystals of galena give increased response with a small e.m.f. applied in the proper direction.

The primary cell detector generates its own local e.m.f. The magnetic detector requires no local battery, as its source of "local energy" is the clock-work for rotating the iron band.

HEAD TELEPHONES.—The telephone magnets should have a great number of turns when they are connected in a detector circuit of high resistance. This demands magnet coils wound with fine wire which necessarily possess a high resistance. So-called "current operated" detectors require the use of low resistance telephones. The magnetic detector, for example, demands telephones of 150 ohms while crystals and vacuum tubes call for telephones of two or three thousand ohms.

Some advantage is gained by the use of light weight telephone diaphragms whose natural period of vibration corresponds to the spark frequency of the transmitter.

The so-called "mica diaphragm" telephones* known as "ampliphones" and by other trade names, are particularly sensitive. Certain types will give 10 times the strength of signals obtained from ordinary telephones. The increase in the strength of signals is very marked and they are recommended to all experimenters who wish the best results from their apparatus.

POTENTIOMETERS.—Wire wound potentiometers of various types are in use. Such a potentiometer can be made by winding a sufficient number of turns of No. 32 resistance wire on a square or round rod to give a total resistance of 400 ohms. A sliding contact bearing on a semicircular piece of *graphite* or *carbon* of the requisite resistance is satisfactory. A number of small buttons of carbon having resistance of 20 ohms each may be mounted on the studs of a circular multi-point switch and connected in series. The blade on the switch is the "potential divider."

BANKED COILS FOR RECEIVING TUNERS.—From the designs for single layer solenoids already given, it is evident that long wave length tuners require very large coils which are undesirable for purely mechanical reasons, such as space considerations, etc. So-called "bank" windings provide large values of inductance with relatively small coils and are widely used particularly for reception at long wave lengths.

In regard to the use of stranded and solid conductors for tuning coils it may be stated that the resistance ratio $\dfrac{R}{R_0}$ of a coil of stranded conductors consisting of a considerable number of very fine insulated wires, is less for moderate frequencies than for a similar coil of solid wire. (R, in the above expression is the resistance of a conductor to a given frequency n, and R_0 is its resistance to direct current.)

For every coil of stranded conductor there is a critical frequency above which the stranded cable has the larger resistance ratio. The critical frequency becomes higher as the strands become finer, but for conductors made up of several dozen strands of say, No. 40 wire, the critical frequency is usually above the range of frequencies ordinarily employed.

---

*For a complete explanation of the operation of these telephones, the reader should consult the author's "Practical Wireless Telegraphy," pages 168 and 169.

Stranded conductors known as "Litzendraht" are widely used as tuning coils. In the case of tuning coils wound with stranded conductor, the resistance ratio is always greater than for a straight conductor. At high frequencies the ratio may be two or three times as great, particularly for very short coils.

Banked coils for *moderate radio frequencies* are preferably wound with stranded conductor. For very high frequencies around the 200-meter wave length a single layer solenoid wound with solid wire will show smaller losses than a coil of stranded wire. Banked coils are not essential at wave lengths below 1000 meters

The reason for banking the turns of a multi-layer coil lies in the fact that the capacitance between layers of the ordinary multi-layered solenoid is excessive. This capacitance gives the coil a tendency to oscillate at some particular frequency and moreover introduces large dielectric losses. In banked coils instead of winding on a complete layer followed by successive similar layers, one turn is wound successively in each of the layers. The maximum voltage between adjacent wires is thus reduced and accordingly the distributed capacitance.

The order of placing the wires on a two-bank coil is as follows:

	3	5	7	9	11	13	15	
1	2	4	6	8	10	12	14	

For a three-bank winding:

	6	9	12	15	18	
5	4	8	11	14	17	
1	2	3	7	10	13	16

For a four-bank winding:

	10	14	18	22	
8	9	13	17	21	
7	6	5	12	16	20
1	2	3	4	11	15	19

The form on which such coils are wound should be threaded so that the wires will stay in place during the winding process. The turns are wound on by hand. This is a painstaking job and in order to do it rapidly requires some experience.

Let us assume that the amateur does not possess "Litzendraht" wire and desires to make up a banked coil for long wave lengths. The inductance of such a coil can be calculated to an accuracy of 5% to 9% by Nagaoka's formula (**40**) making "a" the average radius of the banked coil.

Assume a coil 6″ in length, 5″ mean diameter (including the two layers), wound with two banks of No. 28 d.s.c. wire. Each layer will have $53 \times 6 = 318$ turns, and two layers, $2 \times 318 = 636$ turns. Hence,

$$a = 6.35$$
$$b = 15.24$$
$$n = 636$$
$$\frac{2a}{b} = 0.83$$
$$K = 0.728$$
$$L = 39.47 \times \frac{6.35^2 \times 636^2}{15.24} \times 0.728 = 30{,}778{,}356 \text{ cms.} \qquad (40)$$

Although this coil is but 6″ long, when shunted by a capacitance of 0.0005 mfd. it will resonate at 7500 meters. Hence, if the saving of space is a factor, the amateur should use *two or three bank* windings. The experimenter may now work out for himself the dimensions of banked coils which will permit tuning to wave lengths up to 18,000 meters. Litzendraht wire should be used if it can be conveniently obtained.

Let it be thoroughly understood that the object of banking the turns is to reduce the distributed capacitance of the coil. As already explained a simple multi-layered coil possesses excessive values of distributed capacity.

NOTE TO EXPERIMENTERS

IN the following chapters a number of circuit diagrams for the use of the Marconi V. T. in transmitting and receiving circuits are shown. It is to be noted, that in the majority of simple detection circuits the grid leak of the detector bulb is connected from the grid to the positive side of the "A" battery; whereas, when the valves are used as amplifiers, the grid leak is connected from the grid to the negative side of the filament battery. These methods of connection have been found best for the Marconi V. T., although they may not hold good for other types of bulbs. Connecting the grid leak from the grid to the positive side of the "A" battery gives the grid a slight positive potential which gives better signals when a hard bulb is used as a detector. In amplifying circuits it is usually desirable to hold the grid at a slight negative potential. This can be obtained either by use of the grid condenser with the grid leak connected to the negative side of the "A" battery, or by means of a grid battery of about 1½ volts.

The grid battery provides a slightly easier method of adjusting the grid potential than the grid leak and grid condenser, but the latter method is quite satisfactory. If the experimenter possesses other bulbs than the Marconi V. T., he should try connecting the grid leak to the negative and to the positive side of the "A" battery and then, of course, leave it where it gives the best results.

Marconi V. T.'s of the soft type (known as Class I) require that the grid leak be connected first to the negative side of the "A" battery and then to the positive side, the connection being left where the best signals are secured. (See also note in appendix.)

CHAPTER VII

THE VACUUM TUBE DETECTOR AND AMPLIFIER— CIRCUITS FOR DETECTION OF DAMPED AND UNDAMPED WAVES—CHARACTERISTICS OE THE MARCONI V. T. DETECTOR— PREFERRED CIRCUITS FOR THE MARCONI VALVE

TYPES OF VALVES.—It is safe to say that no instrument or apparatus employed in wireless telegraphy and telephony is susceptible to so many uses as the vacuum tube. It is not possible in the space at hand to cover the subject exhaustively, but an effort will be made to explain the working of the valve circuits considered most feasible for the amateur station. .

The field for experimentation is almost without limit. Hundreds of circuits have been devised. Although many of these appear at first sight to be different from others they are really identical except that the component parts of the system are arranged in a different way in the circuit.

The vacuum tube may be used as a *plain detector*, as a *radio or audio frequency amplifier*, as a *self-amplifier*, as a *"beat" receiver* or as a *generator of continuous oscillations for radio telegraphy or telephony*. The three-electrode valve is also used as an amplifier in land line telephony. It has been employed to control the voltage of a dynamo and there are many other uses to which it may be put.

The first important discovery was made by Fleming whose two-element valve has been used commercially for a great number of years. Deforest inserted a so-called grid element between the filament and plate of Fleming's valve and increased its sensitiveness. Armstrong was the first to show the operating characteristics of the three-electrode valve and his adoption of the regenerative principle marked a distinct advance in the art. Weagant contributed the three-electrode valve with the external grid element, and, finally, the four-electrode valve with an *internal unconnected grid element* and an *outside electrostatic control element*. Donle has recently devised a tube with the grid and filament placed inside the bulb and the plate element on the outside.

All of these valves utilize the *electronic emission* of incandescent lamp filaments in vacua to obtain either rectification or a repeating action. The filament, grid, and plate are made of various metals. One commercial detector tube has a *nickel filament,* a *tungsten grid* and a *molybdenum plate.* Another has a *tungsten filament,* a *nickel grid* and a *nickel plate.* A transmitting tube developed by one manufacturer contains a *tungsten filament,* a *molybdenum grid,* and a *nickel plate.* Still another tube has an *aluminum plate* and a *spiral grid of copper* containing a slight percentage of thorium.

Coated filaments are used to increase the electronic emission. The preparation of these filaments is generally kept secret by the manufacturers. It is known that some manufacturers coat their platinum filaments with a mixture of oxides, such as barium or strontium. Such filaments emit large quantities of electrons at low temperatures, thus tending to prolong their life. Tungsten filaments generally are burned at high temperatures. Most bulbs are exhausted to a very high vacuum, the internal elements being heated to high temperatures during the evacuation process to remove all occluded gases. This prevents ionization under normal conditions of operation, tends toward constancy of action, and gives the tubes definite operating characteristics which can be relied upon throughout their life.

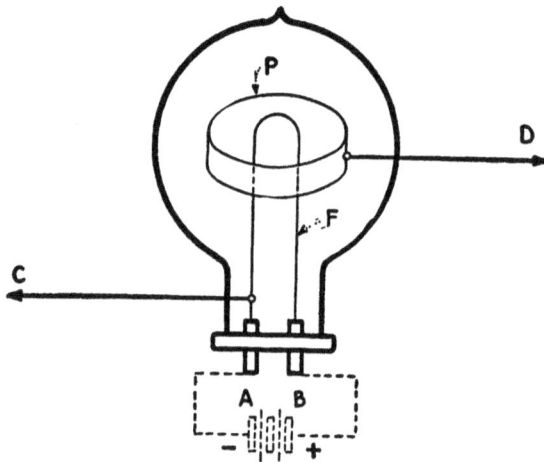

Fɪɢ. 164.   Fleming's two-electrode oscillation valve for the detection of radio frequency currents.

FLEMING'S VALVE.—This detector shown in the drawing Fig. 164 utilizes a *lamp filament* and a *metallic plate* mounted in a glass bulb from which all air or occluded gases are removed. The filament is of carbon or tungsten and the plate of copper or nickel. The filament F is rendered incandescent by a 4- to 12-volt storage battery.

If connection is made to the *negative terminal* of the filament F and the plate P at C and D respectively, it can be shown that the valve conducts best when the plate is connected to the *positive pole* of the current source. If the polarity of the externally applied e.m.f. be reversed little or no current will flow through the valve. It is now clear that the valve possesses *"unilateral conductivity,"* i.e., it is a *rectifier* and as such it may take the place of the carborundum rectifier shown in previous circuits.

The valve not only possesses rectifying properties, but the resistance from the plate to the filament varies with the amount of current flowing through the valve. This can be demonstrated by the circuit of Fig. 165. Here the filament is heated by battery B-1, while the terminals of the battery B-2 are connected to the plate and to the negative terminal of the filament A is a milliammeter. If the e.m.f. of B-2 as read by an appropriate voltmeter is slowly increased from some minimum value, and the reading of the ammeter A noted at intervals, it will be found at first that the current through the valve does not increase directly with the applied e.m.f., and that after a critical e.m.f. is passed, the current increases more rapidly than the applied e.m.f. would appear to call for.

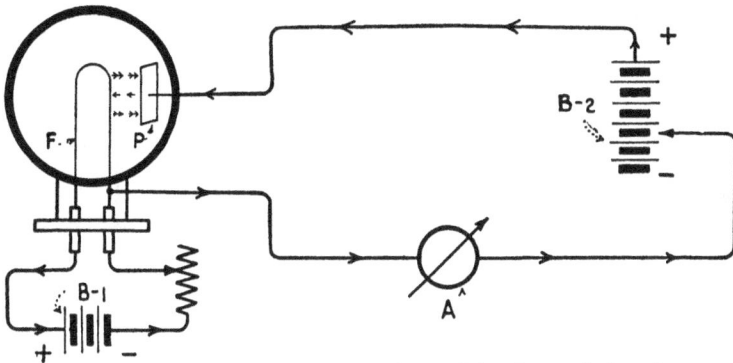

FIG. 165.   Circuit for obtaining the d c characteristics of an oscillation valve.

The data thus obtained may be plotted in the form of a curve as shown in Fig. 166. It is seen that the resistance of the valve falls rapidly at the point B.   At point C further increase of the plate e.m.f. causes no further increase of current.   This is the point of *saturation*.

The current maximum is limited by two things; the *temperature limitation of the filament* and the *space charge*.   Obviously if the temperature be increased above some critical value the filament will burn out.   If the temperature be raised to the maximum permissible value the current maximum between P and F will be that corresponding to the available supply of electrons.   Further increase of the e.m.f. of B-2 will not increase the electron current because all the electrons available are drawn over to the plate.   But before all the electrons are used the *space charge* becomes effective in reducing the plate current.   The explanation of this phenomenon is as follows: Electrons are small charges of *negative electricity* and when a sufficient number have accumulated between the filament and plate they constitute a negatively charged field of such intensity that they repel the electrons behind them so that only a portion of those emitted by the filament reach the plate.   Thus, it is seen that the *space charge* as well as the temperature limitation mentioned above, tend to limit the strength of the electron currents through the tube.

Suppose that the curve $A\ B\ C$ Fig. 166 is obtained with a given filament temperature.   If the filament temperature is increased a new curve $A\ B\ D$ is secured which indicates another point of saturation at D. That is, although the increase in filament temperature has increased the electron emission and, therefore, the plate to filament current, a point of saturation is soon reached because of the space charge and the fact that all the available electrons have been drawn over to the plate.

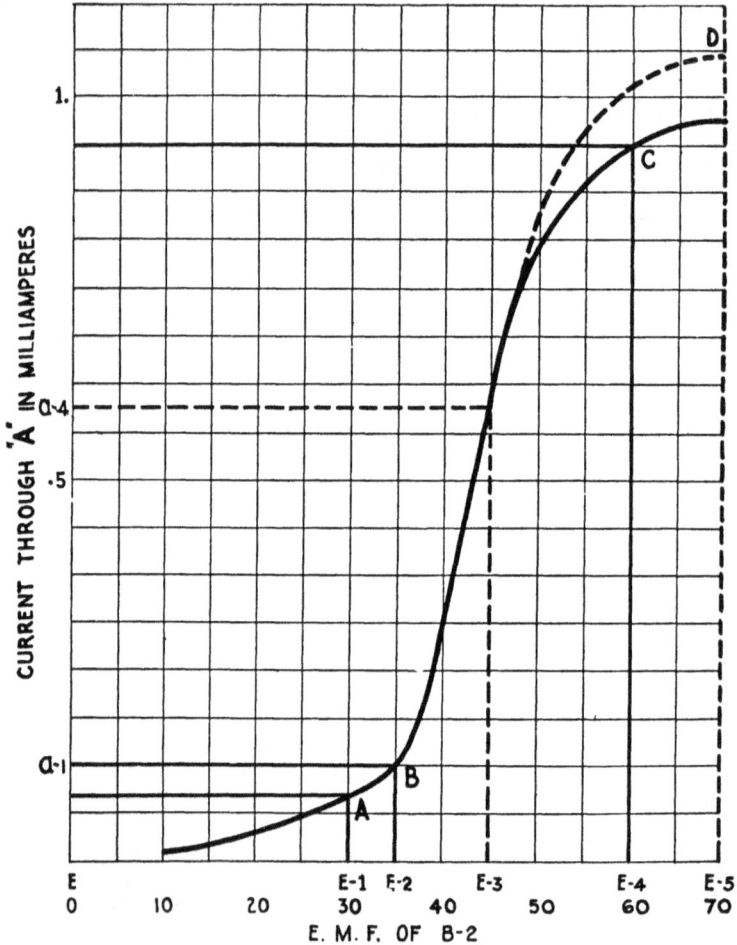

FIG 166   Plate voltage — plate current characteristic of a two-electrode valve showing its non-uniform conductivity.

CIRCUIT FOR FLEMING'S VALVE.—The detection circuit for Fleming's valve which takes advantage of its *non-uniform conducting properties* as indicated in Fig. 166, is shown in Fig. 167. Here the battery *B*-1 not only heats the filament but supplies the local e.m.f. to permit working on a point of the characteristic curve favorable for the detection of radio frequency currents.

The necessary regulation is obtained by potentiometer *P*-1 of 400 ohms shunted across *B*-1. By moving the tap along *P*-1 to the right the plate *P* is charged to an increasing positive potential. The current from B-1 flows through the telephone P-2, through the secondary coil L-2 to the plate P, thence to the filament and back to the negative terminal of the battery.

Assume, then, that the local e.m.f. is adjusted to correspond to point B in Fig. 166. This is the point where the resistance of the valve falls rapidly. If the incoming radio frequency currents be superposed upon the direct current already flowing through the valve, it is clear from the

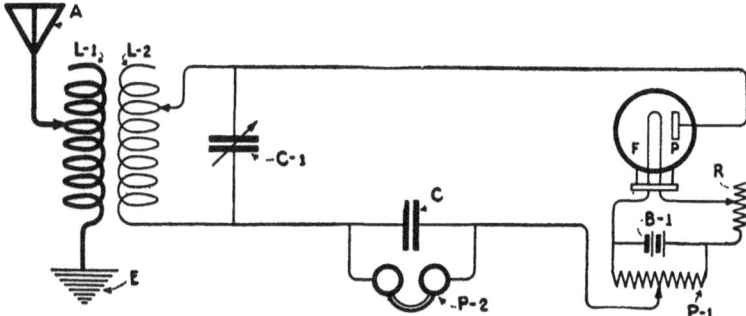

Fig. 167. Circuit for Fleming's valve using a local current to energize the plate circuit. The same source of current is used for the filament and plate circuits.

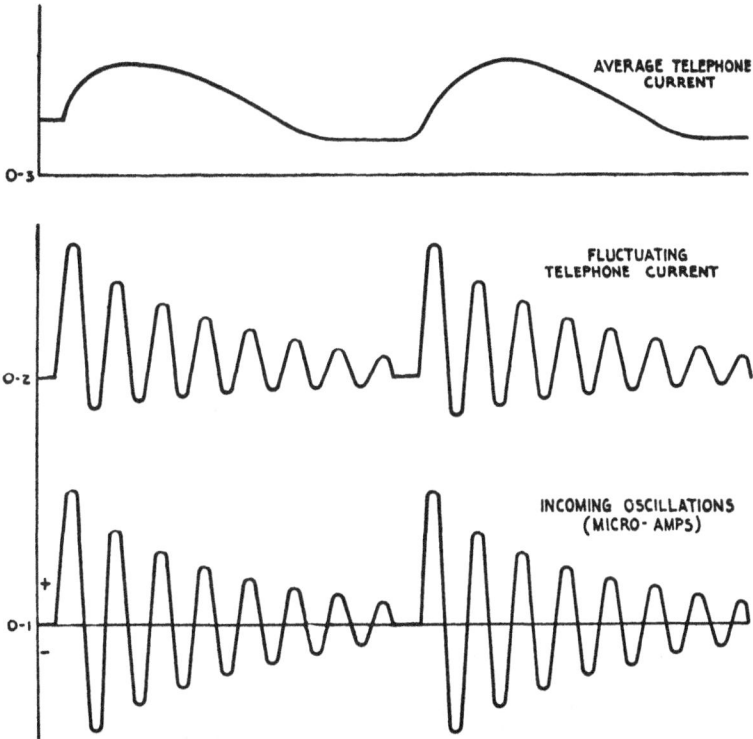

Fig. 168 Graphs showing the phenomena involved in the detection of radio frequency currents by the oscillation valve.

curve that when the incoming e.m.f. and the local e.m.f. are in the same direction, the increase of current through the head telephone will exceed the decrease when these two e.m.f.'s oppose. Since the preponderance of current is in one direction, what amounts to rectified currents flow through the head telephones (during the reception of radio frequency oscillations).

The foregoing actions are depicted graphically in Fig. 168. The lower graph O-1 shows two groups of incoming radio frequency currents. The graph O-2 shows the fluctuations of the telephone current, and the graph O-3 indicates approximately the telephone current which is an average of the amplitudes in O-2. It is now clear that each spark at the transmitter will give one pull at the diaphragm of the receiving telephone.

It is important to note that along the *straight portion of the curve* Fig. 166, the plate current would increase and decrease by equal amounts for any small incoming e.m.f. In that region of operation the telephones would be traversed by a *radio frequency component* to which they would give no response.

If, on the other hand, when the plate e.m.f. is adjusted to the point C on the curve of Fig. 166, and e.m.f.'s generated by incoming signals are impressed upon the filament and grid, it will be seen from the curve that the plate current will decrease to a greater extent below its normal value than it will increase (above normal), but in so far as the head telephones are concerned, the effect will be the same as on the lower bend of the curve; that is, the telephone diaphragm will be impulsed *once* for each spark of the transmitter. The valve is usually more sensitive at the lower bend of the curve.

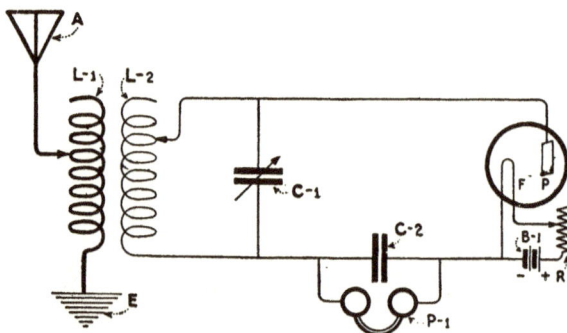

Fig. 169.   Circuit for use of the Fleming valve as a simple rectifier.

The circuit for the Fleming curve as a *simple rectifier* is shown in Fig. 169. Here one terminal of the secondary coil *L-2* is connected to the plate *P* and the other terminal through the telephone condenser *C-2* to the negative side of the filament *F*. *B-1* is a battery of about 6 volts and *R* a rheostat of 10 ohms resistance. The capacity of *C-2* is about 0.002 mfd. The action of the valve during reception of signals is somewhat as follows: When the incoming radio frequency currents charge *P* to a positive potential electrons are drawn over from the filament, which is equivalent to saying that a semi-cycle flows from *P* to *F*. For the negative half cycles, *P* is charged negatively, no electrons are drawn over to the plate and consequently no current passes the valve. Condenser C-2, therefore, receives a uni-directional charge over the duration of a wave train and discharges through the head telephone P-1.

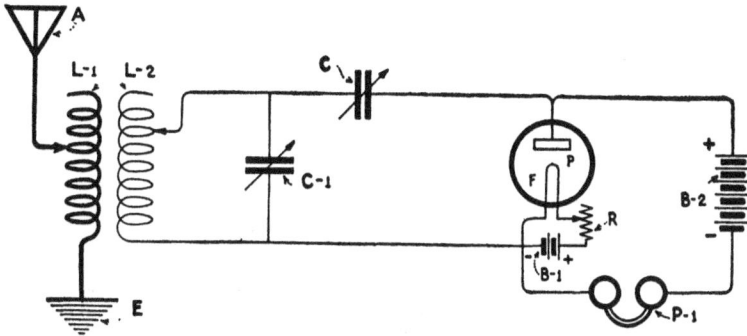

FIG. 170. Circuit for the Fleming valve using separate plate and filament batteries.

Fig. 170 shows another circuit for the Fleming valve, in which a separate battery B-2 of 20 to 40 volts is employed to energize the plate circuit.

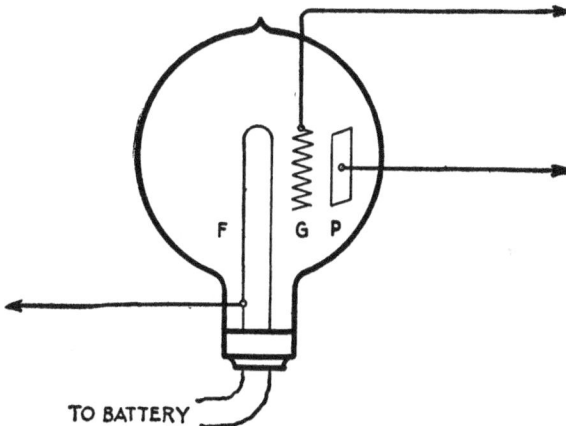

FIG. 171. The three-electrode oscillation valve or audion. F is the filament, G the grid, and P the plate

THE THREE-ELECTRODE VALVE.—The construction of the three-electrode oscillation valve is shown fundamentally in Fig. 171 where F is the *filament*, G the *grid element*, and P the *plate*. The materials commonly employed for the filament, plate and grid have already been mentioned. In this valve, the grid G acts as a *controlling element* of the electron currents between the filament F and the plate P. Such control is effected by *varying the grid to filament potential* by an externally applied e.m.f. Like the two-element tube, the operating characteristics of the three-electrode bulb can best be comprehended by analyzing its characteristic curves. Certain characteristic curves of the popular

Marconi V. T. will follow. This tube is suitable for all-around experimental use, as it possesses universal operating characteristics. It may be used as a *detector* and *amplifier* of radio frequency currents, or as a *generator* of continuous oscillations.

The beginner should understand what is meant by the *input* and *output* circuits of the three-electrode tube. For the bulb circuits shown in Fig. 171, the *input* terminals are the *negative side* of the filament *F* and *grid G*, the *output* terminals are the *plate P* and a *terminal* of the filament, or leads tapped off some part of the plate circuit. Thus, in Fig. 177, *A B* are the input terminals of the valve and *C D* the output terminals.

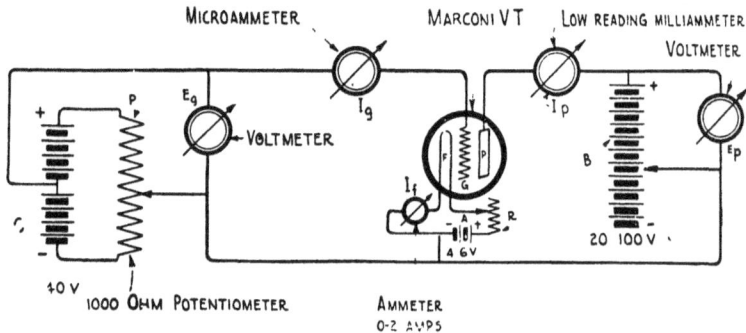

Fɪɢ. 172. Complete circuits for obtaining the various characteristics of the three-electrode valve.

As a preliminary description* of the operating characteristics of the valve, the reader should note the effect of varying grid potentials upon the strength of the plate current. Referring to the diagram, Fig. 172, assume that the filament is incandescent and the plate circuit energized by the plate battery *B*. Current flows from the positive terminal of *B* to the plate *P*, from the plate to the filament and then to the negative pole of the battery. We may then place varying potentials on the grid *G* (in respect to the negative terminal of the filament *F*) by the *grid battery C*. If the grid is charged by *C* to some upper *negative potential*, it will be found that the plate current as read at I_p falls to zero.

If the potential of *C* be slowly changed from some negative value to zero and then gradually increased to some positive value the current through I_p will increase; but as in the two-electrode valve, the resistance from *P* to *F* is not constant. It varies with the current as will be shown by the characteristic curves following. The curve which shows the strength of the plate current corresponding to various grid potentials is called the "*d.c. characteristic*" of the vacuum tube.

To gain a better understanding of the effects of filament, grid and plate potential in a vacuum tube detector, we will consider the curves of *grid potential* vs. *plate current*, *plate potential* vs. *plate current*; and *voltage amplification* vs. *series plate resistance* for the Marconi V. T.

---

*For a simple but more complete explanation of the theory underlying the functioning of the two electrode and three electrode valve, the reader is referred to the author's. Vacuum Tubes in Wireless Communication "

The data for all characteristic phenomena may be obtained by the circuits and apparatus of Fig. 172. C is a battery of about 80 volts shunted by a 2000-ohm potentiometer. A tap is brought from the middle of the cell bank so that moving the slider from one end towards the other end will reverse the polarity of the battery in respect to the input terminals of the tube. E_g is an appropriate voltmeter. I_g is a microammeter (such as the Paul type) for measuring feeble grid currents. I_p is a microammeter or a milliammeter for measuring the plate current. E_p is an appropriate voltmeter for the plate battery. I_f is the filament circuit ammeter whose maximum scale for receiving tubes need not exceed 2 amperes. Several characteristic curves of the Marconi V. T. will follow. The grid-voltage, grid-current curve is not shown for it is relatively unimportant to the experimental tests of the amateur worker.

THE GRID-VOLTAGE, PLATE-CURRENT CHARACTERISTIC.— The data for this characteristic are obtained by adjusting the filament temperature and plate voltage of the tube to normal operating values. The potential of the grid (in respect to the negative side of the filament) is then varied by potentiometer P (Fig. 172) noting at each step the readings of the voltmeter E_g and the ammeter I_p.

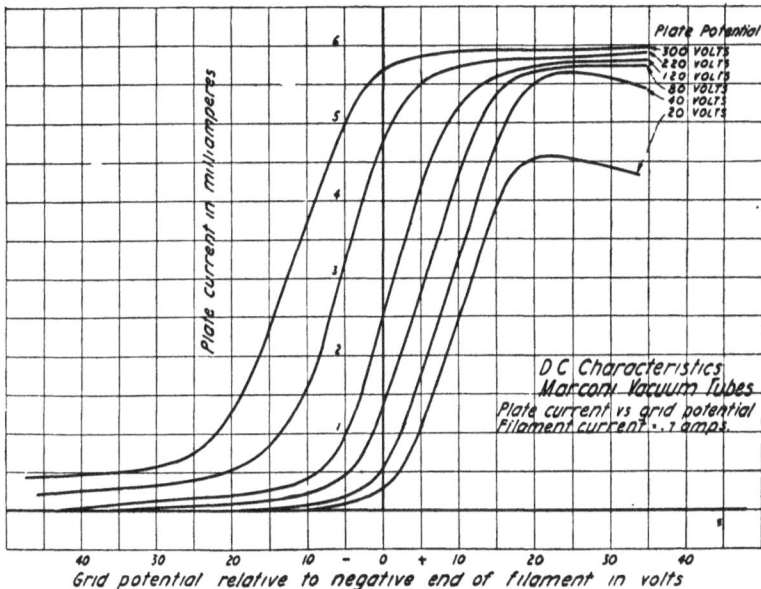

FIG 173. Curves of grid voltage vs. plate current for various plate potentials of the Marconi V T.

Fig. 173 shows a set of curves obtained in this way for the Marconi V. T. with plate voltages varying from 20 to 300 while the filament current is held steadily at 0.7 ampere. (These curves were taken with a Class II tube.)

Noting the curve for 40 volts plate e.m.f., it will be seen that when the grid potential is about —18 volts in respect to the filament the plate current is zero   The peak of the lower bend of this curve corresponds to a grid potential of about —2½ volts.   The plate current corresponding to this is 0.26 milliampere.   When the grid and filament are at the same potential, i e., at zero potential, the plate current is 0.6 milliampere.   The graph is linear from that point on for a considerable range of grid voltages and the point of saturation occurs when the grid potential is +20 volts.

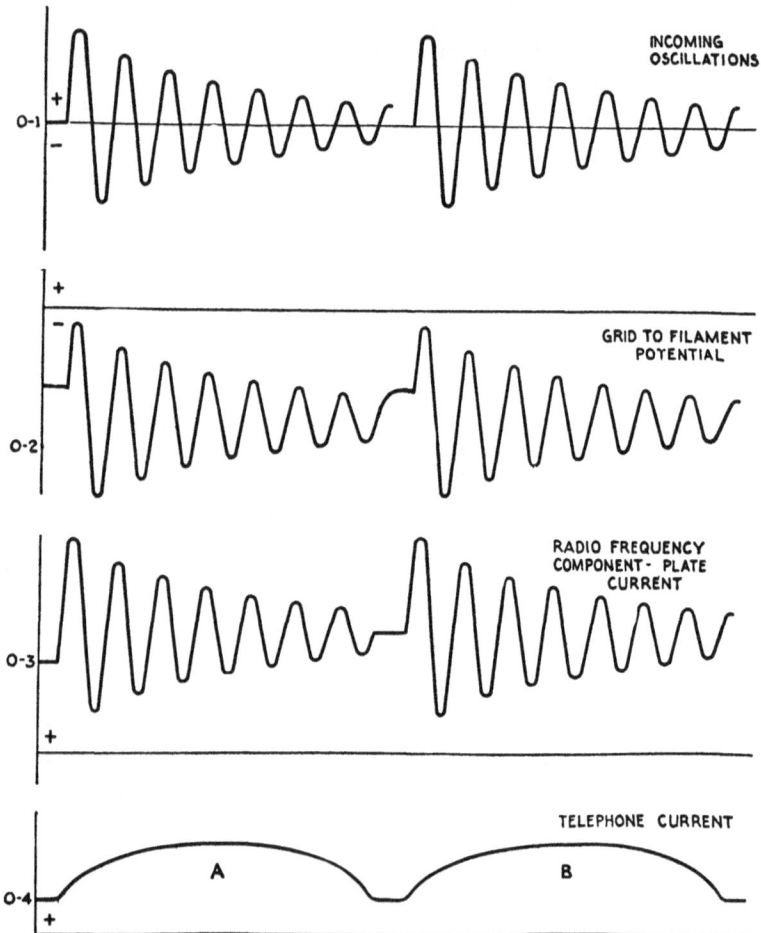

FIG. 173a.   Phenomena of detection with the three-electrode valve.

Referring to the valve circuit in Fig. 179 the point on the characteristic curve most favorable to detection in a circuit of that kind is the *lower bend*.   Say, for example, the plate voltage is 80 volts and the grid potential —5 volts.   Then, from Fig. 173 I_p the plate current = 0.45 milliampere.   If then the potential of the incoming signal for purposes of illustration, is 3 volts, then the potential of the grid to filament will vary between (—5)+(—3) = —8 volts and (—5)+(+3) = —2 volts.   The plate current I_p will accordingly vary between 0.25 milliampere and 0.9 milliampere; that is, it decreases 0.45—0.25 = 0.20 milliampere below normal and increases 0.9—0.45 = 0.45 milliampere above normal (for the first cycle in the wave train).

The increase of current through the head telephone evidently exceeds the decrease. Hence the telephones are energized by uni-directional currents which pull the diaphragm of the receiver once for each spark at the transmitter.

The phenomena of reception are outlined in Fig. 173a. Graph O-1 indicates two groups of incoming radio frequency currents; graph O-2 the fluctuating grid potential, graph O-3 the corresponding plate current and graph O-4 the approximate telephone current.

If the valve is worked at the upper bend of the curve, detection occurs also but the graph O-4 would then show a decrease rather than an increase over the time of a group of oscillations.

Further study of Fig. 173 shows that in the *linear* or *straight portion* of the curve, a small oscillating e.m.f. applied to the grid or input circuit will be reproduced with fidelity in the plate circuit. In this region of the curve incoming radio frequency currents will cause radio frequency variations in the output circuit and if there is no distortion the head telephone will give no response. It is along that portion of the curve that the valve characteristic is correct for cascade radio frequency amplification but not for detection except when the circuit of Fig. 180 is used. In that circuit good detection is often secured when the grid to filament potential is zero.

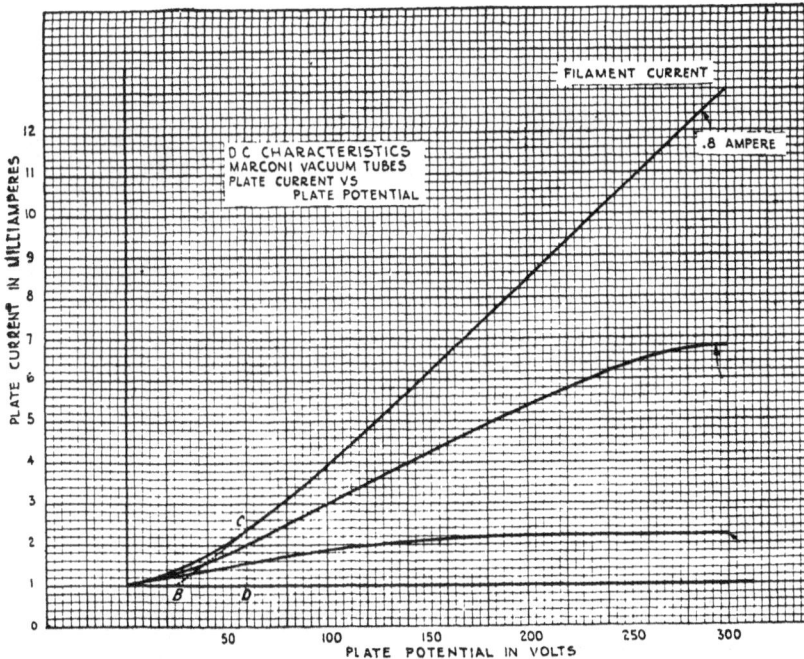

FIG. 173. Curves of plate voltage vs plate current for the Marconi V T

PLATE-POTENTIAL, PLATE-CURRENT CHARACTERISTIC.— The data is secured by connecting the grid to the negative terminal of the filament (Fig. 172), setting the filament current at normal value and varying the plate voltage progressively. Note is made of the readings of E_p and I_p.

Fig. 174 shows *plate-voltage, plate-current curves* of the *Marconi V. T.* for filament currents of 0.6, 0.7 and 0.8 ampere, with the grid at zero potential. These curves indicate how the characteristics of Fig. 173 may be expected to vary with different filament currents. They also permit computation of the internal resistance of the valve as will be explained on page 217.

FILAMENT-CURRENT, FILAMENT-POTENTIAL CHARACTER-ISTIC.—The data for this characteristic is obtained by connecting a voltmeter across the filament F, Fig. 172, and noting the reading of I_f at different voltages.

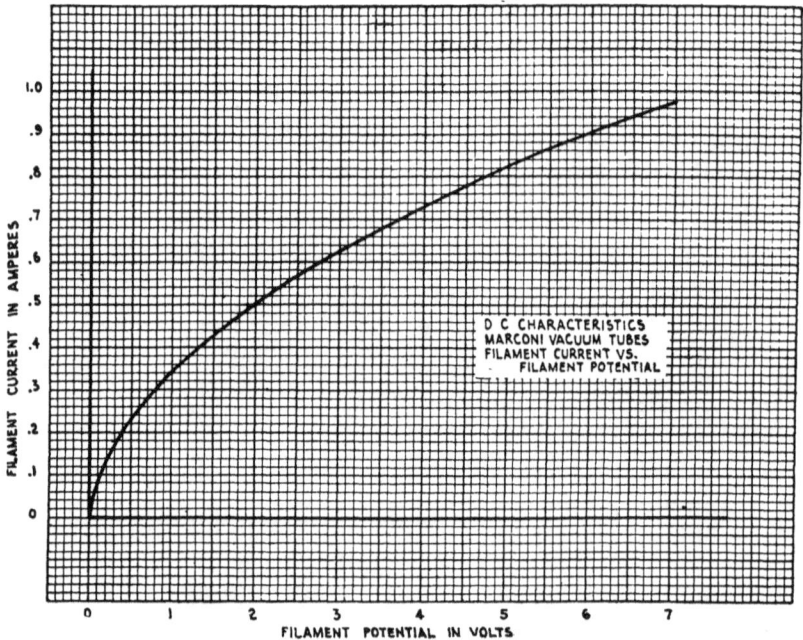

Fig. 175. Curve of filament current vs. filament voltage for the Marconi V. T.

The curve for the Marconi V T is shown in Fig. 175. The value of the curve is that it indicates the battery e.m.f. necessary to maintain the filament current at any desired value within its operating range.

AMPLIFICATION CONSTANT.—The amplification constant of a vacuum tube is defined as the ratio of the *change of plate voltage* necessary to bring about a given *change in plate current*, to the *change in grid potential* required to produce the *same change in plate current*.

For an illustrative example: Suppose that the grid potential is zero, and the plate current corresponding thereto at 80 volts plate e.m.f., is 1.35 milliamperes. Assume then, that the grid is held at +1 volts, and the plate voltage varied until the plate current is again 1.35 milliamperes. If the plate voltage is then 74.5 volts, it is evident that a change in the grid potential of 1 volt has an efiect of 5.5 times that of a similar change in the plate potential. The *amplification constant* is therefore 5.5. It is expressed by K.

The amplification constant of a tube may be obtained roughly by dividing the plate voltage by the grid voltage necessary to reduce the plate current to zero. The amplification constant of the Marconi V. T. is approximately 8.5 at operating adjustments.

A method of obtaining the value of K by direct measurement will be described on page 218.

FILAMENT TO PLATE IMPEDANCE.—In the design of cascade amplifiers knowledge of the internal impedance of a vacuum tube is essential. If this is known the impedance of the load (the output circuit) can then be given the value necessary for maximum power output. A somewhat parallel case is that of the ordinary dynamo where the maximum power is obtained in its output circuit when the resistance of the load is the same as that of the generator. Since the resistance of the load and generator are equal, equal amounts of power are dissipated in each and the efficiency of the load will be 50%. This low efficiency is obviously not permissible in power work, but when only small amounts of power, such as are present in the output circuits of vacuum tubes, are available equality of load impedance and tube impedance is desirable.

The *resistance of the plate to filament*, that is, the *internal impedance of the tube* may be calculated directly from the curve Fig. 174 or measured as in the paragraph following. Take, for instance, the upper curve in Fig. 174; the impedance at point C may be determined by drawing the tangent $C\,B$ to the curve at the operating potential which has been selected in this case as 60 volts. Then draw the line $C\,D$.

The impedance is then equal to the voltage $B\,D$ divided by the current $C\,D$. Since $B\,D = 36$, and $C\,D = 0.00135$ ampere, then R_t, the

$$\text{impedance} = \frac{36}{0.00135} = 26{,}666 \text{ ohms.}$$

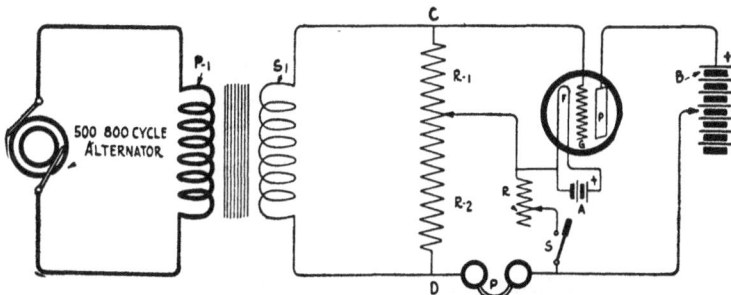

Fig. 176. Miller's method of measuring the internal impedance and amplification constant of a vacuum tube These measurements are of particular importance in the design of cascade vacuum valve amplifiers

MEASUREMENT OF INTERNAL IMPEDANCE AND AMPLIFI-CATION CONSTANT.—The method to be described was presented by J. M. Miller in Vol. 6, Proceedings of the Institute of Radio Engineers (June, 1918). The circuits are shown in Fig. 176. $C D$ is a slide wire resistance of about 7 ohms. A generator of 500 to 800 cycles coupled to $C D$ through a transformer designed so that the current through $C D$ is about 50 milliamperes. R is a dial resistance going up to approximately 10,000 ohms.

The amplification constant K is obtained by adjusting the slider on $C D$ with the switch S open until the telephones are silent, then

$$K = \frac{R_2}{R_1} \tag{59}$$

If $R_2 = 9$ and $R_1 = 1$, then $K = \frac{9}{1} = 9$.

To determine the internal impedance R_t, set the slider on $C D$ at the middle point, close the switch S and vary R until silence is again obtained, then

$$R_t = (K-1) R \tag{60}$$

If, for example,

$K = 9$, $R = 3300$ ohms, then
$R_t = (9-1) \ 3300 = 26,400$ ohms.

For any other ratio of $\frac{R_2}{R_1}$ having a value less than K formula (60) becomes

$$R_t = \left(\frac{R_1}{R_2} K - 1 \right) \ R \tag{61}$$

The tube should be adjusted to normal operating conditions during the measurement.

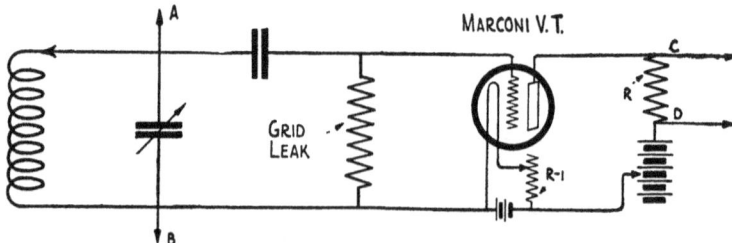

Fig 177. Indicating the "input" terminals A B and "output" terminals C D of the three-electrode valve.

VOLTAGE AMPLIFICATION.—The voltage amplification of a vacuum tube is the ratio of the voltage obtainable across a resistance in series with its plate circuit to that impressed upon the grid circuit. Thus, in Fig. 177, it is the ratio of the voltage between the terminals C and D of R, to the voltage applied between A and B.

If K = the amplification constant of a tube, R_t = its internal resistance, R = the series plate resistance and μ = the voltage amplification, it can be shown that

$$\mu = \frac{K \times R}{R + R_t} \tag{62}$$

Thus if the maximum amplification constant of a tube = 9, its internal impedance = 29,000 ohms and the load or coupling resistance = 1,000,000 ohms, then,

$$\text{voltage amplification } (\mu) = \frac{9 \times 1,000,000}{1,000,000 + 29,000} = 8.7.$$

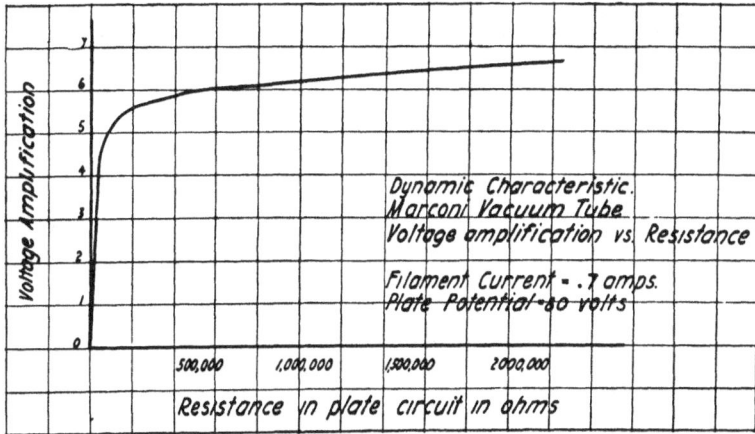

Fig. 178.   Curve of voltage amplification vs. series plate resistance for the Marconi V T

A curve of *voltage amplification* plotted against *series plate resistance* for the *Marconi V. T.* is shown in Fig. 178.   The value of this curve lies in the fact that it shows the correct value of coupling resistance to give the maximum efficiency with cascade amplifiers.   It is seen that the voltage amplification increases rapidly as the load resistance is increased to 200,000 ohms.   At 2,000,000 ohms, the voltage amplification is 6.5 which compares well with the best detector tubes.   A load impedance of this value will be used in the cascade amplification circuits following.

The amplification constant K of a tube is dependent upon tne ratio of the spacing between the grid and filament to the spacing between the grid and plate and upon the construction of the grid.   The amplification increases with a decrease of the ratio and with smaller grid meshes. Since these factors of construction are decided upon by the maker the amateur must design his output circuits to fit the impedance of the tube.

DETECTOR CIRCUIT FOR THE MARCONI V. T.—There are two general methods of detecting damped oscillations by a vacuum valve. The circuit of Fig. 179 utilizes a *grid battery* C to permit operation on the most favorable point of the characteristic curve. The second circuit, Fig. 180, employs a *grid condenser* C-1 to give maximum rectification. This difference may be said to constitute the distinguishing features of the two circuits.

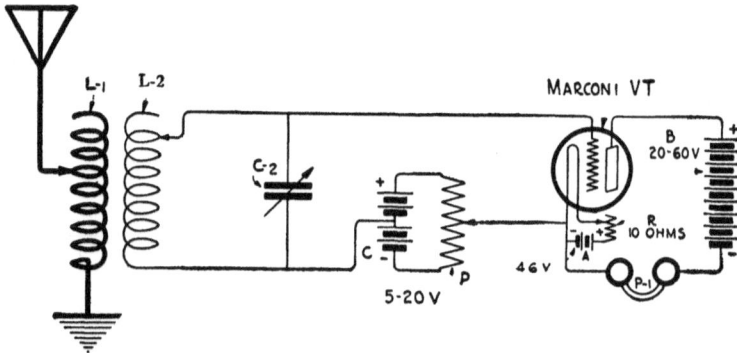

FIG. 179.   Detection circuit for the three-electrode valve using a grid battery to permit operation on a favorable point of the characteristic curve of the valve.

Referring to Fig. 179, L-1 and L-2 are the primary and secondary of a tuning transformer. C-2 is a shunt secondary condenser. C is a grid battery varying up to 20 volts. P is a 400-600 volt potentiometer; B, the plate battery of 60 volts and A, a 4-6 volt storage battery. P-1 are telephones of 2000 ohms. By varying the e.m.f. of C, the grid can be held at any desired negative or positive *potential relative to the filament* within the limits of the tube.

Referring to Fig. 173, noting the "*80-volt plate e.m.f.*" *curve*, it will be seen that the lower bend occurs where the grid potential is —5 volts. The grid may be held at that potential by the battery C. In the region of the bend of the curve, an oscillating e.m.f. impressed upon the grid and filament (the input circuit) will be repeated in the plate circuit with *distortion*, that is, when the incoming e.m.f. impressed upon the grid is positive the resulting increase of the plate current above the normal value exceeds the decrease (below the normal value) when the grid is negative. The preponderance of current therefore flows in one direction, i.e., a rectified current flows in the plate circuit.

The result is the same as that obtained with the two-electrode valve, the graphs of which were shown in Fig. 168. While the signal is passing, the plate current on the average increases, decreasing to normal between sparks at the transmitter. If, on the other hand, the grid potential is made positive in respect to the filament, that is, the valve is worked at the upper bend of one of the curves of Fig. 173 each group of incoming oscillations will cause a decrease in the average telephone current.

The grid battery is not strictly essential in the circuit of Fig. 179. Varying negative grid potentials can be obtained with some bulbs simply by varying the filament temperature and the plate voltage. In other words, the favorable point on the curve for detection may often be located in this way without a grid battery.

PREFERRED DETECTION CIRCUIT FOR THE MARCONI V. T.— Instead of the grid battery of Fig. 179 the detection circuit for the Marconi V.T. in Fig. 180, has a *condenser in series with the grid.* Usually with this connection and no oscillations in the grid circuit, the grid and filament are at zero potential and no current flows in the grid circuit; that is, no electrons pass from the filament to grid.*

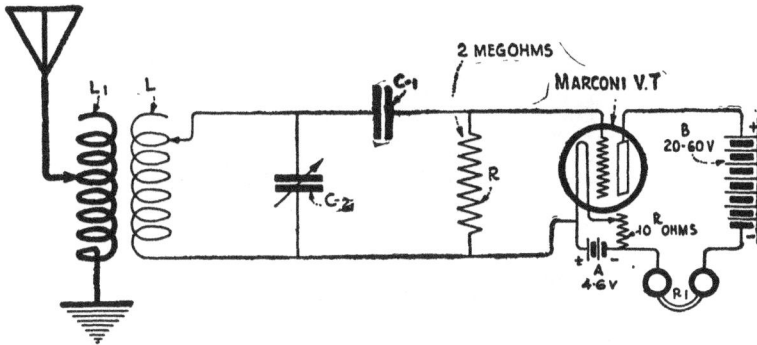

Fig. 180   Detection circuit for the three-electrode valve utilizing a grid condenser to obtain maximum rectification. This circuit is generally preferred for reception from spark transmitters. A grid leak is required to limit the charge accumulating on the grid.

When radio frequency e.m.f.'s are impressed upon the grid circuit of Fig. 180 and the grid to filament potential is approximately zero, the grid is made alternately *positive* and *negative.* When the grid is positive electrons are *drawn over* to it, but when it is negative the electrons are *repelled.* Thus, for each succeeding half oscillation electrons are drawn to the grid, placing a charge in the grid condenser C-1, which is negative on the grid side. As the characteristics of Fig. 173 show, an increasing negative charge on the grid acts to reduce the plate to filament current; and hence, while a group of incoming oscillations undergo rectification, the telephone current is reduced. If means are provided for leaking the charge out of the grid condenser slowly, the grid and likewise, the plate voltage will return to their normal potential between sparks at the transmitter. In tubes which are not highly exhausted the charge in the grid condenser will leak through the bulb if *positive ionization* occurs. Some leakage may take place through the *dielectric* of the grid condenser or through the glass supports of the grid. *Artificial leakage* must be provided with such valves as the Marconi V. T. In practice, leak resistances are called *"grid leaks"* and they vary from, say, 80,000 to 2,000,000 ohms.

Grid leaks may be made by drawing a number of *parallel lead pencil lines* on cardboard between binding posts. If the resistance is too low some of the lines may be rubbed out with an eraser. Certain grades of

---

*By the use of a grid condenser and a variable grid leak the grid can be held at any desired negative potential.

drawing ink if deposited on paper between binding posts will give the necessary leakage. *Carbon paper* has been used. Grid leaks mounted in glass tubes with sealed-in ends ready for use, of the type shown in Fig 180a, may be purchased from the Marconi Company They are positively essential for best operation of the Marconi V. T.

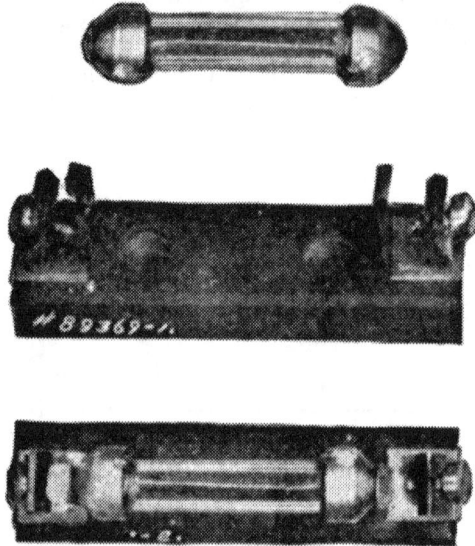

Fig. 180a. Grid leak and base as manufactured by the Marconi Company.

In regard to the complete apparatus called for in Fig. 180: the dimensions of *L*-1 and *L*-2 can be obtained by calculation as explained on pages 169 to 173. For short wave lengths *C*-2 can be dispensed with. For long wave lengths its capacity should not exceed 0.0005 mfd. *C*-1 may be of fixed capacity—about 0.0001 mfd. *R* has resistance of 2 megohms. The e.m.f. of the *A* battery is 4 volts, of the *B* battery about 60 volts.

To obtain the best response from the Marconi V. T., set the plate voltage at some voltage between 25 and 60 and slowly increase the filament temperature by the rheostat *R* until a signal maximum is heard in the telephones.

The life of the bulb with normal usage is about 1500 hours.

CASCADE AMPLIFICATION.—It is clear from the preceding explanations that the vacuum tube is an amplifier of a variable e.m.f of any wave form. A single bulb itself is an amplifier and if used as in Fig. 179 or 180 will give many times the strength of signal obtainable with a *simple rectifier* like carborundum. If either radio or audio frequency voltages are impressed upon the input terminals *A B* of Fig. 177, *amplified voltages* of the same frequency will exist across the terminals of the resistance *R*. These voltages may then be impressed upon the *grid circuit* of a second tube wherein further amplification takes place.

The output and input circuits of as many as 8 tubes may be coupled in cascade for progressive amplification. Generally the limit is reached with about 8 tubes.

In fact the limits of amplification are reached with any combination when the current variation of the last tube has reached the point of saturation. Weagant, of the Marconi Company, and his staff have developed a very successful 8-tube amplifier with which signals may be copied three or four thousand miles on a small frame aerial a few feet square.

In addition to a *pure resistance coupling* of Fig. 177, *capacitive, inductive and conductive coupling* (auto transformer coupling) are employed in cascade circuits.* For instance, in a *cascade radio frequency amplifier* wherein the successive plate and grid circuits are coupled *inductively*, use is made of an *air core radio frequency transformer* as the *coupling element*. Iron core coupling transformers are employed for *audio frequencies*. These may be either of the *open* or *closed core* type. A single impedance coil is sometimes used.

In the earlier types of cascade amplifiers separate filament and plate batteries were employed for each bulb. Circuits have been developed where *one filament battery* and *one plate battery* were used to energize all bulbs. There are few circuits where it is still an advantage to employ separate batteries for each step.

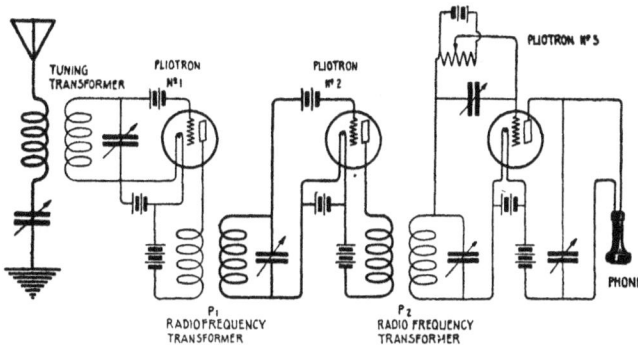

Fig. 181   Circuits of a cascade radio frequency amplifier using inductive intervalve couplings   Separate filament and plate batteries are used for each valve. This is one of the earlier type of circuits which has been improved by using one plate battery and one filament battery for all valves whether used for radio or audio frequency amplification.

The *resistance coupled* amplifier is rapidly gaining preference with experimenters because of the simplicity of the circuits and the ease of obtaining maximum amplification. The advantage of this method of coupling over the method that uses *radio frequency tuning transformers* between valves is obvious. The latter method requires tuned *input* and *output* circuits throughout the system, and any change in tuning in the radio frequency circuits of the first bulb calls for retuning the grid and plate circuits of all bulbs. This is a complex procedure requiring the attention of a skilled manipulator.

*For a more comprehensive description of the many circuits for the vacuum tube, the reader should consult the author's "Vacuum Tubes in Wireless Communication " Only the up-to-date circuits believed to be the most feasible for amateur use are described here

With the resistance coupled amplifier the only tuning required is that usually done at the tuning transformer. A change of wave length does not call for any particular changes in the constants of the amplifier circuits.

A *three-valve circuit* for the amplification of radio frequencies using separate plate and filament batteries for successive steps, is shown in Fig. 181. The output and input circuits of successive bulbs are coupled through radio frequency tuning transformers P-1 and P-2. The plate circuits of all bulbs are untuned but the grid circuits are carefully tuned to the frequency of the incoming signal by shunt variable condensers. The first two bulbs function as radio frequency amplifiers and the last bulb acts as a detector in which the oscillations are rectified and stored up in the grid condenser. Favorable grid potentials, which are generally obtained along the linear portion of the characteristic curve, are given to the first two bulbs by grid batteries which in practice are shunted by potentiometers. The battery and potentiometer shunted around the grid condenser of the last tube are employed to limit definitely the potential difference between the grid and filament. This prevents the grid potential rising to such a high negative value that the tube becomes inoperative.

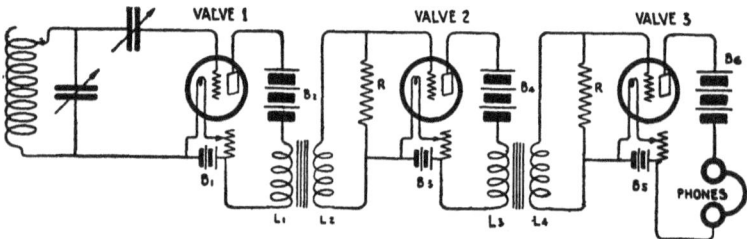

FIG. 182. Circuits of a cascade audio frequency amplifier with iron core intervalve couplings.

The fundamental circuits of an *audio frequency amplifier* are shown in Fig. 182. Here the first bulb is the detector and the second and third bulbs are audio frequency amplifiers. The output circuit of the first bulb is coupled to the input circuit of the second bulb through the transformer L-1 and L-2. Transformer L-3, L-4 performs the same function between the second and third valves. Suitable intervalve transformers for this circuit will be described on page 239. The intervalve transformer is designed to have an impedance at, say, 500 cycles equal to or somewhat greater than the internal impedance of the tube. It is designed with a step-up ratio of turns.

The operation of the circuit is as follows: During reception from spark stations, the incoming oscillations are rectified between the grid filament and, as already explained, the charge and discharge of the grid condenser sets up audio frequency voltage variations across the primary L-1. Similar voltages are induced in L-2, which vary the grid potential of the second bulb wherein further amplification occurs and so on.

Grid batteries ("C" batteries) may be used in the grid circuits of all tubes. With some bulbs control of the grid potential is essential.

CASCADE CIRCUIT FOR THE MARCONI V. T.—There are any number of circuits which can be used. A cascade system for the Marconi

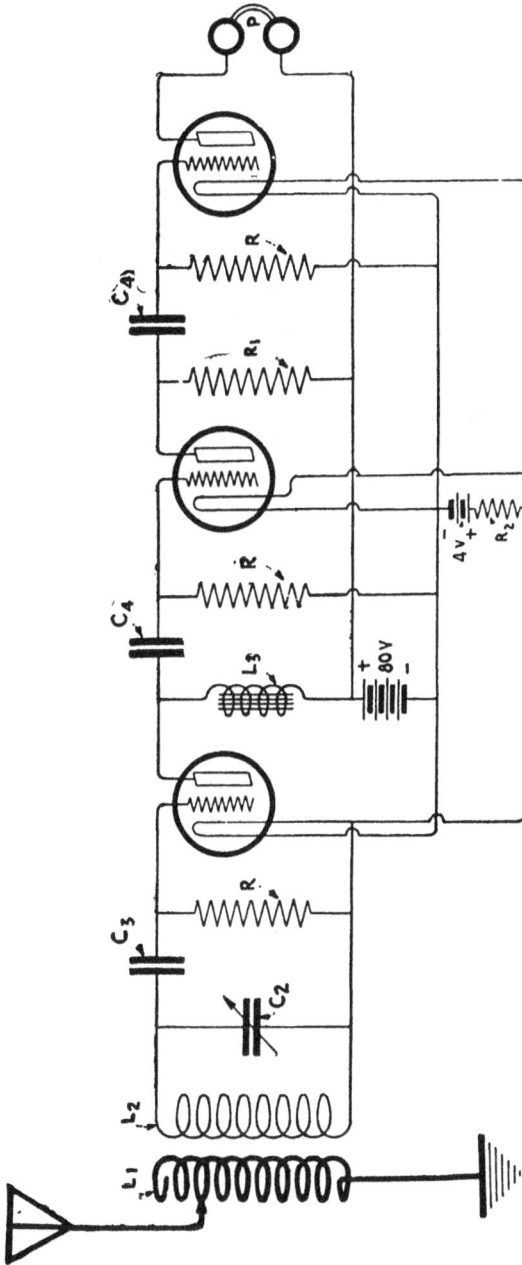

FIG. 183. Cascade amplification circuit worked out especially for the Marconi V. T. The first valve is used as a detector and the second and third bulbs as amplifiers. The first valve is coupled to the second valve through an iron core impedance but the second and the third valves are coupled through a resistance of 2 megohms. This circuit is particularly recommended for amateur communications at 200 meters although it may be used for longer wave lengths.

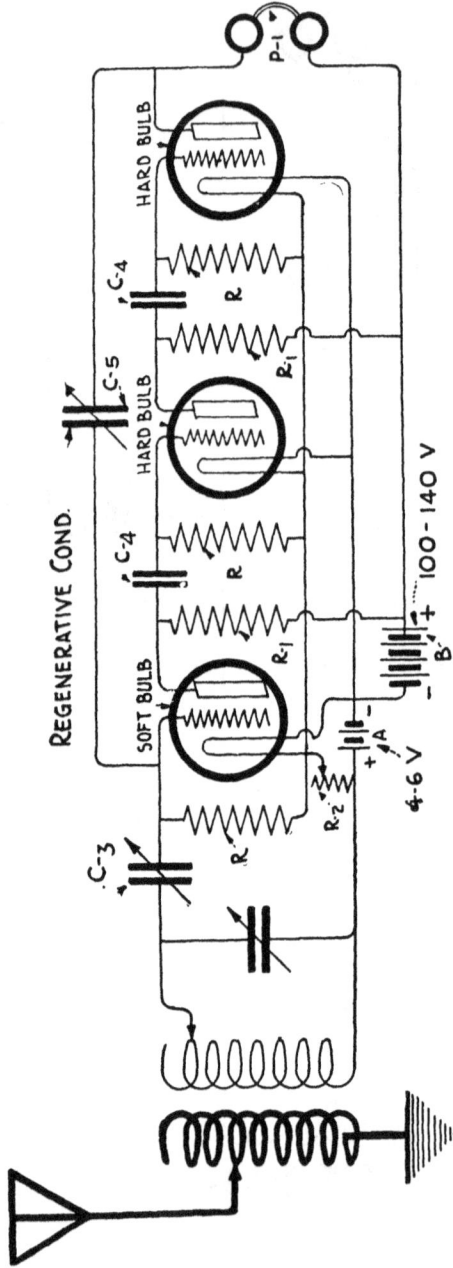

Fig 184   Cascade amplification circuit for the Marconi V T using pure resistance couplings between the output and input circuits of successive valves The resistance of these couplings usually should exceed the internal impedance of the valves by many times. Any number of valves up to 8 may be used in this circuit.

V. T. that has been worked out particularly for amateur use is shown in Fig. 183. Amplifications of 60 are obtained with three (Class II) bulbs. The feature that will appeal to the experimenter is the use of a single plate battery and a single filament battery for all tubes.

Referring now to the diagram, L-2, C-2 is the *secondary circuit* of the tuning transformer. C-3 is a *grid condenser* of 0.0001 mfd. L-3 is an *iron core impedance* of 20 henries which acts as the coupling element between the first and second bulbs. R-1 is a 2 megohm resistance which acts as the *coupling element* between the second and third bulbs. The resistances R, are grid leaks of 2 megohms each. These prevent the grid potentials of successive valves rising to such a negative value that they become inoperative. The blocking condensers C-4 have capacitance of 0.005 mfd each. P is a 2000-ohm telephone.

Note that the filaments are fed by the same A battery. The B battery, which now ha an e.m.f. of 80 volts, energizes the plate circuits of the three tubes.

The operation is as follows: The first tube acts as a detector, the incoming oscillations being rectified by the valve action.

Thus, audio frequency voltages are built up across the impedance L-3 which are impressed upon the grid and filament terminals of the second tube. This tube acts as an amplifier and still greater voltages are built up across the coupling resistance R-1. The output of the second bulb is again amplified by the third tube. A coupling resistance similar to R-1 might have been inserted in the plate circuit of the first bulb instead of the coil L-3. Instead of this the coil was employed to permit the use of lower voltages at the B battery. Had resistance couplings been used throughout the circuits, the e.m.f. of the B battery would have had to be around 110 volts or more.

The impedance L-3 can be made by winding 10,000 turns of No. 36 enamel wire on a core of silicon steel or iron wire $\frac{5}{8}''$ in diameter and $3''$ in length. The inductance of the coil is approximately 20 henries.

As mentioned before, the leak and coupling resistance may be made of lead pencil lines drawn on a piece of cardboard, or appropriate grid leaks may be purchased from the Marconi Company. It is believed that this circuit will satisfy all the requirements of the amateur station and it is particularly recommended for use in long distance relay work.

The circuits of a *resistance coupled amplifier* containing any number of bulbs up to eight, are shown in Fig. 184. This method of coupling gives good results with the Marconi V. T. The output and input circuits of successive tubes are coupled through resistances of 2 megohms each. The resistance* of the coupling units in such a circuit should always exceed the internal impedance of the valve. The B battery should have a potential of 120 volts. The condensers C-4 have capacitance of 0.005 mfd. each.

In this circuit, the first tube rectifies incoming oscillations and the second and third tubes amplify the audio frequency component of the plate current of the first tube. R-2 is a filament rheostat of 10 ohms. If the voltage of the filament battery is four volts or less, the rheostat is not required. Still greater amplification can be obtained by

---

*Note paragraph numbered (2) and (3) page 341 in the appendix.

connecting the condenser C-5 in the circuit as here shown. The condenser provides regenerative coupling between the plate circuit of the last bulb and grid circuit of the first bulb. The condenser will only give regenerative amplification at short wave lengths For long wave lengths a "tickler" circuit is preferable.

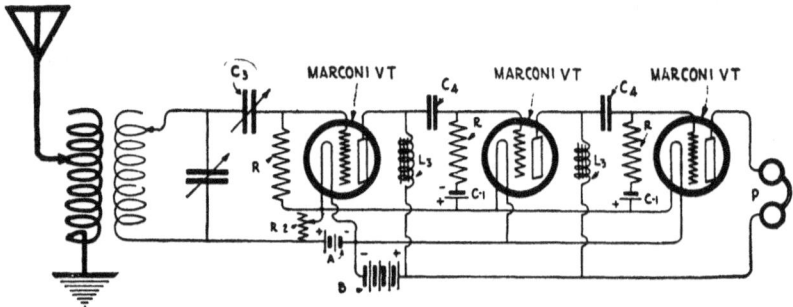

FIG. 185. Cascade amplification circuit using iron core chokes as intervalve couplings. The impedance of these chokes should be somewhat greater than the internal impedance of the tubes.

The circuits of an *impedance coupled amplifier** for use with the Marconi V. T. are shown in Fig. 185. This system has been used in receiving sets in wireless telephony for recording speech signals. The transformers between the amplifier bulbs are designed to give the best response at average "speech frequency" which is approximately 800 cycles. The impedance of the coils is somewhat greater than the internal impedance of the tubes.

The output circuits of the first and second bulbs are coupled to successive input circuits through the iron core impedances L-3, each of which are designed to have an impedance equal to, or two or three times the internal impedance of the corresponding valve. These chokes may consist of an iron core about $2\frac{1}{2}''$ long, $\frac{1}{2}''$ in diameter, wound to a diameter of $1''$ with No. 38 enamel wire. The choke L-3, the dimensions for which are given on page 238 is also satisfactory.

The grid leaks R have resistance between 1 and 2 megohms. The grid batteries C-1, in the second and third bulbs, permit working on favorable points of the characteristic curves of the amplifier tubes. A battery of a few volts will fulfil the requirements. The resistance of the telephones P should be rather high, and if high resistance telephones are not available a series resistance of 10,000 to 30,000 ohms may be connected in series with the telephones.

The first bulb is used for detection purposes. The grid condenser C-3 may be dispensed with in that circuit and a grid battery of a few volts substituted as shown in preceding diagrams. The grid battery may be shunted by a potentiometer to permit close regulation of the grid voltage. If the condenser C-3 is employed, it should have a capacitance of about 0.0001 mfd. The condensers C-4 should have capacitance of 0.005 to 0.5 mfd.

---

*For audio frequency amplification.

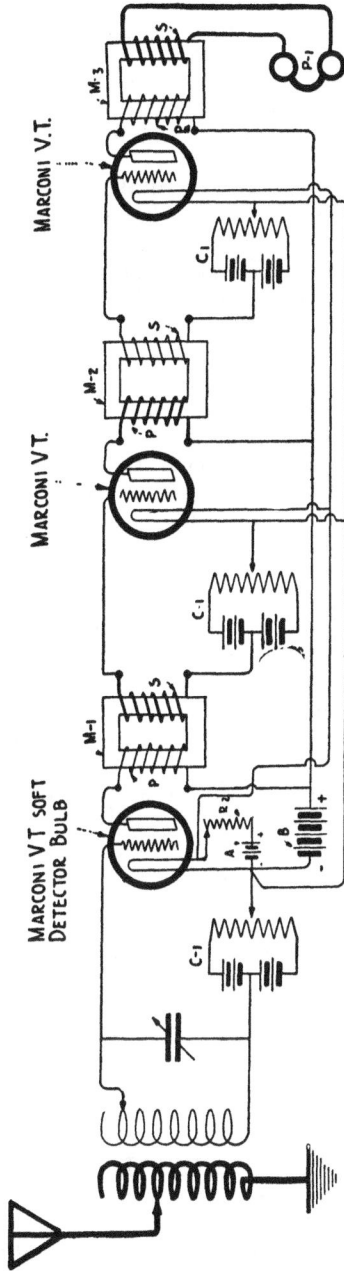

Fig. 186 Cascade amplification circuit using closed core step-up transformers between valves. These transformers are desirable for tubes having low amplification constants and low internal resistances. The impedance of the transformer with the secondary on load should equal the interal impedance of the tube. The dimensions of a suitable transformer are given in Fig. 199. The use of M-3 is not always essential.

The circuit in Fig. 186 is particularly applicable for cascade amplification by tubes having low amplification constants and low internal resistances. The tubes in this circuit are coupled through *closed core* transformers. If the amplifier is to be used for wireless telephony the impedances of the transformers should be approximately equal to the internal impedances of the tubes at 800 cycles. The requisite impedance may be calculated for lower frequencies (500 cycles) and the design of the transformer changed accordingly. The ratio of transformation should range from 3 to 6 according to the design of the valve.

The grid batteries *C*-1 permit operation on favorable points of the characteristic curve which generally are obtained with negative values of grid potential. The telephones *P* should have an impedance equal to the impedance of the transformer *M*-3. Transformers *M*-1, *M*-2 and *M*-3 are of identical design and one suitable for the Marconi V. T. is shown in Fig. 199.

Summing up the foregoing circuits Fig. 183 is particularly applicable for cascade amplification with the Marconi V. T. at short wave lengths, assuming the reception of damped oscillations. Fig. 184 is suitable for the same purpose. Fig. 185 is a circuit applicable to telephonic and telegraphic reception, provided the second and third bulbs are designed for high amplification constants and high internal impedances; but difficulty is apt to be encountered in practical operation with several stages due to the tendency of the circuits to oscillate at audio frequencies. Fig. 186 is suitable for amplification using bulbs having low amplification constants and low internal impedances. This circuit functions well with bulbs requiring plate voltages around 25 volts.

It may prove of interest to state here that the impedance of a telephone receiver is often sufficient for an intervalve coupling. One standard make of telephone has an impedance of 22,000 ohms at the frequency of 500 cycles.

Regarding the correct grid potential for amplification or detection the experimenter should understand that if the valves are to be used simply for amplification they should be worked along the linear or straight portion of the grid-potential, plate-current curve. This holds good for radio or audio frequency amplification. But for the detection of radio frequency currents in telegraphy, a grid potential should be selected corresponding to some point on the lower bend on the curve where the loudest signals are obtained. Assume a detecting circuit made up of a single bulb with a three-stage amplifier, the first, the detection bulb, should be adjusted to operate on the bend of the characteristic curve (if no grid is employed) but the grid potential of successive bulbs is preferably such as will permit working along the straight portion of the d.c. characteristic curve.

SELF-AMPLIFIERS OR REGENERATIVE RECEIVERS.—The regenerative principle for self-amplification was first disclosed by Armstrong. The underlying idea may be understood by referring to the circuit Fig. 177. If radio frequency e.m.f.'s are impressed across the input terminals *A B*, *amplified radio frequency voltages* will be set up across the terminals *C D* of the resistance *R*. If some of the energy in the plate circuit is then fed back to the grid circuit by *inductive, conductive* or *capacitance* coupling, the amplitude of the original grid oscillations will be *increased* and still greater variations in the plate current will result. The incoming signals may thus be amplified enormously. Either the *radio or audio frequency component* of the plate current can be amplified in this way. For *radio frequency amplification*, the grid and plate circuits are coupled through a *condenser* or a *radio frequency* transformer; for *audio frequency amplification*, an *iron core* transformer is employed. In all regenerative circuits caie must be taken to have the plate oscillations in the correct phase relation so as not to oppose the grid oscillations. The correct relation between the grid and plate currents is readily found by experiment. When regeneratively coupled circuits are used for the reception of damped oscillations, the coupling of the regenerative transformer *should not exceed a certain critical value* (determined by trial), as otherwise the valve will be set into self-oscillation at a radio frequency and *beat reception* will be obtained. Although *beat receivers will amplify spark signals* to a very great degree, the note of the spark station is distorted and a note of different pitch than the normal tone obtained. As a matter of fact any regenerative receiver designed for the amplification of spark frequency signals will permit the reception of undamped waves if the coupling between the grid and plate circuits is *sufficiently close*. All of the circuits which follow will receive either damped or undamped waves. In the case of undamped wave reception, the inductance and capacitance of the tuning elements must be of the correct magnitude for obtaining resonance with the longer waves usually employed by high power stations using undamped wave generators.

There are many ways of connecting vacuum tubes for regenerative amplification. Some of the more important circuits suitable for the amateur station will be described.

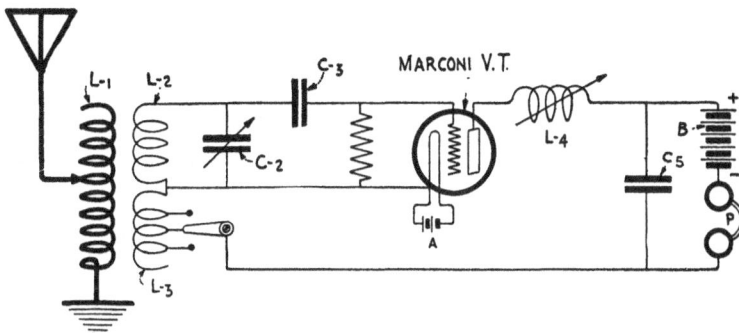

FIG. 187. Armstrong's original regenerative circuit for amplification of radio frequencies.

PRACTICAL REGENERATIVE CIRCUITS.—Fig. 187 is Armstrong's original regenerative circuit. The plate circuit is coupled to the grid circuit at *L-2*, *L-3*. *L-3* consists of several turns of wire in inductive relation to *L-2* through which the amplified plate voltages are impressed upon the grid circuit for further amplification. If an untuned plate circuit is used and the coil *L-4* eliminated, the coil *L-3* should have from 35% to 75% of the inductance of *L-2* for the longer wave.

Condenser *C*-5 provides a path around the head telephone for the radio frequency fluctuations of the plate current. *L*-4 is a large tuning inductance which is not required at short wave lengths. In practical operation the signal is first tuned in with loose coupling between *L*-2, *L*-3, after which the coupling is made closer for maximum amplification.

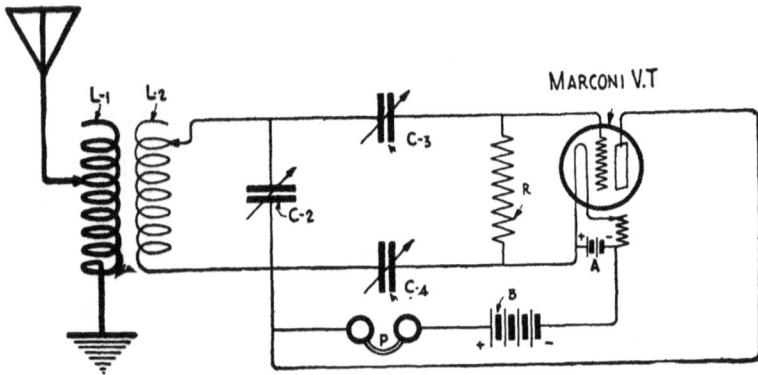

FIGURE 188-A

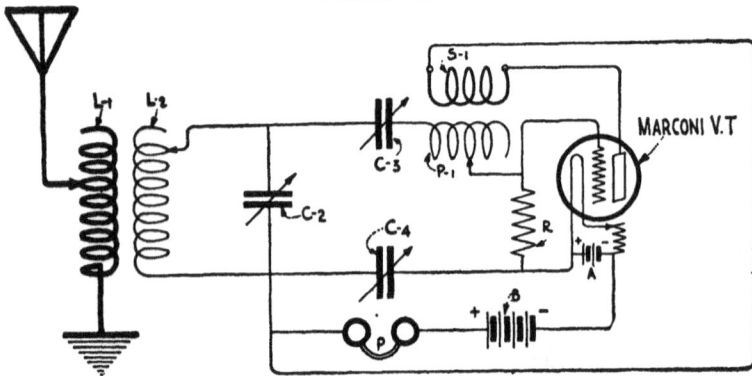

FIGURE 188-B

Fig. 188a.   Ultra-audion circuits providing electrostatic coupling for regenerative amplification.
Fig. 188b.   Modified circuit providing inductive and electrostatic regenerative coupling.

Deforest's ultra-audion circuits are shown in Figs. 188a and 188b. In the first diagram the secondary terminals of the receiving transformer are connected to the grid and plate of the tube rather than to the grid and filament, which is the more common connection. The condenser *C*-4, shunted by the head telephone *P* and battery *B*, provides *electrostatic regenerative coupling;* that is, the condenser *C*-4 is charged by the fluctuating voltages across the plate battery (and head telephone). The condenser discharges into the grid circuit amplifying the oscillations therein.

The modified circuit in Fig. 188b has the *"tickler" coil S*-1 which is in series with the plate circuit. Inductive regenerative coupling in addition to the electrostatic coupling afforded by the condenser is thus provided. These circuits may be used for the reception of *arc* or *spark signals*. Fig. 188*b* is applicable to long wave lengths provided *P*-1 and *S*-1 have sufficient mutual inductance to keep the valve in oscillation at the lower frequencies.

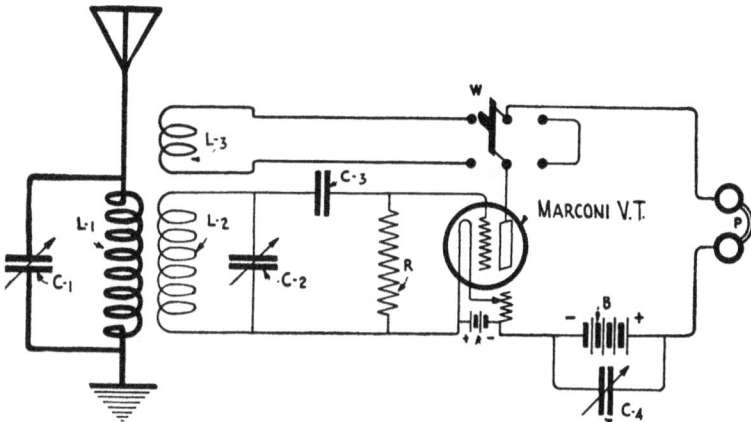

FIG. 189.   Inductive regenerative coupling using a so-called "tickler" coil in the plate circuit, which is in inductive relation to the grid circuit   In practice the tickler coil has from 35% to 75% of the inductance of the secondary circuit for wave lengths from 1000 to 18,000 meters.   For very short wave lengths the inductance of the tickler should be equal to that of the secondary.

Another way of using Armstrong's regenerative principle is shown in Fig. 189, where a "tickler" coil *L*-3 connected in series with the plate circuit is placed in inductive relation to the high potential end of the secondary coil *L*-2.   The "tickler" may be cut in or out of the circuit by the switch *W*.   When the switch is placed to the right, a "plain detector circuit" for the Marconi V. T. results, but when thrown to the left, *regenerative amplification* is obtained.   The regenerative connection will amplify the incoming signal about six times.   It is possible to design

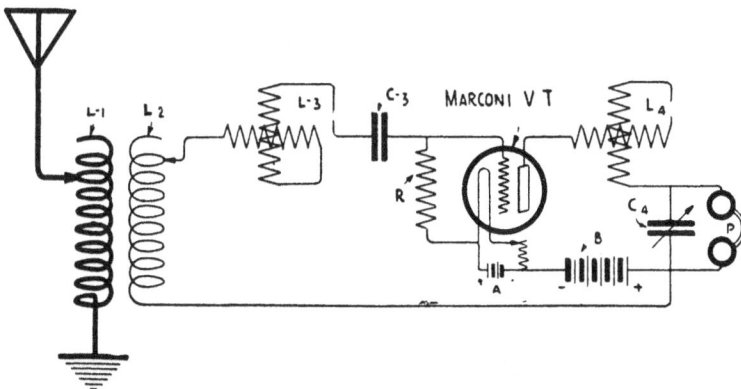

FIG. 190.   A system of regenerative coupling suitable for the amplification of signals at short wave lengths.

the primary and secondary coils *L*-1 and *L*-2 of the tuning transformer so that for the range of wave lengths, 200 to 600 meters, tuning can be accomplished by the condensers *C*-1 and *C*-2 alone. This simplifies the mechanical construction of the receiving transformer. This circuit is recommended for *long or short wave reception* and is much used in government work.

The *regenerative circuit* shown in Fig. 190 is widely used and seems to be preferred by the majority of amateur wireless experimenters for *short wave lengths*. Regenerative coupling is here obtained by connecting a terminal of the secondary coil *L*-2 to the positive terminal of the plate battery *B*. Accordingly, any fluctuations of e.m.f. across the battery, during reception of signals, react upon the grid circuit and thus provide regenerative amplification. *L*-3 and *L*-4 are small *variometers*, one in the grid circuit and the other in the plate circuit. They are used for the purpose of tuning. Amplification is obtained in any vacuum tube circuit by tuning the plate circuit as well as the grid circuit to the frequency of the incoming signal. The telephones *P*, in Fig. 190, are shunted by a small variable condenser *C*-4 which generally need not exceed 0.0005 mfd. This circuit is recommended for the Marconi V. T.

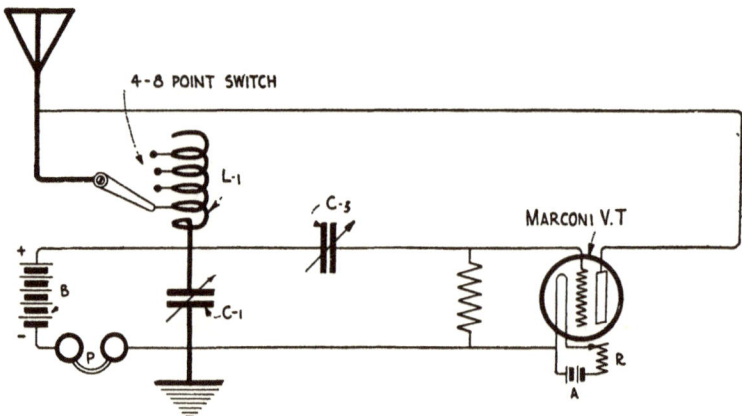

Fig 191   Simple regenerative circuit feasible for transmission and reception.   The plate circuit inductance and a series condenser are used as the tuning elements of the antenna, grid and plate circuits

The circuit shown in Fig. 191 was suggested by M. W. Sterns and has been found good for reception of long or short waves. The plate inductance *L*-1 and the series condenser *C*-1 act as the tuning elements of the antenna circuit and provide regenerative coupling. The path of the plate current is from the positive terminal of the battery *B*, through the coil *L*-1, to the plate of the tube; thence from the plate to the filament, through the head telephones *P*, and back to the negative side of the battery. Condenser *C*-1 should have a maximum capacitance of 0.0005 to 0.002 mfd. For the Marconi V. T. the capacitance of the grid condenser *C*-3 should be about 0.0001 mfd. The grid leak *R* should have a resistance of 2 megohms. The dimensions of the coil *L*-1 will vary with the wave length to be received.

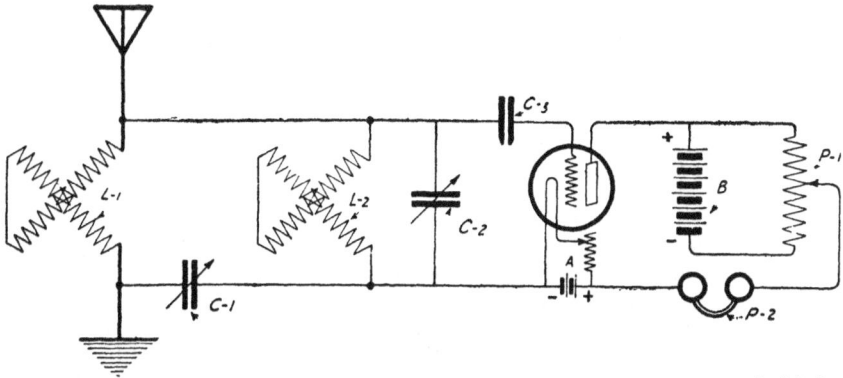

FIG. 192. Circuits of a short wave receiver preferred by many amateurs. Tuning is effected solely by variometers.

For short wave lengths many amateurs use the tuning circuits shown in Fig. 192 in which the primary and secondary inductances of the tuning transformer are the variometers L-1 and L-2. C-1 is a coupling condenser of variable capacitance. Although the secondary coil L-2 is shunted by a variable condenser C-2, the use of the condenser is not advisable at short wave lengths, for the distributed capacitance of the variometer L-2 is sufficient to establish resonance with the antenna circuit.

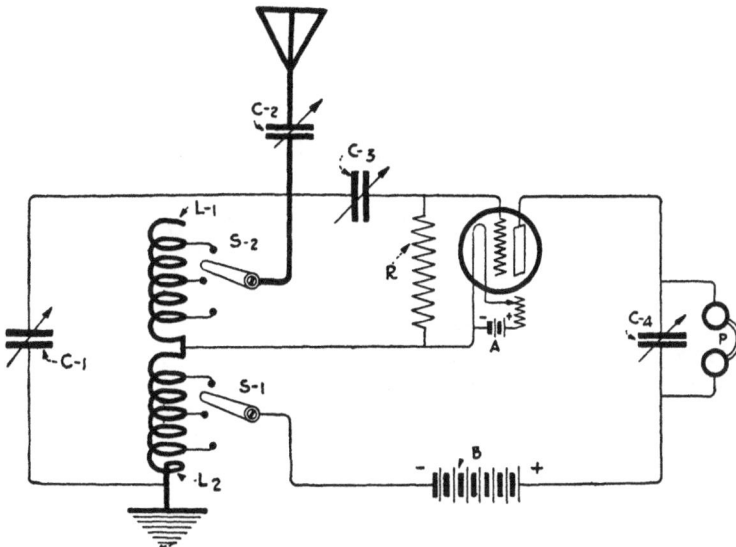

FIG 193 Simplified regenerative circuit feasible for reception of long or short waves

The plate battery B is shunted by a 10,000-ohm potentiometer P-1 through which the plate voltage can be very carefully regulated. Although this circuit does not have a regenerative coupler, feed-back amplification can be obtained by *disconnecting the lower wire of the secondary from the filament, and attaching it to the positive pole of the B battery.* A grid leak (not shown) should be employed. The circuit of Fig. 192 is recommended for short wave lengths. For long wave lengths variometers are not practical as tuning elements.

A modified regenerative circuit is shown in Fig. 193 which is good for the reception of long or short wave lengths. The coils L-1 and L-2 may be one long coil tapped at the center as shown. The magnitude of the regenerative coupling is varied by the switch S-1, the studs of which are tapped off L-2 say every 2″. L-1 is varied by a sliding contact or a multi-point switch. C-1 is a variable condenser of say, 0.0005 mfd. C-3 is a grid condenser of the usual dimensions and R a grid leak. C-2 is a series variable condenser which will be found useful with long aerials. This circuit has been tested with the Marconi V. T. and has been found particularly suitable for short waves because of its simplicity. A series condenser should be placed in the antenna circuit, with aerials over 100 feet in length.

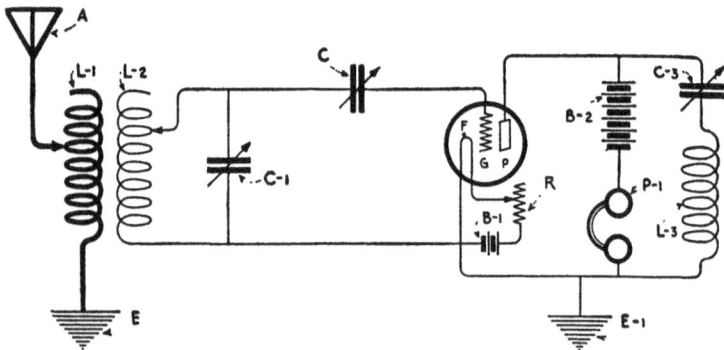

Fig. 194.   Weagant's "X" circuit receiver suitable for the reception of damped or undamped oscillations.

In regenerative circuits for the reception of *long wave lengths* the author prefers *Weagant's X circuit* receiver shown in Fig. 194. The principal point of departure from other systems is the method of connecting up the tuning elements of the plate circuit.

The plate battery B-2 and head telephone P-1 are shunted by a variable condenser C-3 connected in series with a variable inductance L-3. This circuit will oscillate without coupling between the grid and plate circuits other than that furnished by the internal capacity of the valve itself; but in some cases, the coil L-3 should be near enough L-2 to provide external regenerative coupling. L-2 and L-3 may be separated 6″ or 8″ in practice. For very long wave lengths a loading inductance should be connected in series with L-2, both this coil and L-3 having large values of inductance. This circuit has been found to give very good signals and it is recommended particularly for the reception of undamped wave signals because of the readiness by which stations can be "brought in." The dimensions of the coils for long waves will be given in the chapter following. If a grid leak is provided the circuit will be found very suitable for the Marconi V. T. detector.

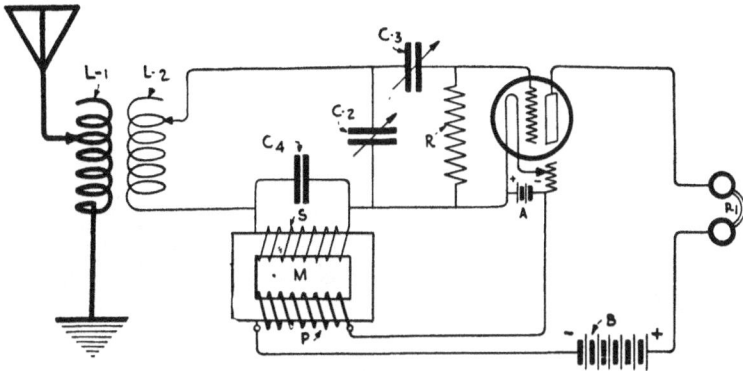

FIG 195   Regenerative circuit for the amplification of the audio frequency component of the plate current.

Audio frequency regenerative coupling may be obtained in the circuit of Fig. 195. The audio frequency voltages set up in the plate circuit during reception of spark signals are impressed upon the grid circuit through the transformer *M*. The ratio of transformation of this transformer varies between 4 and 6. The impedance of the transformer should be given a value equal to, or slightly greater than, the internal resistance of the tube. An open core step-up transformer may be used instead. To by-pass the radio frequency currents of the incoming signals around the secondary winding *S*, the transformer must be shunted by a fixed condenser *C*-4 of about 0.002 mfd.

Regenerative circuits for the simultaneous amplification of the *radio and audio frequency component* of the plate current have been devised. Such circuits have been used in practice but have been found somewhat unstable in operation. Some experimenters prefer radio frequency regenerative coupling for the detector tube, further amplification being obtained by a two- or three-stage cascade audio frequency amplifier coupled to the plate circuit of the detecting bulb. Audio frequency regenerative circuits are not widely used in practice.

A cascade amplification circuit of 7 stages now in common use was developed during the war. The first three tubes are connected in cascade for radio frequency amplification. The fourth tube is then connected for detection and the resulting audio frequency pulses in its plate circuit are amplified by a three-stage cascade audio frequency amplifier. Needless to say, this circuit gives enormous amplifications.

Any experimenter possessing the required number of valves can make up a similar circuit by studying the diagrams for radio and audio frequency amplifiers already given. It will perhaps prove an expensive experiment.

The several regenerative circuits just shown represent only a small proportion of those proposed. Obviously many variations of the principle are possible. The question as to whether a *cascade amplifier* or a single bulb *regenerative receiver* is preferable is not difficult to answer. The cascade amplifier will be used when very great amplifications are required. But if lesser degrees of amplifications will satisfy conditions the single bulb regenerative receiver will be employed.

DIMENSIONS OF TUNING ELEMENTS FOR SOME REGEN-ERATIVE RECEIVERS.—The method of computing the required induct-ance and capacitance in a receiving circuit for any wave length has been explained on pages 169 to 173.

FIG. 195a   Grebe regenerative receiver and two-stage amplifier.   Range 170 to 580 meters.

Some suggestions and helpful hints regarding the regenerative circuits in Figs. 188 to 195 will be given. Suitable dimensions for Fig. 187 will be given in the following chapter. Figs. 188a and 188b represent nothing unusual. The dimensions of the primary and secondary coils may be computed by the method already given. C-4 in Fig. 188a has capacitance

FIG. 195b   Showing the coupling transformer, vacuum tube sockets and variometer in the Grebe short wave amplifying receiver

of about 0.001 mfd. The dimensions of P-1, S-1 vary with the wave length and can easily be found by experiment.

Arno Kluge of California gives the following dimensions for the circuit of Fig. 189, if operated with an aerial whose natural period is about 175 meters. The primary L-1 is a tube 3″ in diameter wound with 30 turns of No. 22 d.s.c. wire. This coil will raise the wave length of the antenna mentioned above to 600 meters. The secondary L-2 is a tube $2\frac{1}{2}$″ in diameter wound with 50 turns of No. 24 d.s.c. The "tickler" coil L-3 is mounted on a ring that will fit over the end of L-2 and is wound with 12 turns of No. 18 annunciator wire in a groove $\frac{3}{8}$″ wide. The winding consists of 3 layers of 4 turns each. The capacitance of C-4 is 0.0001 to 0.001 mfd.

In connection with the circuit of Fig. 190, F. V. Bremer gives the following dimensions for 180- to 580-meter reception. The primary L-1 is a cardboard or bakelite tube $3\frac{3}{4}$″ in diameter and $2\frac{1}{2}$″ long wound with 56 turns of No. 22 s.c.c. wire. The secondary is shaped like a *variometer ball*, is mounted in one end of the primary and contains 36 turns of No. 22 s.c.c.

The ball for the variometer in the grid circuit is $4\frac{1}{2}$″ maximum diameter at the center and tapers to 3″ at the ends. It is wound with 32 turns of No. 20 s.c.c. The stationary winding is $4\frac{7}{8}$″ inside diameter wound with 30 turns of No. 20 s.c.c.

The plate circuit variometer L-4 has the same dimensions as that for the secondary but it has only 27 turns of No. 18 s.c.c. on the ball and 25 turns on the stationary frame. The construction of variometers is treated on page 184.

The circuit of Fig. 191 may be used for *transmission* and *reception*. Its range* is about ten miles, with one valve. The coil L-1 as constructed by M. W. Sterns, is a *banked winding* split into two sections. The wire is a Litzendraht cable composed of 42 separate strands of No. 36 B. & S. enameled wire with a double silk covering. A four-bank coil is employed. The first section has 50 turns wound on a tube 3″ in outside diameter. A space† of $\frac{3}{8}$″ is left between this and the next section which is wound with 220 turns, with a tap at 55 turns, and another tap at the end of the coil. Sterns reports that the circuit, with this coil and a small variable C-1, will function at wave lengths between 130 and 1850 meters with an antenna of 0.0005 mfd.

For reception up to 18,000 meters, the coil L-1 in Fig. 191 should be 18″ in length, 6″ in diameter and wound full with No. 30 d.s.c. wire.

For the variometers in Fig. 192, Carleton Howiler has given the following dimensions: Each variometer consists of 2 cardboard cylinders, the inner one 4″ in diameter, and $1\frac{1}{2}$″ long; the outer one $4\frac{1}{2}$″ in diameter, and $1\frac{1}{2}$″ long. Both are wound full with No. 26 s.c.c. wire. With the amount of inductance afforded by the primary variometer, an antenna with a natural wave length less than 200 meters will respond to wave lengths up to 600 meters.

The dimensions of the coils L-1 and L-2 in Fig. 193 can be calculated by the method described on pages 169 to 173. As mentioned before, the inductance may be one straight coil tapped at the center. For very close tuning the portion L-2, should be tapped every $\frac{1}{2}$″ or so, while the re-

---

*Transmitting range.
† This space is left to reduce the distributed capacitance at the lower wave lengths.

maining portion *L-1*, preferably is fitted with a "units" and "tens" switch.   Condenser *C-1* may vary in capacity up to 0.001 mfd.   For the shorter wave lengths capacities around 0.0003 mfd. are preferred.

The dimensions of the coils in Fig. 194 will be given in the chapter on undamped wave reception.

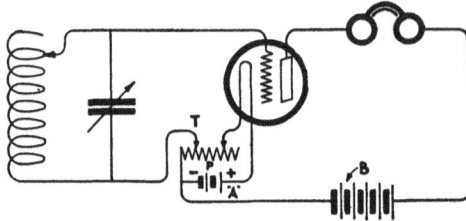

FIG. 196. Method of obtaining negative grid potentials without the use of a special grid battery or a grid condenser.

NEGATIVE GRID POTENTIAL.—The use of the "C" or grid battery has been shown in the diagram, Fig. 179.   The requisite negative grid potential can be supplied by the "A" battery as shown in Fig. 196.   A resistance *P* is cut in the negative side of the *A* battery and by moving the tap *T* along the resistance, the grid can be held at various negative potentials.   A resistance of 1 ohm is generally sufficient.

INTERVALVE TRANSFORMERS.—The design of *intervalve couplings* for audio frequency amplification is a puzzling matter to some experimenters.   As already mentioned the *impedance of these couplings* must at least be equal to or greater than the internal impedance of the valve.   To save the experimenter the trouble of carrying out the necessary computations dimensions will be given of transformers that have been found suitable for commercial types of valves.

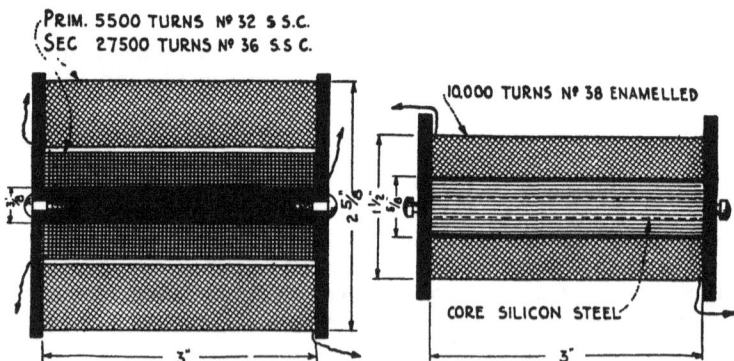

FIG. 197. Air core intervalve coupling transformer for audio frequency amplification. A step-up ratio of turns (1 to 6) is employed
FIG. 198 Dimensions of a single coil coupling impedance suitable for a cascade connection of Marconi V T.'s.

Fig. 197 gives the dimensions of an air core step-up transformer suitable for detector and amplifier bulbs such as used by the Government.

The transformer is suitable for any bulb whose internal impedance is about 25,000 ohms. The core is a hard rubber rod $\frac{3}{8}''$ in diameter and $3\frac{1}{2}''$ long. The primary is wound with 5500 turns of No. 32 s.s.c.; the secondary with 27,500 turns of No. 36 s.s.c. It is a good transformer for amplifiers in radio telephony and may be used in the circuit of Fig. 186.

The dimensions of the *coupling impedance* L-3 in Fig. 183 are given in Fig. 198. This coil is also suitable for Fig. 185 when using the Marconi V. T.

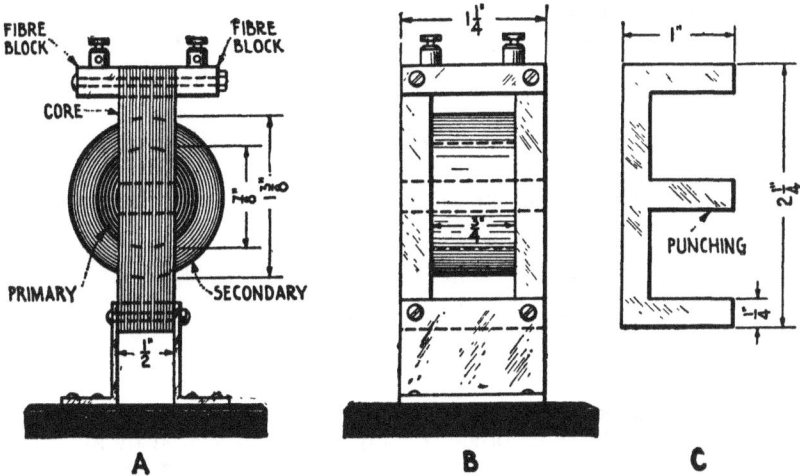

FIG. 199. Dimensions and constructional details of the Federal Telegraph and Telephone Co.'s shell type of transformer suitable for intervalve couplings in a cascade circuit using Marconi V. T.'s.

Fig. 199 gives the dimensions of a *closed core intervalve transformer designed especially for the Marconi V. T. by the Federal Telegraph and Telephone Company.* Because of the amplification obtainable and the small space it occupies this transformer is recommended to amateur experimenters.

The transformer is of the shell type and is wound with a *step-up* ratio of turns. The primary and secondary coils are mounted on the middle leg of the core. The core is built up of "E" shaped silicon steel punchings 14 mils in thickness. Thirty-eight pieces as shown in the detail C are required.

The primary consists of 3900 turns of No. 44 enamelled wire and the completed coil has a diameter of about $\frac{7}{8}''$. The secondary, which is wound immediately over the primary, has 12,000 turns of No. 44 enamelled wire. The d.c. resistance of the primary is 2200 ohms, and of the secondary 9150 ohms.

The core punchings are compressed tightly by machine bolts passed through the corners. Small fiber blocks are mounted on the top of the core to support the binding posts for the primary and secondary terminals.

A side view of the completed transformer is shown in the detail B, while A gives the general coil dimensions. These transformers give excellent amplifications at audio frequencies and the experimenter desiring a two- or three-stage amplifier giving high magnifications should use them.

All valves may be energized by one filament and one plate battery as in the diagrams previously shown.

The experimenter will find it an extremely difficult matter to wind these coils with such fine wire and he is therefore urged to purchase them from the manufacturer. The Federal Company also makes an excellent radio frequency intervalve transformer which has been designed especially for the Marconi V. T.

HOW TO DETERMINE THE IMPEDANCE OF INTERVALVE COUPLINGS.

—The impedance of a circuit or coil is equal to the square root of the sum of the resistance squared and the reactance squared or,

$$Z = \sqrt{R^2 \times X^2} \tag{63}$$

The reactance of a circuit in ohms is obtained from the expression,

$$X = 2\pi N L \tag{64}$$

where N = frequency in cycles

and L = inductance in henries

If the resistance of an intervalve coupling is measured on a Wheatstone bridge and the inductance is computed or measured by any of the well known methods, then the impedance may be computed from formula (**36**) above.

Take for example the single-layer solenoid of Fig. 198. R as measured = 900 ohms. L as computed or measured = 20 henries (approx.). Hence at the frequency of 800 cycles,

$$X = 628 + 809 + 20$$

$$= 100,500 \text{ ohms (approx.)}$$

$$\text{Then } Z = \sqrt{900^2 \times 100,500^2}$$

$$= 104,024 \text{ ohms.}$$

It is difficult to separate the reactance and resistance components of the impedance of an intervalve coupling, except by a laborious bridge measurement. The effective impedance of the primary, when the secondary is loaded or unloaded, may be determined by the simpler method shown in Fig. 199a. Here the grid circuit of the vacuum tube is connected to the secondary S of the intervalve transformer, the primary of which is coupled to an 800-cycle alternator. A pair of telephones P-1 is connected in series with the input circuit. A double-throw switch S-1 is provided This, when thrown to the right connects the input current to the primary P, and to the left substitutes a calibrated

resistance box going up to 100,000 ohms or more. The intensity of
signals is noted with the switch S-1 thrown to P, and then the switch
is thrown the other way and the resistance varied until a signal of equal
intensity is obtained. The value of this resistance evidently is equal to
the effective impedance of the transformer primary. The effective
primary impedance may then be measured with the filament of the valve
burning and the plate voltage and grid voltage adjusted to normal operating conditions.

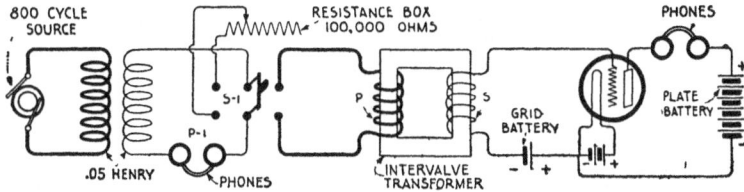

FIG. 199a. Method of determining the effective impedance of an inter-valve transformer.

It will be found that the effective primary impedance of certain
valves varies widely for different grid potentials. It may vary from a
value of 80,000 ohms with one or two volts negative grid potential down
to 10,000 ohms with a half volt or so positive potential. The experimenter may utilize this method in determining the impedance of any
inter-valve transformer he may happen to have at hand.

CHAPTER VIII

UNDAMPED WAVE RECEIVERS—AUTODYNE AND HETERODYNE CIRCUITS—PREFERRED CIRCUITS—ROTARY CONDENSER

RECEPTION OF CONTINUOUS OSCILLATIONS.—In amateur transmitters, the group frequencies of the antenna currents lie between 60 and 400 per second. In commercial wireless telegraph transmitters, group frequencies up to 1000 per second are employed. A special airplane transmitter, developed during the war, is operated off a 1000-cycle generator. The group frequency, in this case, is 2000 per second. The *amplitudes of the antenna oscillations* in *damped* wave transmitters are *modulated at an audio frequency* by the audio frequency charging current supplied by the high voltage transformer making it unnecessary to do anything more than rectify the receiver currents for response at the receiving station.

In the case of undamped wave transmitters, the successive amplitudes of the antenna oscillations are uniform and to obtain response in the telephone they must be modulated at an audible frequency either by a "chopper" at the transmitter or by similar or other means at the receiver. The modulation of continuous oscillations is most conveniently effected at the receiver. The *"tikker,"* which is a *current interrupter* connected either in the antenna circuit or in the secondary circuit of a receiving transformer, is the simplest form of undamped wave detector. It takes the place of the crystal detector in the circuit for damped waves.

Since the average telephone receiver gives best response when it is impulsed between 500 and 1000 times per second the "tikker" should be designed to interrupt the incoming radio frequency oscillations 500 to 1000 times per second. The telephones will then be energized periodically at a suitable audio frequency as in spark telegraphy.

The phenomena of detection with a rectifier placed in series with a "tikker," are shown in Fig. 200. The graph O-1 indicates incoming oscillations of continuous amplitude, graph O-2 the resulting groups when the incoming cycles are interrupted by a "tikker," graph O-3 the rectified groups, and graph O-4 the approximate telephone current. If the "tikker" interrupts the circuit 1000 times per second, 1000 groups of damped oscillations will flow in the receiving circuits where they are rectified by the crystal. The telephone will be impulsed 1000 times.*

---

*For more detailed explanation of the problems involved in the transmission and detection of damped and undamped oscillations, the reader should read the introduction of the author's "Vacuum Tubes in Wireless Communication."

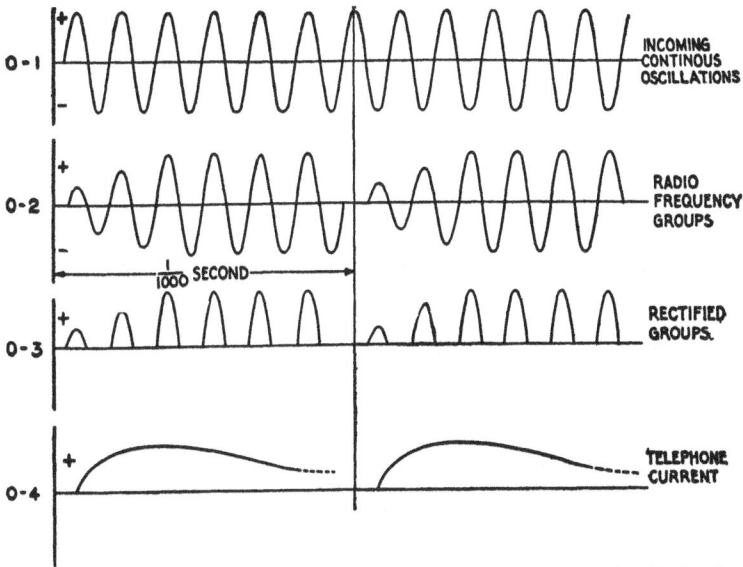

FIG. 200. Graphs showing the phenomena involved in the detection of undamped oscillations by means of a "tikker" with a rectifier in series.

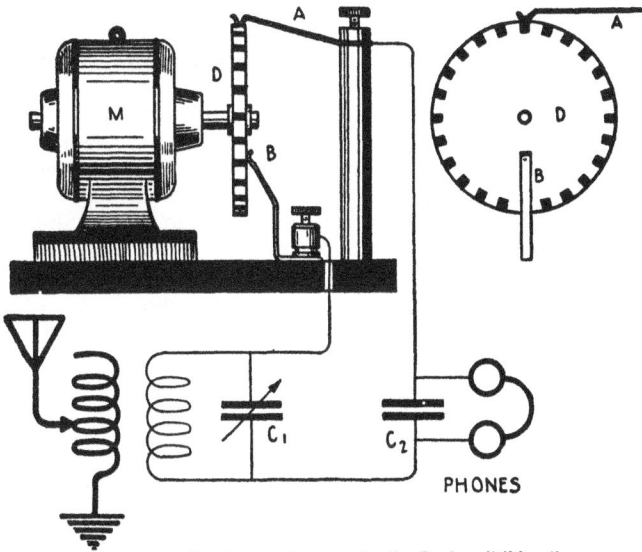

FIG. 201.   Fundamental circuit for the Poulsen "tikker."

A simple circuit for the Poulsen "tikker" is shown in Fig. 201 which also gives a rough idea of its construction.   In this case, it is a form of a *commutator interrupter*.   Connection to the disc D is made by the brush B and the second brush A makes contact with the copper segments. The graphs in Fig. 200 do not show the phenomena involved when the "tikker" is employed without a series rectifier. The only value of the rectifier is that it tends to smooth out the note in the head telephone.

The primary and secondary coils of a receiving tuner to be suitable for the tikker should be made of low resistance conductor such as Litzendraht. The telephone condenser C-2 should have rather high capacity. Values of 0.01 mfd. are often used in practice.

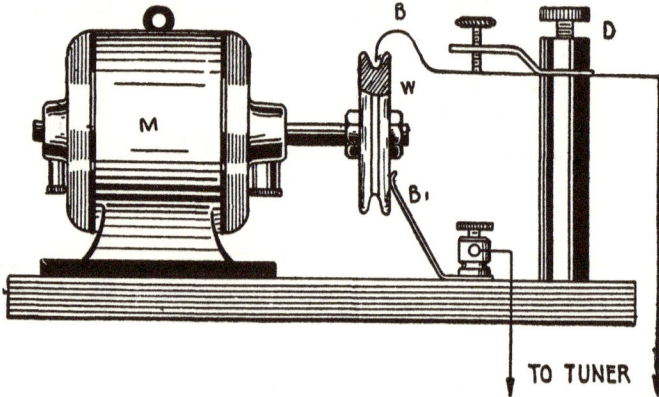

Fig. 202. The "slipping contact" detector—a modified type of "tikker."

Another type of tikker or undamped wave detector is the *"slipping contact"* detector shown in Fig. 202. While the drawing shows the essentials of construction the design may be varied in many ways. The disc or wheel W is mounted on the shaft of a small motor. One terminal of the secondary circuit is connected to the disc through the brush B-1 and the second terminal to the binding post D which supports an elastic piece of steel or copper wire B bearing lightly on the wheel W. When the motor is in operation, the wire B forms a contact of variable resistance which, during reception, varies the telephone current accordingly.

The slipping contact detector gives a slightly clearer note than the tikker of Fig. 201. Both types are efficient and will permit undamped wave reception over comparatively long distances. The chief objection to the tikker is the poor note which it produces in the head telephone.

HETERODYNE RECEIVER.—As already explained, in continuous wave reception it is necessary to modulate the amplitudes of the incoming oscillations at audible frequencies at a rate suitable to the head telephone. Although the tikker provides a simple means of obtaining such modulation, it is done much more effectively by the so-called *heterodyne* receiver. The operation of this receiver is based upon the interaction of two alternating currents of different frequencies. It can be shown that if an alternating current of 50,000 cycles per second, for example, flows through a given circuit and if there is superimposed on the circuit another current of a frequency of 49,000 cycles per second, the result will be a so-called *"beat"* current with a frequency equal to the numerical difference of the two applied frequencies or 1000 per second. This principle is extensively used in undamped wave reception.

Take for example, the ordinary receiving circuit for damped wave reception; if the frequency of an undamped signal is 50,000 cycles and a detector circuit with a simple rectifier be employed, uni-directional pulses of radio frequency will pass through the telephone receiver without emitting a sound.   But if the incoming oscillations interact with a locally generated current of say 49,000 cycles 1000 beat currents will be set up in the receiver circuits and the amplitude of the d.c. pulses flowing through the head telephone will rise and fall 1000 times per second.

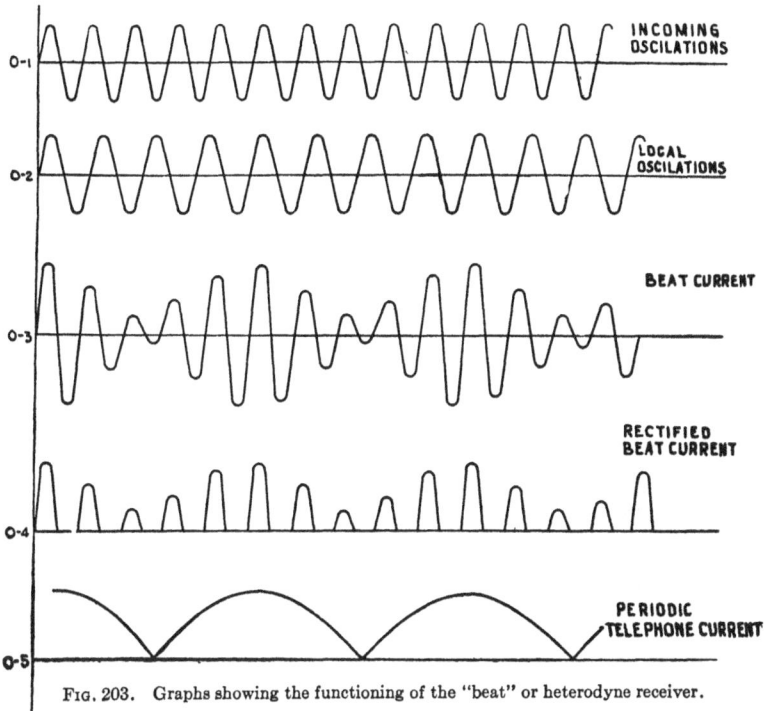

FIG. 203.   Graphs showing the functioning of the "beat" or heterodyne receiver.

The characteristic phenomena of the beat receiver are best explained by the graphs of Fig. 203.   Graph 0-1 represents a few cycles of *incoming continuous oscillations*, and graph 0-2 indicates the cycles of a *locally generated radio frequency current* differing slightly in frequency from the current of the incoming signal.   The graph 0-3 indicates the resulting *beat currents* which are radio frequency cycles with amplitudes varied at audible frequencies.   The reader must bear in mind that the individual cycles of each group of beat currents occur at radio frequencies and that in order to obtain audible response they must be rectified as shown by the graph 0-4.   The periodic telephone current is shown by the graph 0-5.

In an earlier form of the beat receiver the local radio frequency currents were generated by a Poulsen arc generator or a high frequency alternator, but the vacuum tube generator is employed in modern systems.

The *vacuum tube generator* is more suitable for this purpose than the Poulsen arc because it is less difficult to obtain oscillations of any desired frequency with the tube than with an arc generator. A tube generator will oscillate steadily at frequencies from one cycle to 10 million per second whereas the oscillations of the arc are apt to be rather unstable at the higher frequencies.

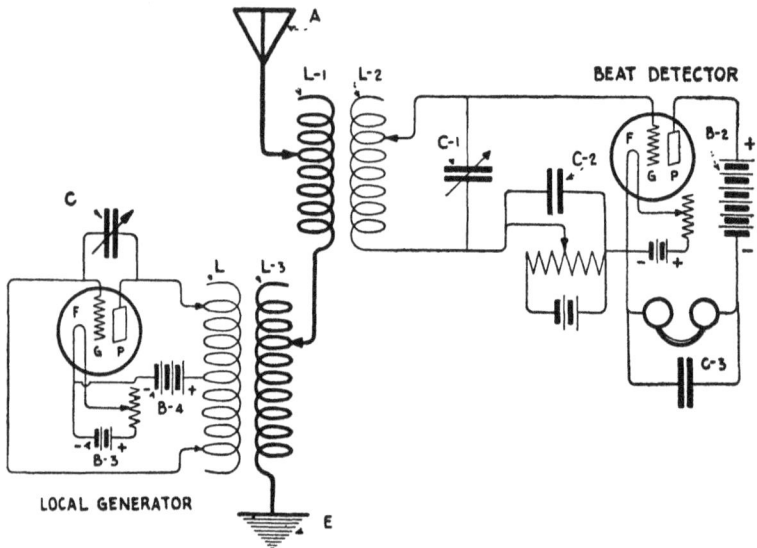

Fig. 204. Circuits of the external heterodyne using a vacuum tube generator as the source of local oscillations.

A practical circuit for the heterodyne receiver is shown in Fig. 204. It is often called the *external heterodyne* because the local oscillations are generated by a source external to the receiving circuits. The tube generator is shown in the left hand part of the drawing. The right hand part of the drawing shows the circuits of a vacuum tube detector. L-1, L-2 is a tuning transformer and C-1 a shunt variable condenser. The grid battery and potentiometer are connected in series with the grid circuit and are shunted by a fixed condenser C-2. The grid battery is employed to maintain the grid at some negative potential most suitable for the detection of beat currents. If the Marconi V. T. is used in this circuit a *grid condenser* and *grid leak* may be employed instead of the grid battery. The remainder of the circuit is no different from the detection circuits previously outlined.

It may prove helpful to explain how the vacuum tube (to the left of the drawing) generates radio frequency currents. It will be noted that the grid and plate circuits are coupled through the coil L which has a tap at the center. The upper half of L is the plate circuit inductance and the lower half of L the grid circuit inductance. C is a variable condenser which in reality is shunted around the coil L; that is, it is common to both the grid and plate circuit inductance. The operation of the tube circuit is practically as follows: Assume that the potential of the battery B-4 is varied by any means: the intensity of the field surrounding the coil L is changed and an e.m.f. will be generated in the coil which will tend to increase the charge in the condenser C. C will then discharge through the coil L setting up radio frequency oscillations. Since the grid element G is tapped off a portion of the coil radio frequency potentials will be impressed upon the grid. The fluctuating grid potential will in turn vary the plate current and amplify the plate oscillations. Through the coupling afforded by the coil L, a still greater oscillating voltage is impressed upon the grid which will increase the amplitude of the plate oscillations until the limits of the tube and circuits are reached.

Practically all tubes will oscillate in a circuit of this kind, but the best results are obtained from tubes whose characteristic curves show a *steep slope*. With such tubes small changes in grid voltage will cause relatively large changes in the plate current.

A convenient voltage for generating powerful oscillations with the Marconi V. T. is about +300 volts on the plate, and a grid voltage of about —18. The grid may be held at this negative potential by the grid battery, or the same result will be obtained by the use of a condenser and an adjustable grid leak. The Marconi V. T. will oscillate at much lower plate voltage with sufficient intensity for beat reception.

When the taps on the coil L (one leading to the plate and the other to the grid) are set at some value of inductance and the capacitance of the condenser C is varied some upper value of capacity will be found at which the tube will cease to oscillate. The inductance in the plate and grid circuits must then be increased until oscillation is obtained. The correct proportions of inductance and capacitance are readily found by trial.

The best results are obtained with the circuit in Fig. 204, when the primary and secondary circuits are *tuned* to the desired wave length by a *wavemeter*. The wavemeter should be excited by a buzzer and its coil placed in inductive relation to the earth lead of the receivers say at the coil L-3. The inductance and capacitance in both the primary and secondary circuits of the detector tube should then be varied until maximum response is obtained from the buzzer indicating that the receiving set is tuned to the desired wave length.

When adjusting the receiver by trial the operator should first ascertain if possible whether the station desired is actually sending. He should then light the filament of the *generator tube* and adjust the voltage of the plate battery to normal operating values. The taps on the coil L should be placed where the tube will oscillate at a radio frequency *slightly off the frequency of the incoming signal*. The capacitance of condenser C may then be varied until a musical note is heard in the head telephone. From the heterodyne principle outlined above it is obvious that the frequency of the beat note can be varied over a considerable range. By varying the local frequency the operator can then produce a pitch suitable to his ears. Not only does the beat receiver permit the pitch of the note to be varied over a wide range of tones, but the heterodyne principle itself affords a marked degree of selectivity.

Assume, for example, that when the frequency of the incoming signal is 50,000-cycles per second, and the local generator is set at 49,000-cycles, an interfering signal is received whose frequency is 49,500-cycles.

The 50,000-cycle signal will produce a 1000-cycle beat note, and the 49,500-cycle signal a 500-cycle beat note. If the frequency of the local generator be then set at 49,500-cycles, a zero beat note will result from the interfering station, and a 500-cycle beat note from the station desired.

The circuit for the external heterodyne in Fig. 204 is recommended for experimental use. If an 8-stage, resistance-coupled amplifier patterned after that shown in Fig. 184 is employed instead of the single detection bulb shown in that diagram, undamped wave signals can be received over several thousand miles on small frame aerials consisting of a few turns of wire wound on a frame 4 or 5 feet square.

The *external heterodyne* has a distinct advantage over the *autodyne receiver* which we will now describe. In the former the frequency of the beat note can be varied over a wide range *without changing the tuning of the circuits of the detector bulbs;* whereas with the autodyne circuits, the pitch of the beat note can *only* be changed by changing the tuning of the detector circuits.

To make up a practical circuit like the one in Fig. 204 the experimenter may select primary and secondary coils for any desired range of wave lengths from the table Fig. 134. For very long wave lengths, say 18,000 meters, loading coils should be connected in series with L-2 and L-1. The requisite inductance for the coil L (for long wave lengths) will be about 75% of that required in the secondary circuit (including the loading inductance). For stable oscillation the capacitance of the condenser C should not exceed 0.001 mfd. Suitable dimensions for a long wave tuner are given in design No. 5, Fig. 134.

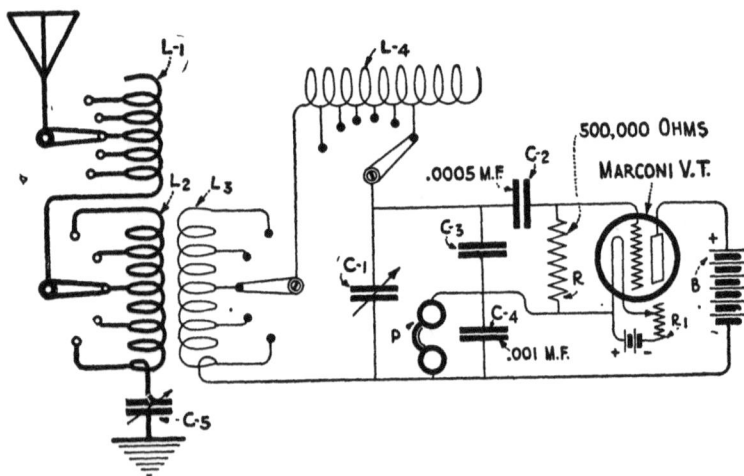

FIG. 205. Modified ultra-audion circuit for the reception and detection of undamped oscillations*.

*This circuit sometimes functions with increased sensitiveness when a tickler coil having approximately 75% of the secondary inductance is placed in series with the plate circuit and in inductive relation with the secondary coils L-3 or L 4 In one type of Navy tuner using substantially this circuit the condensers C-4 and C-5 are made up into one condenser having two sets of fixed plates and one set of moving plates. The plates are so mounted that as the capacity on one side increases, the capacity on the other decreases The rotating element is also mechanically connected with the tickler so that the ratio of the grid side capacity to the plate side capacity is varied in such a way as to maintain oscillations. Since an increase of capacity on one side is accompanied by a decrease on the other side the tuning corresponding to any particular setting of the condenser C-1 is unaffected

CIRCUITS FOR UNDAMPED WAVE RECEPTION.—As we have said, the circuits in Fig. 187 to 195 are applicable to undamped wave reception. The circuits following, Figs. 205, 206 and 207, are particularly suitable for such reception at long wave lengths.

Fig. 205 is a *modified ultra-audion circuit* employing electrostatic coupling between the grid and plate circuits. It is claimed that with this circuit no difficulty is encountered in obtaining *beat reception* over a wide range of wave lengths. Fig. 206 is a simple *Armstrong regenerative circuit* with inductive coupling between the grid and plate circuits. Regenerative coupling is provided by the tickler coil *L*-5 which is in inductive relation to the secondary coil *L*-4. Fig. 207 is *Weagant's X circuit*, recommended to the beginner because it is easily adjusted and very stable. Although an auto-transformer *L*-2 is used as a coupler, inductive coupling may be used as well.

Coming now to the dimensions of the various coils. Design No. 5, Fig. 134 gives the dimensions of inductances suitable for the three circuits shown. It is not essential that these dimensions be duplicated. Coils of smaller diameter and greater length will do as well provided the necessary inductance is obtained. We find from Fig. 134 that the antenna coil *L*-1 in Fig. 205 is 7″ in diameter, 24″ long wound with 923 turns of No. 24 d.s.c. The primary coil *L*-2 is 7″ in diameter, 14″ long, wound with 532 turns of No. 24 d.s.c. The secondary coil *L*-3 is 6″ in diameter, 14″ in length wound with 742 turns of No. 28 d.s.c. Coil *L*-4 is 6″ in diameter, 24″ long, wound with 1856 turns of No. 32 d.s.c.

The maximum capacitance of the variable *C*-1 is 0.0005 mfd. *C*-2 varies between 0.0001 and 0.0005 mfd. *C*-3 is of fixed capacity—about 0.0005 mfd. *C*-4 is fixed and its capacity about 0.001 mfd.

L-1 is tapped every two inches, *L*-2 every half inch, or the latter may be fitted with a "tens" and "units" switch. The *units* switch for example, should have 20 taps, with each tap connected to a single turn. The *tens* switch should have 25 taps each taking in 20 turns. The last tap obvi-

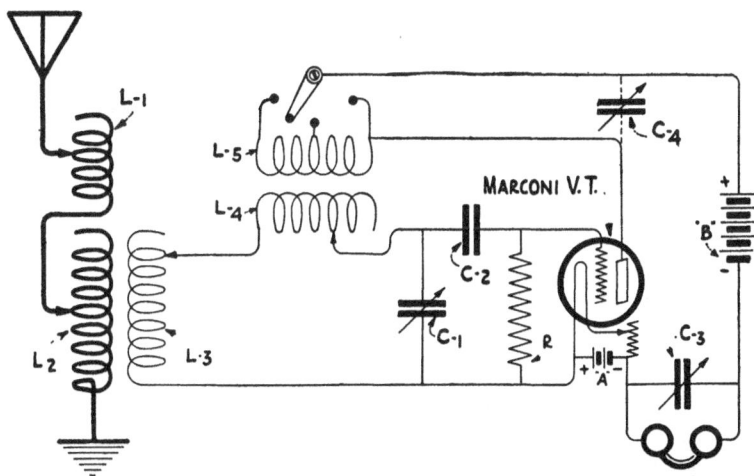

FIG. 206. Regenerative receiver with "tickler" coupling for the reception and detection of undamped oscillations In some cases the condenser *C*-3 should shunt the plate battery and the head telephone. The condenser *C*-4 is of value with some designs

ously will need to have $20+12=32$ turns. Coil L-3 may be tapped every $2''$; similarly coil L-4. For wave lengths above 15,000 meters a small variable condenser should be shunted around L-1 and L-2. For the Marconi V.T. the grid leak R has 2 megohms resistance. The A battery is of 4 to 6 volts, the B battery 60 volts.

For the circuit in Fig. 206, L-1, L-2, L-3, L-4 have the same dimensions as in Fig. 205. The tickler coil L-5 should have approximately 75% of the total inductance of the secondary circuit, for long wave lengths. It may be shunted by a small variable C-4 of 0.0005 mfd. Some designers wind the turns of L-5 in grooves on a narrow drum which just fits over the end of L-4. The turns are wound in multilayers. Three taps should be provided for L-5. These will be sufficient for the range of the longer waves used commercially.

The coils L-4 and L-5 need not necessarily be mounted coaxially. Sufficient coupling is generally provided if the coils are stood on end near one another. The connections must be made so that the current alternates through L-5 in the direction to assist the oscillations in the grid circuit, and not to oppose them.

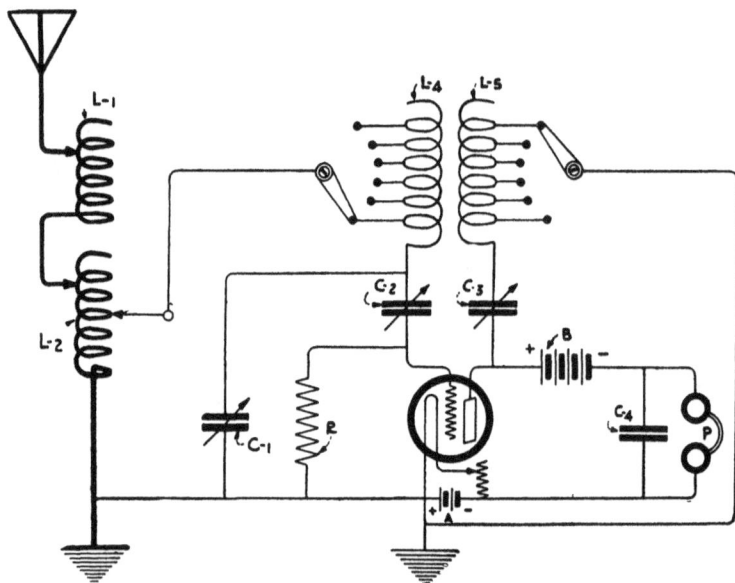

Fig. 207. Weagant's "X" circuit for undamped wave reception. This is an excellent circuit for reception at long wave lengths.

For the "X" *circuit* of Fig. 207 the coil L-2 has the same dimensions as the primary in Figs. 206 and 207. L-5 should have the same dimensions as L-4, viz., $6''$ in diameter, $24''$ long, wound with No. 32 d.s.c. L-1 has the same dimensions as in Fig. 206.

Beat reception may be obtained in the "X" circuit without coupling between L-4 and L-5 but it is an advantage with some valves to place these coils in mutual relation. The principal tuning in the circuit of

Fig. 207 is done at the variable condensers C-1 and C-3. The antenna circuit is tuned at L-1 and L-2.

Fig. 207a   Grebe type $C\,R\,7$ long wave receiving set.   Range 500 to 20,000 meters.

If the beginner is not well versed in the manipulation of beat receivers, he should use the circuit of Fig. 191, for in order to establish resonance with a transmitter it is only necessary to change the inductance of L-1 and the capacitance of C-1.

Fig. 207b.   Showing the arrangement of the apparatus in the Grebe long wave set.

DEFOREST HONEYCOMB INDUCTANCES.—These coils were developed by R. F. Gowen and mark a distinct advance in the design of tuning inductances. They are exceedingly compact and portable, and the losses are lower than in any other multilayer coil yet devised. They are suitable as *primary, secondary* and *"tickler" coils* and make excellent inductance elements for wavemeters.

FIG. 208. Deforest honeycomb coils attached to a special three-coil mounting which permits them to be used as a receiving set These coils may be obtained in various sizes suitable for reception at all wave lengths between 200 and 25,000 meters.

The winding of the honeycomb coil is peculiar in that it approximates a banked winding in one direction. The finished coil is *cellular*, the turns of one layer crossing the preceding layer always at an angle. This construction reduces the distributed capacitance to a minimum. One of these coils made of standard "litz," with a natural period of 1725 meters, had a direct current resistance of 20.4 ohms and a high frequency resistance at 23,500 cycles of only 33.5 ohms. This particular coil had an

TABLE XIV

	Wire	Millihenries Inductance Approx.	Approx. wave length range in meters with ordinary 0.001 mfds. variable air condenser
L- 25		.040	170- 375
L- 35		.075	200- 515
L- 50	No. 24 S. C.	.15	240- 730
L- 75		.3	330- 1030
L- 100		.6	450- 1460
L- 150		1.3	660- 2200
L- 200		2.3	860- 2850
L- 250		4.5	1120- 4000
L- 300	No.25 S. C.	6.5	1340- 4800
L- 400		11.	1860- 6300
L- 500		20.	2340- 8500
L- 600		40.	2940-12000
L- 750		65.	3100-15000
L-1000	No. 28 S. S.	100.	5700-19000
L-1250		125.	5900-21000
L-1500		175.	7200-25000

All coils have an inside diameter of 2", a width of 1", an outside diameter varying from 2¼" to 4½" These coils may be obtained wound with either litzendraht or solid wire.

FIG 208a Giving the catalogue numbers and other data for honeycomb coils to cover definite ranges of wave length.

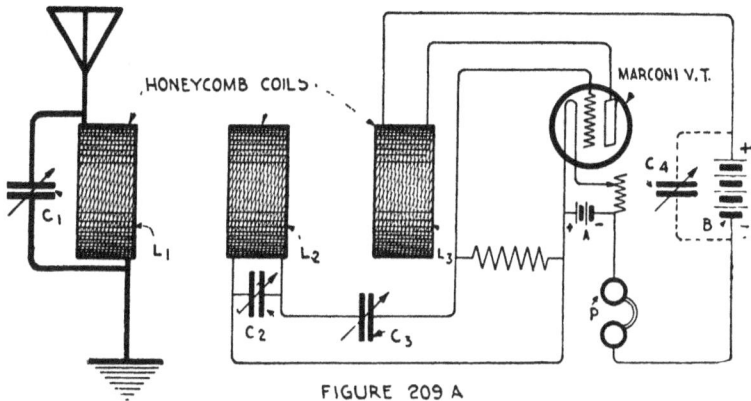

FIGURE 209 A

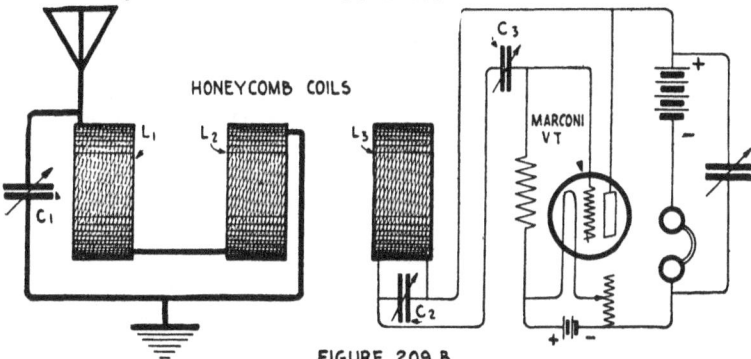

FIGURE 209 B

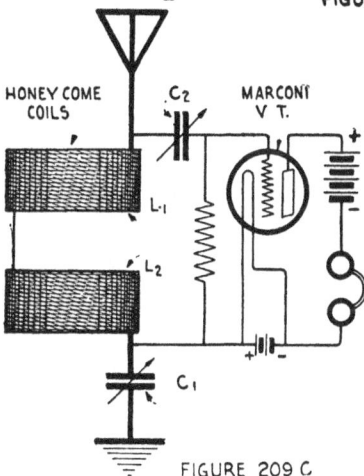

FIGURE 209 C

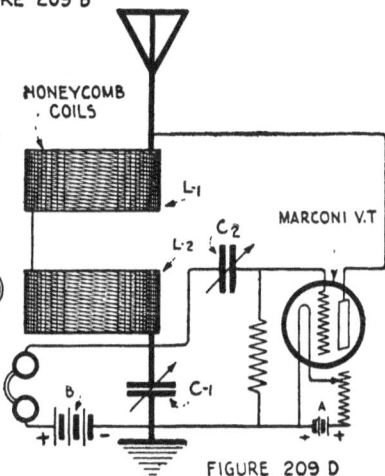

FIGURE 209 D

FIG. 209a.   Three honeycomb coils used as an inductively coupled tuner with inductive regenerative
       coupling between the grid and plate circuits.
FIG. 209b.   Modified circuit for the honeycomb coils wherein two coils are used as an antenna variometer.
FIG. 209c.   A circuit providing direct coupling between the antenna circuit and detector circuit.   Two
       honeycomb coils are used as a variometer.
FIG. 209d.   Simple regenerative circuit for reception of short or long waves.   Two honeycomb coils used
       as a variometer provide regenerative coupling and are employed to vary the wave length.

inductance of 64.5 millihenries and oscillated at 15,250 meters when shunted by 0.001 mfd. The advantage of such coils over the more common types will be readily understood from the following.

A single-layer solenoid for long wave lengths, as is apparent from Fig. 134, is a cumbersome affair and requires a great deal of space for mounting. A plain multi-layer winding is out of the question for radio reception at long wave lengths because of the excessive self-capacity of such coils. To reduce the self-capacity of multi-layered coils the turns are *banked* as explained on page 201. The construction of a banked coil presents many difficulties to the unskilled builder and hence he is apt to resort to the more readily constructed single layer solenoid.

The former chief objections to multi-layered coils as regards losses, etc., have been done away with in the Deforest cellular coil. In addition the honeycomb inductance provides a notoriously small coil even for the longest waves.

Fig. 208 will give the reader an idea of the assembled coil and also a special mounting which has been devised so that three coils may be used as a receiving coupler for damped or undamped wave reception. Representative circuits for their use will follow. It is to be noted that the coils are of fixed inductance and wave length variation is secured by either a shunt or a series condenser.

From the table Fig. 208a the amateur is enabled to select coils suitable for the range of the wave lengths which he desires to cover.

The experimenter may devise numerous circuits for either the two-coil or three-coil mounting. Fig. 209a is a regenerative "tickler" circuit to which the honeycomb coils are admirably adapted. Coil L-1 is the primary which is shunted by the variable condenser C-1, L-2 is the secondary shunted by C-2. The tickler coil L-3 is connected in series with the plate circuit. The standard three-coil mounting permits the coupling between L-1 and L-2, and between L-2 and L-3 to be readily changed.

For best results the tickler L-3 should have from 35% to 75% of the inductance of L-2. The variable condenser C-4 (in dotted lines) may be shunted around B alone or around B and P.

The proper coils for definite ranges of wave length with any circuit can be selected from the foregoing table and the table following. For reception from high power stations operating between 8,000 and 15,000 meters, with an antenna of 0.001 mfd. the coil in the table marked L-600 has about the correct inductance for the primary L-1 of Fig. 209a. Either L-750 or L-1000 is about right for the secondary L-2, and L-300, L-400, or L-500, for the tickler coil L-3.

To ascertain the size of an antenna coil for a definite range of wave lengths, the amateur may first measure the capacitance of his antenna according to the process described on pages 169 to 173. He may then select a suitable loading inductance for any wave length from the foregoing table,

using the relation $\lambda = 59.6 \sqrt{L\,C}$ or $L = \dfrac{\lambda^2}{3552 \times C}$. In these formulae

L is expressed in centimeters and C in microfarads. The formulae, for all practical purposes, apply to the antenna circuit as well as to the closed circuit.

If desired the identical coils can be used for the primary and secondary.

If by this selection the primary proves too large for resonance with a given transmitter, the wave length can be lowered by a series variable condenser. The best method is to connect a switch in the circuit so that the condenser C-1 can be placed either in series or in shunt with the primary L-1.

In the modified circuit Fig. 209b, L-1 and L-2 are used as a variometer and the antenna wave length is varied by changing their mutual inductive relation. When the coils are close together and the current circulates through them in opposite directions their total self-induction will be a minimum. If the connections to one coil are reversed their self-induction will be a maximum. When the coils are placed at any angle the inductance of the variometer varies accordingly. This differential mounting of the coils plus the frequency variation afforded by the shunt condenser C-1 gives a considerable range of wave lengths.

The secondary circuit of Fig. 209b is the *ultra-audion* circuit. The terminals of the secondary tuning coil are connected to the *grid and plate* of the Marconi V. T. rather than to the *grid and filament*. For 2500 meter reception, L-3 may be the coil marked L-300 for a shunt condenser of 0.0005 mfd., and L-1 and L-2 may each be the coil marked L-100 or L-150. L-1 and L-2 each has less inductance than L-3, for the maximum inductance of L-1 and L-2 used in this way is very much greater than that of either coil alone.

Fig. 209c is a simple detection circuit in which the honeycomb coils L-1 and L-2, are used again as a variometer. A single coil, with a series antenna condenser will suffice for some ranges of wave length.

Fig. 209d is the Stern's circuit which has been shown before. The coils L-1 and L-2 are used as an antenna variometer. For very long waves each may be the coil marked L-1000. For long wave reception with very small aerials, the coils L-1500 or L-1250 should be used.

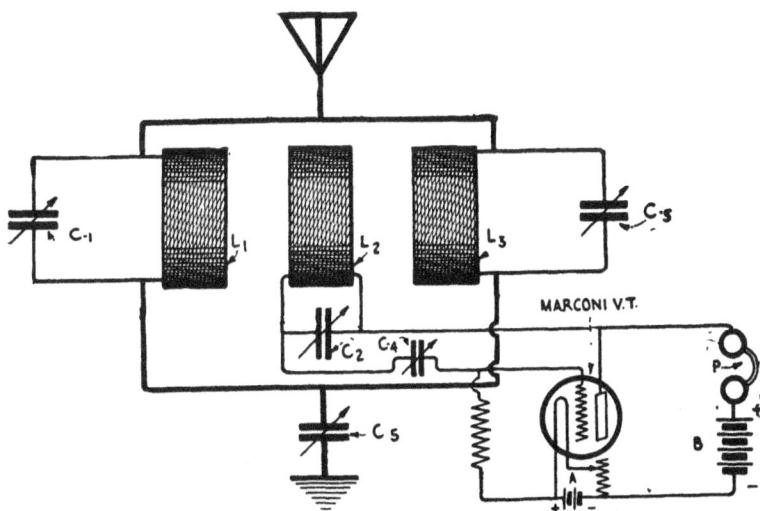

Fig 209e Differential circuit for interference prevention utilizing three honeycomb coils

Fig. 209e is a differential circuit for interference prevention which has been found of some value. For moderate wave lengths, coils L-1, L-2 and L-3 may be identical: The primaries should at least have the same inductance. The experimenter can effect numerous other combinations

It is not difficult for the experimenter to select the proper secondary coil for any range of wave lengths as all the necessary data appears in the foregoing table; but he should use wisdom in selecting the primary coil as is shown in the following brief discussion.

Assume, as an illustration, that with an inductively coupled set the maximum wave length desired is 15,000 meters: Coil L-750 will be satisfactory for the secondary, although if a secondary condenser smaller than 0.001 mfd. be available, coil L-1000 or L-1250 should be used. To point out the possibility of error in the selection of a primary coil assume that L-750 has been selected. If the capacity of the amateur's aerial is 0.001 *mfd. the shortest possible wave length in the antenna circuit without a series condenser* with this coil will be 15,000 meters. In that case it would be preferable to select a smaller coil for the primary, say L-600 or one even smaller, so that a shunt primary condenser may be used for tuning purposes. Consideration of the facts presented in the chapter on receiver design will show the amateur how to make the necessary computations for these circuits and enable him to select the correct coil for any conditions of service.

The table following, Fig. 209f, will be of particular interest to the amateur. It shows the proper coils for the circuit of Fig. 209a, to cover all existing wave lengths. The capacitance of the antenna is assumed to be 0.0006 mfd. the capacitance of the primary condenser 0.001 mfd. and of the secondary condenser 0.0005 mfd.

TABLE XV

Type of Signals	Wave Lengths	Number of coil		
		Primary Coil	Secondary Coil	Tickler Coil
Amateur..................	140– 240	L– 25	L– 25	L–35
Commercial .... {	550– 700 900– 1400	L– 75 L– 100	L– 100 L– 150	L–50 or L–75 L–75 or L–100
Arlington Time Signals ...	1650– 2750	L– 300	L– 300	L–100
High Power............ {	8000–15000 10000–20000	L– 600 L–1000	L– 750 L–1250	L–300, or L–400, or L–500 L–300, or L–400, or L–500

FIG 209f   Giving the catalogue number of honeycomb coils for use in the circuit of Fig 209a for definite ranges of wave length.

ROTARY CONDENSER—The method devised by the author utilizes a specially constructed rotary condenser mounted on the shaft of a small d.c. or a.c. motor. The condenser should be designed to throw the circuits in and out of resonance, say 350 to 600 times per second.

Connected in the circuit of Fig. 210a, the rotary condenser functions as follows: When the condenser C-2 passes through the value of capacity necessary to establish resonance, currents flow through the rectifier D and the head telephone P. When the condenser passes through values of capacity off resonance the amplitude of the currents in the detector circuit is decreased. Hence the telephone is energized periodically at an audible frequency, which is determined by the number of resonance points passed per second.

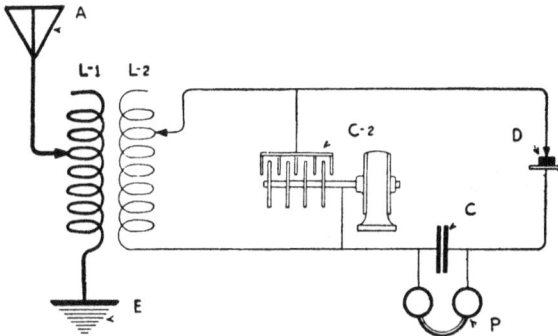

Fɪɢ. 210a   A rotary condenser used as a detector of undamped oscillations as first employed by the author

In practice, the rotary condenser is shunted by a *small stationary variable condenser*. The capacity of the latter is adjusted near to resonance and the rotary condenser will give the required increase of capacity to establish resonance.* The correct proportions between the two capacities are found by trial.

The condenser C-2 must be insulated from the shaft of the driving motor and the motor enclosed in a metal box grounded to earth, for otherwise inductive interference will be obtained in the head telephones.

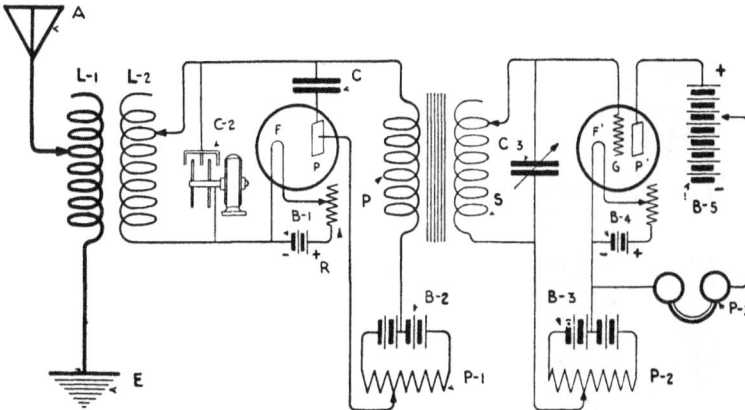

Fɪɢ. 210b   Amplification circuit for the rotary condenser devised by the author.

Fig. 210b shows one circuit devised by the author using a 2-electrode and a 3-electrode valve for detection and amplification. The 2-electrode valve rectifies the groups of radio frequency currents which are produced

*The stationary condenser may be set for resonance and the rotating condenser employed to throw the circuit out of resonance at an audible frequency.

by the rotating condenser and audio frequency voltages are therefore impressed across P, and transferred to S after which they are amplified by the 3-electrode valve. The batteries B-2 and B-3 are employed to adjust the grid potential and plate voltage of the 3-electrode and 2-electrode valve, respectively, for maximum amplification.

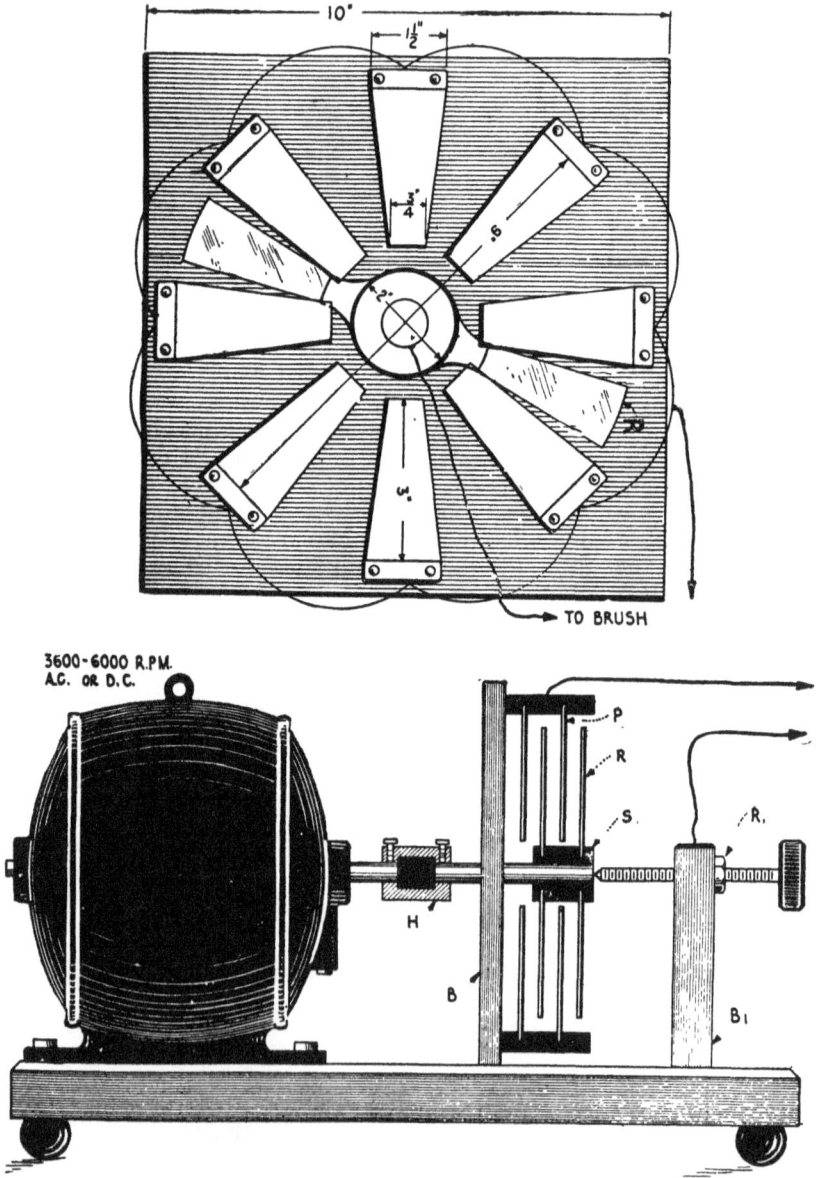

Fig. 210c. Showing the construction of a rotary condenser suitable for undamped wave reception.

Fig. 210c is a suggested plan of construction for the rotary condenser and Fig. 210d a detail of the motor shaft mounting. Two stationary condenser plates connected in parallel are arranged at 8 points around a circle and mounted on a fibre base. They are connected together with a wire leading to a binding post for connection to one terminal of the secondary of the tuning circuit. The rotor R, consisting of two plates of substantially the same dimensions as the stationary plates, is mounted on the shaft of the motor by the sleeve S. Contact with the rotor is made by the screw point R_1 which engages the end of the rotor shaft. R_1 is supported by the brass binding post B-1 and the stationary plates by the fibre base and support B (see side view). The remaining details are left to the builder. The motor should rotate at least 4000 r. p. m.

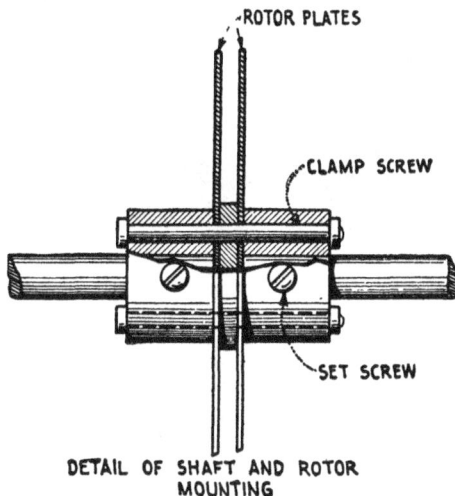

DETAIL OF SHAFT AND ROTOR
MOUNTING

Fig. 210d.   Details of the rotor shaft and clamps.

Because of the high speed of the rotor the construction should be rigid. The plates on the rotor should by no means be soldered to the sleeve S. They are preferably clamped between two rings as shown in the detail Fig. 210d. An insulating coupling between the motor and the rotary condenser is shown at H, Fig. 210c.

The preliminary adjustment is less difficult if one variable condenser is connected in series with the rotary condenser and another in shunt. This permits a closer regulation of the variation of capacity around the point of resonance than can be obtained by the rotary condenser alone.

A rotary condenser of modified construction is shown in Fig. 210e and it will be seen from this that the construction is similar to a rotary spark gap. The rotor B is a solid brass casting. The face of the tips of both the stationary electrodes A and the rotating electrodes on B are about 1″ square. They should not be separated more than 0.01″. The disc must be insulated from the motor shaft. The maximum capacitance between all opposing surfaces is about 20 mmfds. which gives sufficient variation around the point of resonance to make undamped signals audible.

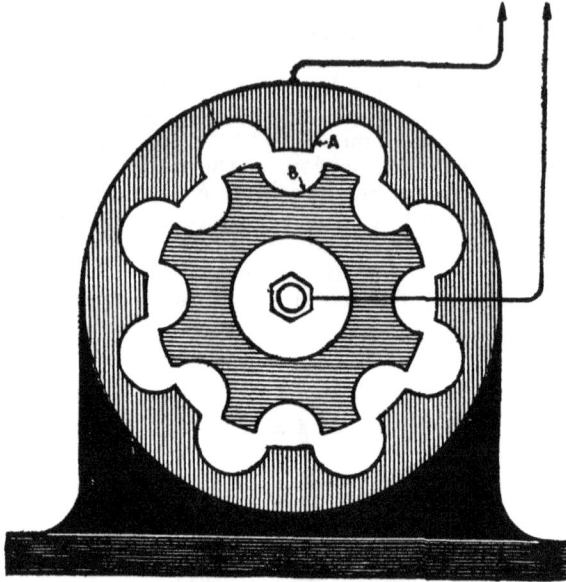

FIG. 210e.   A rotary condenser of modified design.

The design shown in Fig. 210*e* is substantially the author's idea as worked out by the U. S. Navy and used on their radio compass sets. As used by them it is called a *tone condenser* but it is employed as shown in the author's original patent specification.   The rotor is mounted on a jack shaft having a variable speed pulley which in turn is belted to another variable speed pulley which is driven by a motor through a flexible cable such as used on speedometers.   There are ten opposing surfaces and the rotor is driven 6000 r.p.m.   The use of this condenser is absolutely essential in undamped wave reception at short wave lengths between 50 and 400 meters where the heterodyne ordinarily would be employed.   The operation of the beat receiver at very high frequencies is attended by great difficulties for the slightest change of capacitance in the circuits will throw the beat note beyond the limits of audibility.   In the rotary condenser method the incoming oscillations are modulated at an audible frequency, mechanically, making reception independent of any slight changes of frequency.   For short wave lengths a capacity variation of 20 micromicrofarads is sufficient to produce signals, but for the longer waves used by high power stations a rotary condenser of higher capacity such as shown in Fig. 210*c* is essential.   The design in Fig. 210*e* is appropriate for undamped reception at 200 meters, using, say, a vacuum tube transmitter.

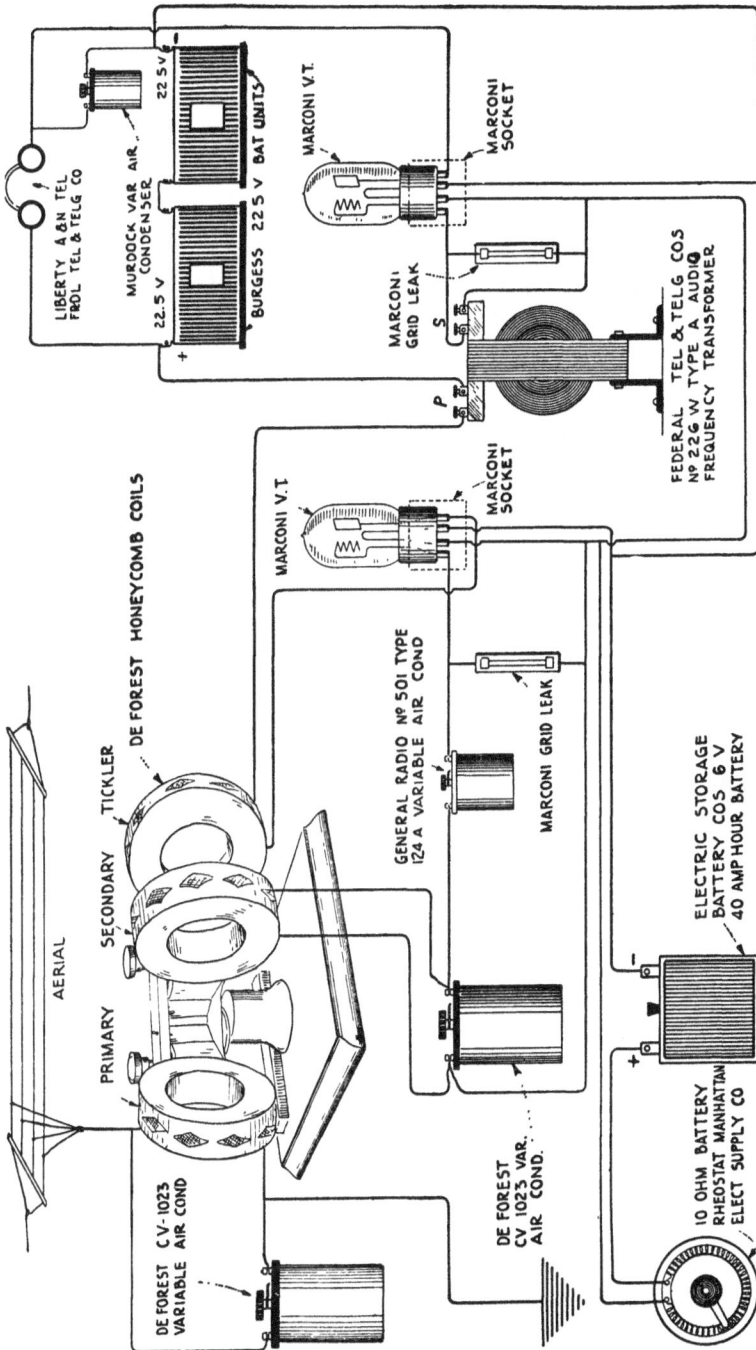

Fig. 210f. Composite receiving set for long and short waves using one Marconi V. T. as a detector and a second V. T. as an audio frequency amplifier. The intervalve coupling is a small step-up transformer. Regenerative coupling is provided for all wave lengths. It is desirable to shunt the primary P of the audio frequency transformer with a small variable condenser.

CHAPTER IX

UNDAMPED WAVE TRANSMITTERS—WIRELESS TELEPHONE TRANSMITTERS—MODULA-TION CIRCUITS—COMBINED DAMPED AND UNDAMPED WAVE BULB SETS

The principle of the vacuum tube generator has been explained in connection with the heterodyne receiver. The fundamental circuits of a tube generator are shown in Fig. 211. Any variation of voltage at the B battery will cause the circuit L-2, C to oscillate at a radio frequency governed by the product of L-2, C. L-1 is then subjected to radio frequency potentials which vary the grid potential and amplify the plate oscillations. This process will continue as long as conditions in the circuit remain constant. It sometimes aids in securing stable oscillation to place one of the taps so that it will include a few turns of L-1.

The purpose of the grid condenser in this circuit is to maintain the grid at a favorable negative potential, the magnitude of which is governed by the variable resistance leak R. Values between 15,000 and 25,000 ohms will be suitable for the average valve.

It should now be clear that if the condenser C be replaced by an aerial and earth wire, radio frequency currents will flow in the antenna system, and the tube may be used for the transmission of telegraphic and telephonic signals.

For *telephony* the antenna oscillations are modulated by a *microphone*. For *telegraphy by damped oscillations*, the antenna oscillations are modulated at audible frequencies by a buzzer which may be connected in the earth lead. For *telegraphy by undamped oscillations*, a key may be placed in the earth lead or in shunt to a few turns of wire connected in the earth lead. Signaling is then effected by varying the wave length as in arc systems.

A tube generator constructed in accordance with the diagram in Fig. 211 will be found extremely useful around the laboratory as a source of steady oscillations of any desired frequency for measuring purposes or for beat reception. (Note also the slightly modified circuit, Fig. 218.)

A single Marconi V. T.* operating on 20 to 80 volts plate battery is suitable for work which requires small amounts of energy, as in radio receiving investigations, or for the production of local oscillations in the external heterodyne system. For use in calibration work where sufficient

---

*Use Class II tubes for this work

262

energy is required to actuate a thermo-couple and a reflecting galvanometer, three tubes in parallel, operating on about 500 volts plate battery
are desirable.   As an indicator in such work, a heater type thermo-couple
galvanometer of about 1 ohm heater resistance with reflecting galvanometer of about 12 ohms resistance (and a 4 to 5 second period) are recommended.

Very accurate radio measurements can be carried on with such an
outfit.   Where it is desired to actuate a portable instrument such as a
hot wire milliammeter or a Weston thermogalvanometer, four tubes in
parallel and plate voltage of about 200 are recommended.

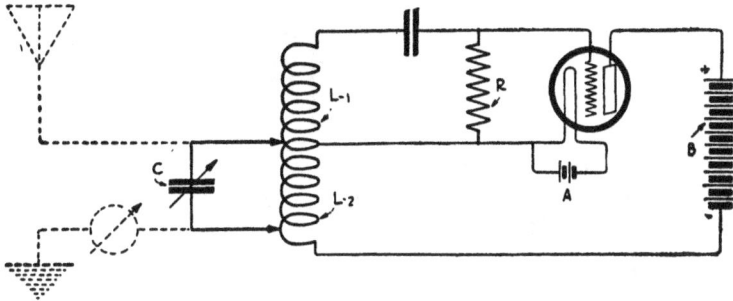

FIG. 211   A Marconi V T connected up for the production of radio frequency oscillations for measurement
purposes or for other uses around the laboratory.

AMATEUR WIRELESS TELEPHONE TRANSMITTER—A radiophone set designed for use with an antenna of about 0.0004 mfd. and
utilizing three Marconi V. T.'s,in parallel is shown in Fig. 212.   Although
three tubes in parallel are indicated it is possible to use one or two
tubes in the same circuit as well.   This circuit is perhaps the most
efficient one that can be devised for the Marconi V. T. for transmission
at short wave lengths.   The amateur experimenter can easily construct
all of the component parts.   His chief difficulty will be to obtain the
necessary plate voltage.

In the diagram, G is a d.c. generator of from 100 to 600 volts.   *Three
hundred volts* is a good working value for amateur transmission but
increased antenna current will be obtained with higher plate voltages.
The generator G is shunted by the condenser C-3 of one mfd. capacitance,
which tends to smooth out the commutator ripple of the generator.   Two
standard two mfd. telephone condensers connected in series will prove
satisfactory.

The condenser C-1 connected in series with the earth lead and the
grid condenser C-2 each have a capacitance of 0.0003 mfd.   Either mica
or air condensers may be employed.   For condensers with air spacings
between plates there should be a separation of $\frac{1}{8}''$ between the plates
to prevent sparking.   The *leak* R_3 should have resistance of approximately
12,000 ohms.   A wire resistance unit of this type may be purchased
from the Ward-Leonard Company.   Carbon rods are not very suitable
for this purpose as they have a tendency to vary their resistance.

Continuing the description of the circuit, M is a standard *low resistance microphone* such as used in wire telephony, which is shunted by coil

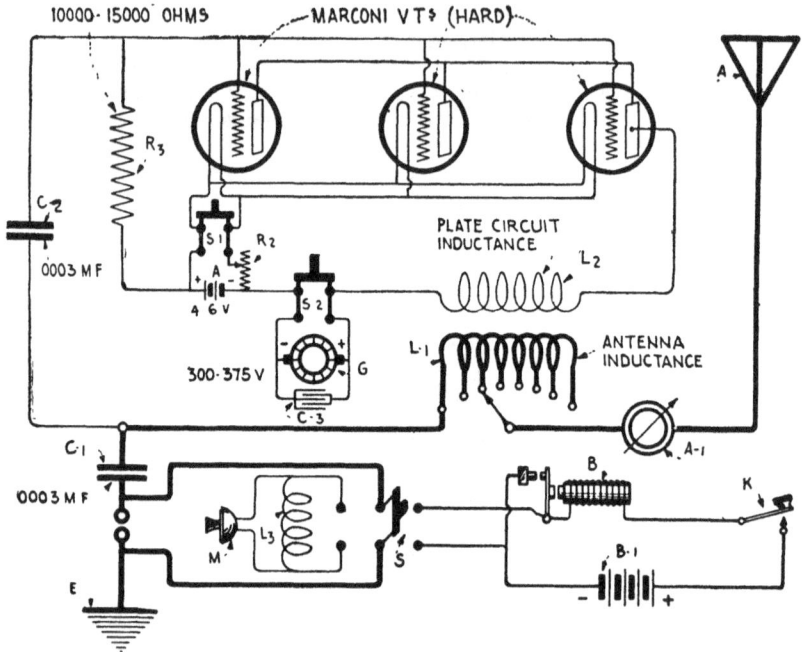

Fɪɢ. 212. Wireless telephone transmitter designed for use of the Marconi V. T. and to be operated at the wave length of 200 meters   Ranges up to 50 miles are obtainable with a detector and two-stage amplifier at the receiving station. (Note.—The positive side of the A battery should be grounded to earth)

L-3 consisting of 10 turns of No. 18 bare wire, wound on a tube 2″ in diameter. The turns are spaced ½″. The function of the coil *L*-3 is to shunt a portion of the antenna current around the microphone as it is not necessary to pass the entire current through it.

For telegraphic transmission by undamped waves a key may be placed directly in series with the antenna circuit. For damped wave transmission, switch *S* is thrown to the right, connecting the breaker contacts of the buzzer *B* in series with the antenna circuit. If the buzzer interrupts the antenna currents 500 times per second, 500 groups of damped waves will be radiated from the antenna. This makes the set of more practical use for communication, which by this arrangement can be established with receiving stations fitted with detectors which are only suitable for the reception of *damped waves*.

The coils of the transformer *L*-1 and *L*-2, which are in variable inductive relation, have been designed to "adapt" the antenna circuit to the output circuit of the tubes, thus giving the maximum degree of efficiency. Maximum antenna current will be obtained for some definite coupling which may be found by trial.

The coil *L*-1 is wound with 20 turns of No. 18 annunciator wire on a cardboard tube 5½″ in diameter and it is tapped every second turn. Coil *L*-2 is wound with 60 turns of No. 18 annunciator wire on a tube 5″ in diameter. Maximum antenna current is usually found with the coil *L*-2 placed inside of *L*-1, i.e., with close coupling.

The preferred process of tuning this set is to vary the turns on *L*-1 until the desired wave length is obtained. Then adjust the coupling between *L*-1 and *L*-2 until maximum antenna current is obtained, as noted by the aerial ammenter *A*-1. The meter should have a maximum scale reading of 1 ampere. A simple hot wire ammeter will do, but a thermo-couple type of ammeter is preferred. As a preliminary test of

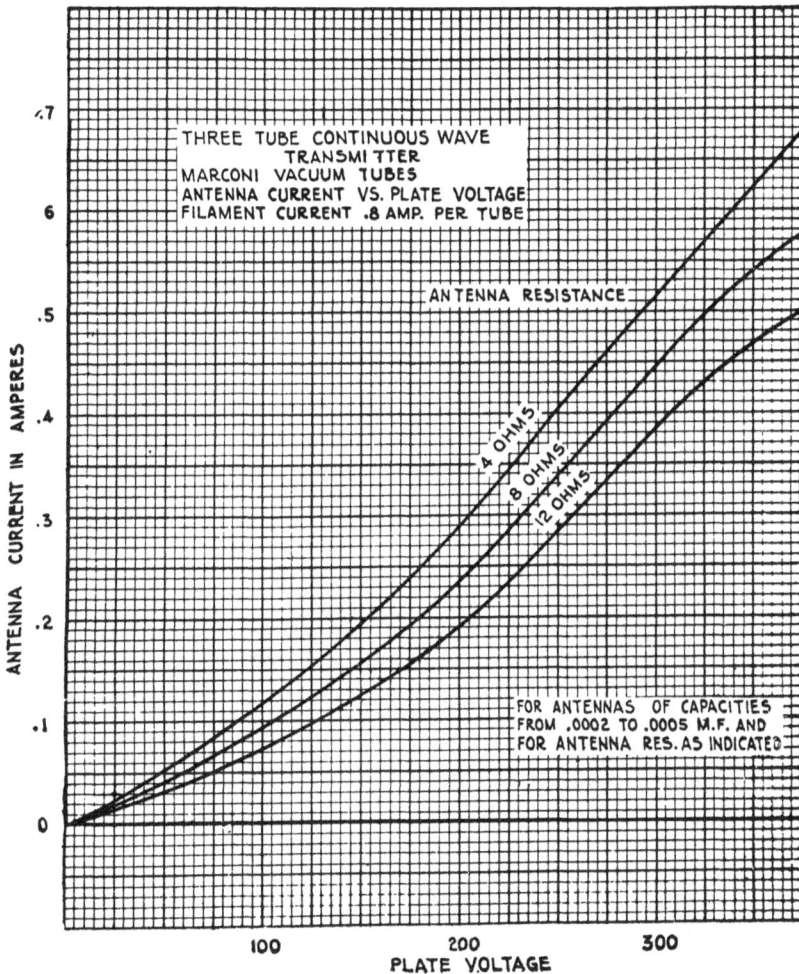

Fig 213 Showing the antenna currents to be expected from the set in Fig 212 with antennae of several capacities and resistances.

the ability of the transmitter, to observe when good modulation is obtained, a receiving set may be set up within a few feet of the transmitter. At the wave length of 200 meters fifty miles can be covered in favorable locations.

The antenna currents to be expected with aerials having capacitances between 0.0002 and 0.0005 mfd. and antenna resistances of 4, 8, and 12 ohms are shown in the curves of Fig. 213. It will be noted that with a plate voltage of 375 volts and an antenna of 4 ohms, the antenna current is about 0.65 ampere.

The data for these curves was obtained with the microphone short-circuited and they in reality indicate the currents available for *telegraphy*. Because of the resistance of the microphone (10 ohms) somewhat lower currents are obtained when it is connected in the circuit.

The plate current for three tubes at 300 volts lies between 20 and 30 milliamperes. The power in the plate circuit is therefore approximately $300 \times 0.025 = 7.5$ watts. The filament current is $3 \times 0.8 = 2.4$ amperes.

If this three-bulb transmitter be used in connection with the three-bulb cascade amplifier of Fig. 183, the telegraph range will be 100 miles and the telephonic range about 50 miles. Greater distances will probably be covered under favorable conditions.

Flash light batteries are recommended as a source of plate voltage when the Marconi V. T. is used as an oscillation detector. They are also suitable when the tube is used as a low power generator for heterodyne reception. These batteries can be purchased in 25 volt units. The cells should be placed well separated from one another in a *moisture proof container*. It is a good plan to purchase fairly large size flash light batteries and place them in a wooden box well separated from one another. *Molten paraffine* should then be poured into the box to cover the cells completely. Leads may be brought from each cell to a set of binding posts on the top of the box. These posts should be mounted on porcelain knobs to insure against surface leakage. When a particular cell dies its terminals can be short circuited, permitting the use of the remainder of the cells.

Batteries are suitable as a source of plate e.m.f. in transmitting sets, but in the long run it will probably be cheaper to make up a motor generator set for this purpose. A simple and inexpensive means to obtain a source of plate voltage is by driving a small 220-volt d.c. motor with ⅛ h.p. motor, such as the amateur uses to drive rotary spark gaps. The motor is, of course, used as a generator, and if the speed is increased above normal 300 volts may be readily obtained. If a source of 220 volts d.c. is available it will be found quite suitable both for the filament and for the plate voltage. The filament circuit should have a resistance in series of sufficient value to reduce the filament current to normal value. Caution should be exercised in handling a circuit of this kind. *Since the potential of the generator is 300 volts, a fatal shock may be obtained.*

The advantage of the tube transmitter over the spark transmitter is as follows: Forced to use the wave length of 200 meters, with spark transmitters, the amateur is compelled to adopt circuit constants to which spark transmitters do not readily lend themselves. The bulb sets, however, will work efficiently at wave lengths even below 50 meters.

During the war considerable transmission was carried on with waves of the order of 50 to 150 meters. With the comparative absence of static interference on these short wave lengths and with tube transmitters and amplifying receivers a new field for investigation in wireless transmission is opened up to the experimenter who has a scientific turn of mind. The results obtained with such transmitters should be recorded and published in magazines for the benefit of other experimenters.

COMBINED TRANSMITTING AND RECEIVING RADIO TELEPHONE SET.—As already suggested, the cascade amplifying set of Fig. 183 may be combined with the three-bulb transmitter of Fig. 212 to make a complete radiophone. When separate bulbs are used for the transmitter and receiver a simple single-blade, double-throw antenna switch or a *double-contact strap key* will suffice to change from transmitting to receiving. The same bulbs may, however, be used for either purpose. This demands a complicated change-over switch which will transfer the bulbs from one circuit to the other.

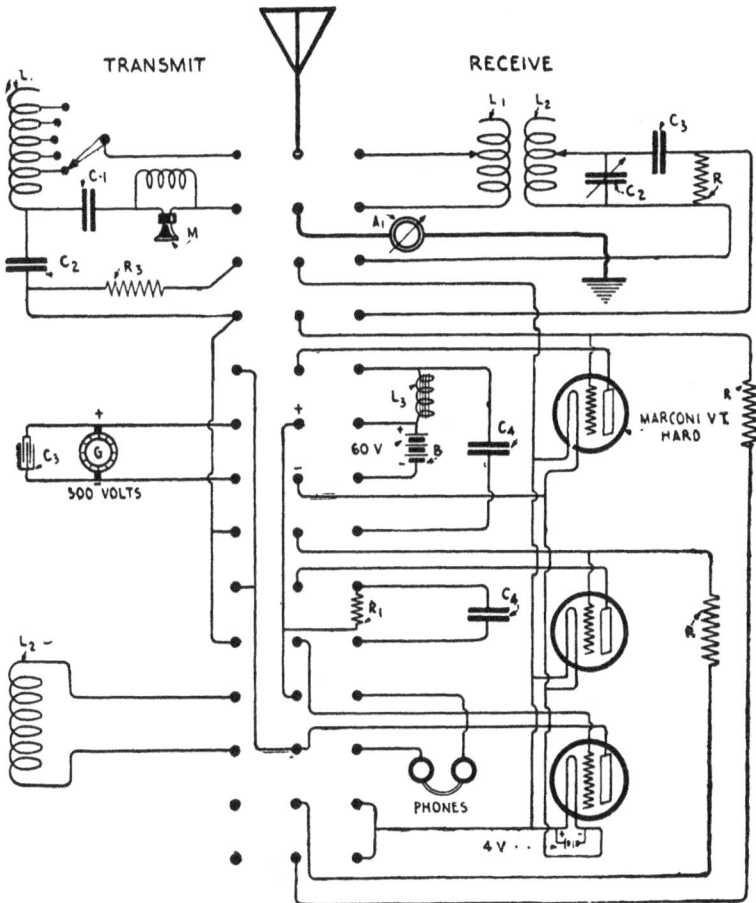

FIG 214   A circuit combining the cascade amplifier of Fig 183 with the transmitter circuits of Fig. 212. The same bulbs are used for transmission and reception.

The essential connections of such a switch to combine the circuits of Figs. 183 and 212 are shown in Fig. 214. A 14-blade double-throw switch is required.   When this is thrown to the right the detector and two-step amplifier circuit of Fig. 183 results; when thrown to the left the transmitter circuits of Fig. 212 result.   The notations in Fig. 214 correspond to those in Figs. 183 and 212.

It is even necessary to transfer the grid leaks.   Thus when receiving, the two lower right hand blades of the switch connect the grid leaks of the two amplifier bulbs to earth.   When transmitting, these leaks are removed from the circuit as the leakage for all tubes is provided by the 10,000 ohm leak R-3 shown in the upper left hand part of the drawing.

It was not thought that a switch of this type would be adopted.   The ingenious and resourceful experimenter will design some form of a rotary switch providing a more rapid change from the transmitting to receiving positions than that which the common d.p.d.t. switch affords.

If the experimenter has not the tools or materials with which to construct a switch of that type he can purchase 7, d.p.d.t. switches for the purpose.   He should remount them in such a way that a small throw of the lever will permit the use of either circuit.   The base of this switch must be made of extremely *good insulating material to insure against surface leakage.*

If the entire set is mounted on a panel the leads connecting the various instruments should not be bunched but kept fairly well separated.

It is a far more practical scheme to provide three bulbs for the transmitter and three for the receiver.   The chief trouble which may be met with in the circuit of Fig. 214, is the fouling of the switch contacts. A well built switch with positive contacts must be provided.

A serviceable switch for shifting these connections would be one of the magnetic type fitted with platinum contacts like those used on switches in telephone work.   The throw of the armature need be no more than $\frac{1}{4}''$ to $\frac{1}{2}''$.   The switch could be operated by a push button and a small battery providing a very rapid shift of circuits.

MODIFIED RADIO TELEPHONE TRANSMITTER.—The circuit of Fig. 215 is substantially a duplicate of that employed by the U. S. army in the European war.   It will give good results with small tubes of high vacuum.   The first tube marked V. T. No. 1 is the oscillator which is connected up regeneratively for the production of radio frequency currents.   The second tube marked V. T. No. 2 is termed the *modulator*.   This *amplifies* the *speech currents* generated by the microphone M and *modulates the plate current* of the oscillator bulb at *speech frequencies*. L-1 is a direct coupled oscillation transformer fitted with primary and secondary taps for tuning purposes. C-3 is the plate blocking condenser to keep the high voltage of the generator off the antenna circuit.   Its capacity is about 0.01 mfd. although 0.1 mfd. has been used. C-2 is the grid condenser which, because of rectification between the grid and filament, maintains a *negative potential* on the grid. The magnitude of the grid potential is limited by the leak R which varies from 10,000 to 20,000 ohms. C-1 is a coupling condenser of the rotary plate type with sufficient spacing between plates to prevent sparking. Its maximum capacitance need not exceed 0.002 mfd. L-2 is a radio frequency choke of about 13 millihenries, which prevents the radio frequency oscillations flowing through the d.c. generator.

The series plate choke *L-3* varies from $3\frac{1}{2}$ to 15 henries, depending upon the type of tube employed. Dimensions for *L-3* will be given later. *P-1, S-1* is a telephone induction coil of either the open or closed core type. For experimental work one of the type used commonly in land line telephony is serviceable. *B* is a 4 to 6 volt microphone battery. The microphone is one of the type used in wire telephony.

C-4 is a shunt protective condenser of 0.1 mfd. capacitance. *B-1* is a grid battery to hold the grid of the modulator bulb at a negative potential. The required negative potential can be obtained by tapping a resistance connected in the *negative side* of the line leading from the generator.

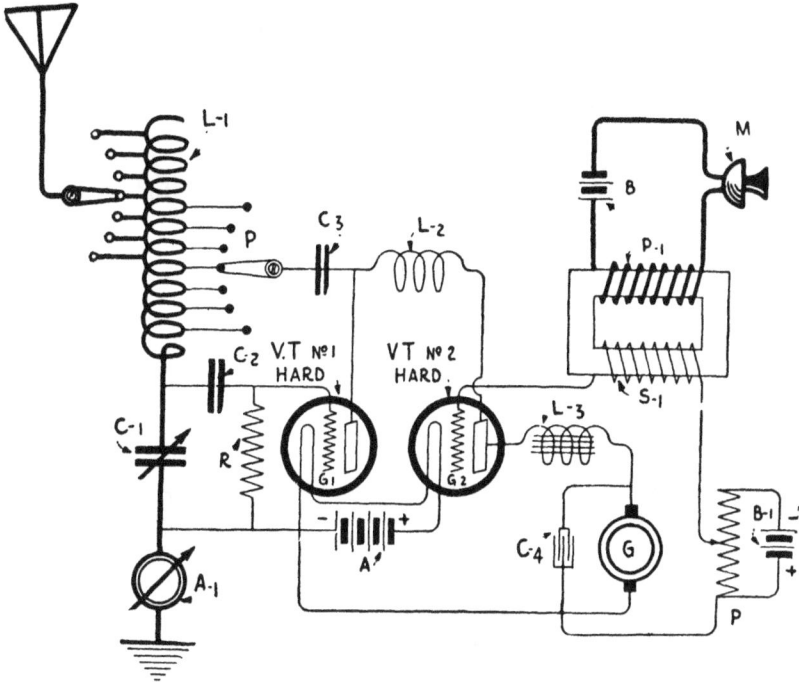

Fig 215. Circuits of a radio telephone transmitter wherein one bulb is used as the "oscillator" and the other bulb as the "modulator." This system has been used extensively by the U. S Government.

For transmission at amateur wave lengths the antenna coupling coil *L-1* may be wound on a cardboard tube (previously soaked in molten paraffine) 5″ in diameter. The coil consists of 40 turns of No. 18 annunciator wire or an equivalent Litzendraht conductor. The antenna side of the coil should be tapped every third turn after the 20th turn from the bottom. The primary side (the right hand tap) should be tapped every 2 or 3 turns for about 25 turns. Bulbs having 4-volt filaments operate satisfactorily in this circuit. The voltage of the generator *G* may be anything from 220 to 600 volts depending upon the "hardness" of the bulbs.

The circuit functions as follows: When both bulbs are lit to incandescence and the plate current is turned on the greater amount of current flows from the plate to the filament of the oscillator bulb. The plate circuit of the modulator bulb is in parallel with the oscillator, but its grid is held at some negative potential by the grid battery B-1.

When the microphone is spoken into, alternating currents of speech frequency flow in the secondary S-1, varying the charge on the grid G-2. This, in turn, varies the plate to filament current in the modulator bulb. An increase of current through the modulator bulb causes a decrease in the plate current of the oscillator bulb and vice versa. These variations occur at speech frequencies causing the amplitudes of the antenna oscillations to be modulated at the same frequencies.

The adjustment of a set like this is not difficult. First, turn out the filament of the modulator bulb, set the generator in operation, and vary the antenna inductance with the primary tap set at some point. Change the inductance at P until maximum antenna current is obtained. Then, vary the resistance of the grid leak between 10,000 and 20,000 ohms. Now, light the filament of the modulator tube and then re-tune the plate and antenna circuits for maximum current. Next, adjust the potential of B-1, and adjust the tap on potentiometer P (400 ohms). Speak into the microphone M trying various grid potentials. Have an observer listen in on a receiving set to note when the best adjustments are secured.

It may be well to vary the potential of the microphone battery B. When the correct adjustments are once found they are not difficult to maintain.

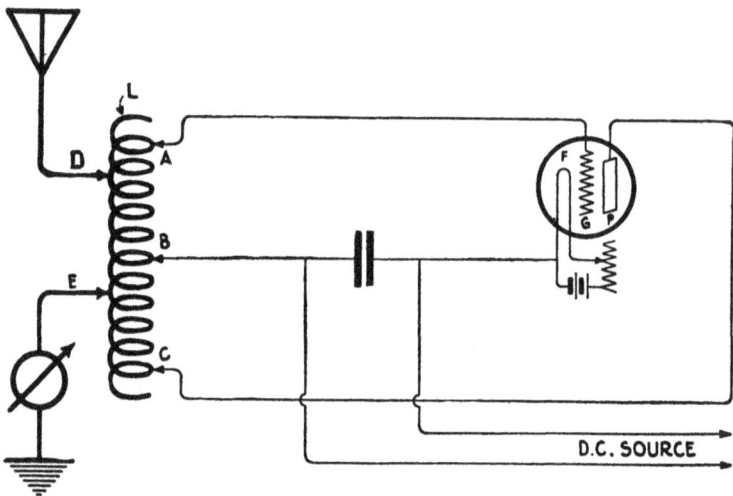

FIG. 216 Circuits of a vacuum tube generator suitable for radio telegraphy and telephony Direct coupling between the plate and grid circuits is shown.

In regard to the choke L-3: Its dimensions vary somewhat with tubes of different characteristics. One that has been found suitable for use with transmitting tubes has the following dimensions. The core is $1\frac{3}{4}''$ x $1\frac{3}{4}''$ x $8\frac{1}{8}''$ and is covered, first, with a sheet of 10 mil ropetag paper, then with a layer of 10 mil empire cloth, and finally with another layer of ropetag paper. Paper insulation of 5 mils is placed between layers of the winding. The total winding consists of 8450 turns of No 29 enamel wire, and there are approximately 615 turns per layer.

The core is made of strips of iron $^{15}/_{16}$ of an inch wide, $8\frac{1}{2}''$ long and 14 mils in thickness. About 62 pieces will be required to make a core of the dimensions given above. The resistance of this coil is 150 ohms and its reactance about 11,000 ohms at the frequency of 500 cycles. The circuit of Fig. 215 is suitable for bulbs of greater power, but the same number of bulbs are required for the modulator circuit as for the oscillating circuit.

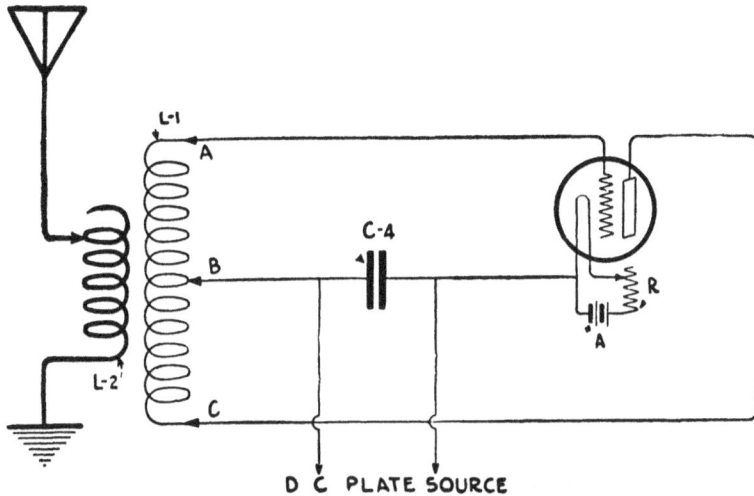

Fig. 217. The circuit of Fig. 216 using inductive coupling between the valve and antenna circuits.

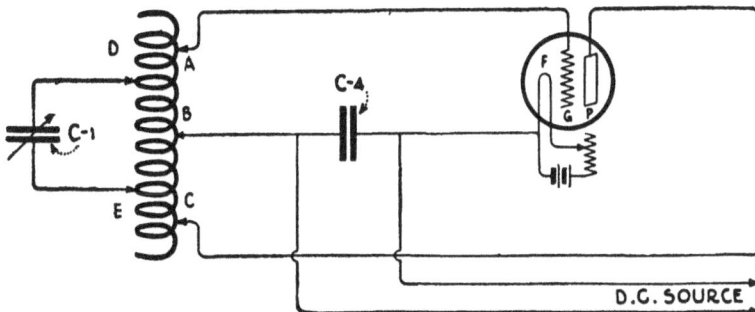

Fig. 218. Vacuum tube oscillator for use in experimental work. By varying the inductance of the coil and the capacity of the shunt condenser C-1 oscillations of any desired frequency may be obtained

Additional *valve transmitting circuits* of practical importance are shown in Figs. 216 to 218. In Fig. 216 the source of plate e.m.f. is connected in series with the plate circuit whereas, in Fig. 215 the terminals of the plate circuit generator go direct to the plate and filament.

The antenna circuit in Fig. 216 is tapped off the coil L by taps D and E. Taps A, B and C permit the most favorable coupling for maximum antenna current. Fig. 217 is the same circuit with inductive coupling.

Fig. 218 shows the circuits of a good *laboratory oscillator* which may be designed to generate currents of any frequency. In practice it will be found that the oscillations cease when the capacity of C-1 for a given inductance is increased beyond some value. It is then necessary to select a larger value of inductance in order to maintain oscillation—and this explains the use of taps D and E.

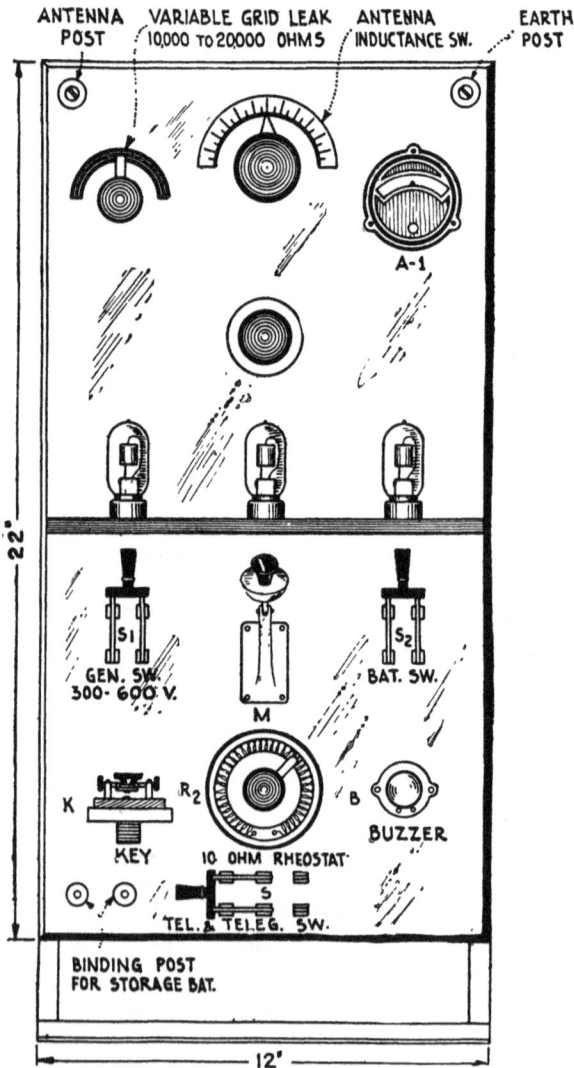

ANTENNA   VARIABLE GRID LEAK   ANTENNA   EARTH
POST   10,000 TO 20000 OHMS   INDUCTANCE SW.   POST

A-1

S1   S2
GEN. SW.   BAT. SW.
300-600 V.
M

K   R2   B
KEY   10 OHM RHEOSTAT   BUZZER

S
TEL. & TELEG. SW.

BINDING POST
FOR STORAGE BAT.

12'

22'

Fig. 219.   Panel wireless telephone set using the circuits of Fig. 212.

With the Marconi V. T. a grid condenser and a grid leak should be used.   With this circuit a small ammeter with a maximum scale of 0.5 to 1 ampere should be connected in series with C-1 to determine the strength of the oscillations. C-1 has capacitance of 0.0005 to 0.001 mfd. For wave lengths up to 2500 meters, the coil in Fig. 218 may consist of a single layer of No. 26 s.c.c. wound on a tube 4″ long, 4″ in diameter. The winding space is $3\frac{1}{2}$″ with 175 turns.   A proportionately greater number of turns must be provided for longer wave lengths.   The capacitance of C-4 is 1 mfd.

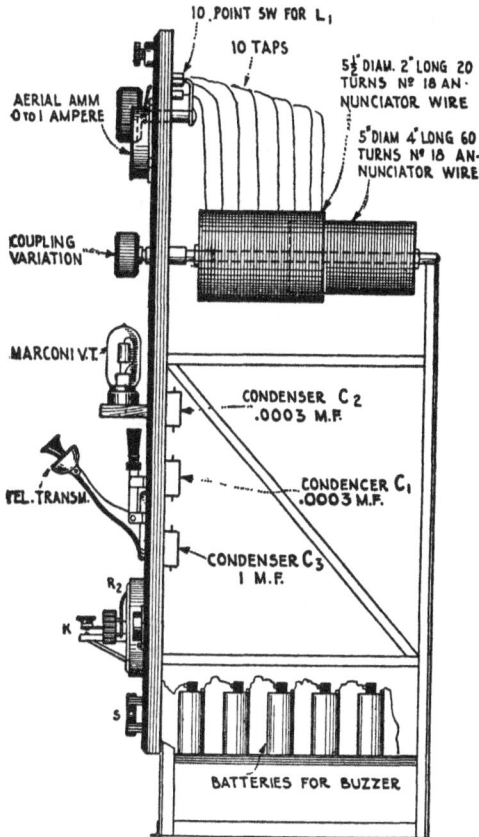

FIG. 220.   Side view of the panel set of Fig. 219.

PANEL RADIO TELEPHONE SET.—A suggested plan of panel transmitter using three Marconi V. T.'s in accordance with the circuit of Fig. 212, is shown in Figs. 219 and 220.   The former is a front view, the latter a side view.   It is possible to vary these approximate overall dimensions for the panel without sacrifice of efficiency.   The position of the instruments may be changed to suit the builder.   The panel board should be of some first class insulating material such as bakelite, dilecto or transite asbestos wood.

The insulating tubes or forms for the oscillation transformer should be made of bakelite or some similar insulating material, but cardboard tubes soaked in molten paraffine will do as well.   The rod which varies the transformer coupling should not be of metal but of some insulating material such as hard rubber.

The Marconi V. T.'s are mounted on the front of the panel so that the filament temperature can be judged by sight, if an ammeter is lacking. The *inductance variation switch* on the top of the panel has 10 taps leading to every second turn of the antenna coil   Switch S-1 is the generator switch and across it there is a 1 mfd. condenser for smoothing out the commutator ripple of the generator. S-2 is a small 10-ampere switch to break the filament circuit.   The *grid leak* at the top of the panel is an adjustable Ward-Leonard resistance 10,000 to 20,000 ohms.   The microphone M is a common 10-ohm telephone transmitter.   The *buzzer B*, for *telegraphy by damped oscillations*, is mounted on the lower right hand part of the panel with the key to the left.   The buzzer battery is in a box behind the panel.   The 10-ohm rheostat R-2 is not required if the filament battery e.m.f. does not exceed 4 volts.   The condensers C-1 and C-2 are of fixed capacitance.   They may be made of sheets of *copper foil* separated by sheets of *mica* and their capacity may be calculated by the formula on page 45.   Air condensers can be used if the plates are separated by $\frac{1}{8}$".   Two binding posts for the aerial and ground are mounted on top of the panel.

The connecting wires on the panel should not be bunched but spaced as far apart as possible.   Particular attention should be paid to the connections.   Use copper lugs and solder the wires wherever possible.

As was stated on page 266 the set should be tuned with the microphone on short circuit.   A switch, which is not shown on the panel, can be shunted around it.   When transmitting by damped oscillations with the buzzer, do not use too much battery on the buzzer; usually one cell is sufficient.   The current for the buzzer may be taken from the filament battery.

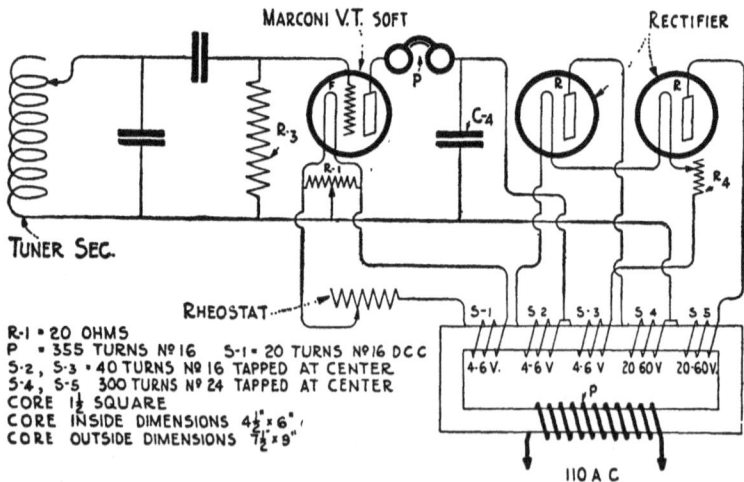

FIG. 221.   Circuits using a special alternating current transformer and two Tungar rectifiers to supply the plate and filament currents of a vacuum tube detector set

HOW TO OPERATE VACUUM TUBES FROM ALTERNATING CURRENT.

—The method was first disclosed by W. C. White (G. E. Co.). An adaption of his circuit suitable for transmitting and receiving bulbs is shown in Fig. 221. The plate current is supplied by two Tungar rectifiers R connected in the secondaries of a closed core transformer, of which the primary is fed by 110 volts a.c. Both halves of a cycle are utilized by this arrangement giving a fairly smooth plate current. The condenser C-4 of from 2 to 6 mfds. capacitance somewhat compensates for the voltage fluctuations.

The transformer is provided with 3 secondaries. Two of these (S-1 and S-2, S-3) light the valve filaments. The other secondary, S-4, S-5, provides the rectified a.c. for the plate circuit.

The filament of the detection tube is shunted by a 20 ohm rheostat R-1 with a sliding contact. The variations of the *potential gradient* which would otherwise exist between the grid and filament, are reduced by properly locating the slider. Then the characteristic "hum" of the alternator is reduced almost to complete silence.

Note that secondary S-2, S-3 is tapped at the center; as is also secondary S-4, S-5. The former provides 4–6 volts for the rectifier filaments and the latter 20–60 volts for the plate circuit.

The dimensions of the transformer core to secure step-down transformation may be calculated by the process outlined on pages 65 to 69. The description of a transformer of suitable dimensions follows. The core is $1\frac{1}{2}''$ square (in cross-section). The inside dimensions are $4\frac{1}{2}''$ x 6″, the outside dimensions $7\frac{1}{2}''$ x 9″. The primary is wound with 355 turns of No. 16 d.c.c. The secondary S-1 has 20 turns of No. 16 d.c.c. The secondary S-2, S-3 has 40 turns of No. 16 d.c.c., with the winding tapped in the middle. Secondary S-4, S-5 has 300 turns of No. 24 d.c.c. tapped at the center. R-4 is a 10-ohm rheostat. The core of the transformer is, of course, laminated. The various secondaries are not necessarily placed on one end of the core but they may be distributed over the three sides of it.

Before using this transformer the builder should test the voltages of the secondaries with a voltmeter. If the voltage of any particular section is too low, he should add more turns; if too high unwind a few turns. If taps are brought from the secondaries, the number of turns for a given voltage may be found by trial.

Three-electrode tubes with their grids connected to the filament or two-electrode tubes of the requisite current carrying capacity, may be substituted for the *G. E. Tungar rectifiers R.*

"HARD" AND "SOFT" BULBS.

—These terms are used to distinguish bulbs with *high vacuum* from those of *low vacuum*. *Hard* bulbs will not ionize (give the characteristic blue glow) at very high voltages. *Soft* bulbs will ionize at low voltages. Soft bulbs are suitable as detectors while hard bulbs are preferable as amplifiers and oscillators. Soft bulbs will give oscillations but their output is very limited. The power in the plate circuit cannot exceed $I \times E$. Hence, the higher E is without ionization, the greater the output.

Soft bulbs will often give better response as oscillation detectors than hard bulbs. Soft bulbs may be used in cascade amplification circuits if the primary and secondary of the inter-valve transformers are variable. A few trial experiments will show the amateur how to obtain good signals with soft bulbs.

MODULATION OF ANTENNA CURRENTS IN RADIO TELEPH-ONY.—The subject of modulation of radio frequency currents in radio telephony has been mentioned in several places.   In radio telephony the amplitudes of the antenna oscillations must be varied at the speech frequencies of the human voice and particular care must be taken to design the apparatus so that these effects are obtained without distortion.

There are two general systems of modulation that may be applied to vacuum valve circuits.   In one system a *shunt circuit*, which absorbs energy according to the impulses of the voice, is connected around some portion of the circuit in which radio frequency currents are flowing.   In the other system a similar circuit and apparatus is connected *in series* with the source of oscillations.

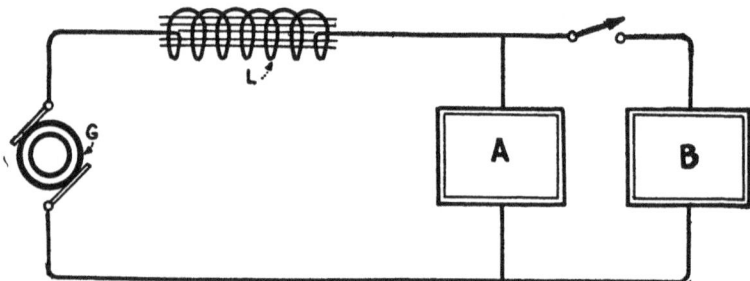

Fig. 221a.   Illustrating the principles of a shunt modulation circuit in wireless telephony.

In the first method shown in Fig. 221*a*, assume that the load *A* is connected to a d.c. dynamo and a large inductance *L* is placed in the supply line.   This inductance will tend to keep the current in the supply mains constant throughout any series of rapid changes of the resistance of the load.   Closing the switch affords a path *B* around *A* and for an instant the current in *A* is decreased.   The current in the line, however, will soon readjust itself to conform with that required by the new load.   If the switch is opened and closed rapidly and the inductance *L* is relatively large, the current in *A*, and (if we assume the resistance of *A* is constant) the voltage across *A* will follow the variations of the shunt resistance *B*.   This effect is amplified by the reactance voltage of the coil *L* which adds to or decreases the potential across *A* (depending on whether the load *B* is disconnected or connected).   It is clear that the load *B* and the switch can be replaced by any variable resistance such as a carbon grain telephone transmitter, and that the source *A* can be an oscillating valve circuit supplied with energy from the d.c. generator *G*.   The condition necessary for satisfactory operation is that the energy taken from the source be kept constant.   As an antenna circuit energized by coupling to an oscillating circuit receives practically a constant amount of energy, a rapidly varying resistance shunted around any portion of this circuit will modulate the current in that portion.

A circuit using this method of modulation (Fig. 221a) was used on wireless telephones supplied to the Army and Navy and is shown fundamentally in Fig. 215.

For greater power several valves may be connected in parallel with each of the valves shown in this circuit. For the best operation there should be as many valves in the modulator circuit as in the oscillating circuit. It is necessary to put a grid battery in the modulator circuit in order that the valve will operate on the linear portion of the characteristic curve.

In the circuit, Fig. 215, the inductance coil L-2 is just large enough to keep the high frequency currents from surging through the modulator system and through the generator, for the inductance coil L-3 may often have enough distributed capacitance to allow a considerable amount of the high frequency energy to pass through the generator. The coil L-2 which offsets the possibility of this may have an inductance of 6 to 13 millihenries.

In the second system of modulation the modulator (either a simple transmitter, or else a three-electrode valve or magnetic amplifier to increase the effect of a telephone transmitter) is connected in series with the circuit. The circuit shown on page 264, Fig. 212, makes use of this principle.

A carbon grain transmitter can be connected almost anywhere in an oscillating valve circuit with some degree of modulation. The optimum position will depend upon the resistance change ratio, that is the ratio between the minimum and maximum resistances of the transmitter, and upon the type of circuit used. The experimenter should try inserting the transmitter in the antenna circuit, in the filament circuit, and in series with the grid circuit and determine the optimum position by listening in at a receiving set placed a short distance away.

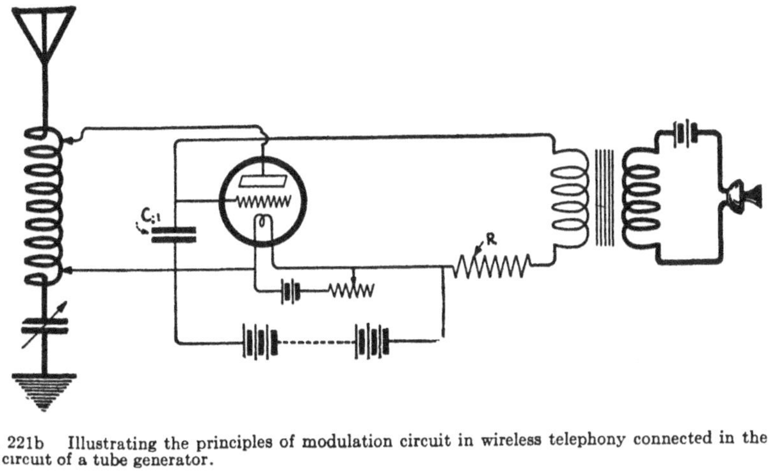

FIG. 221b   Illustrating the principles of modulation circuit in wireless telephony connected in the grid circuit of a tube generator.

Many wireless telephones operate with the transmitter connected in series in some part of the grid circuit of the oscillator valve. A circuit of this type is shown in Fig. 221b, which was used by the Deforest Co.

and has the advantage that only one bulb is necessary. However in a circuit of this kind considerable difficulty may be experienced in getting the valve to oscillate and modulate properly. By experimenting with a grid condensers of various capacities, and with different values of the resistance R the correct values may be found for any type of vacuum valve.

It is impractical with any system of modulation using V. T. oscillators to secure 100% modulation for that would stop oscillation. It generally requires about 5% of the total current to maintain oscillations. It is rarely possible to secure 95% modulation with any modulation circuit system except possibly the circuit shown in Fig. 215.

It is well to remember that it is only the modulated component of the emitted wave that is effective in wireless telephony. Thus a system putting 50 watts into the antenna with 20% of the energy modulated would be no more effective than a 10-watt set completely modulated. This explains the statement made concerning the modulation system of Fig. 215, viz, that as many vacuum tubes are required in the modulation system as are in the oscillating system.

In any system the modulator bulb should operate on the straight portion of the characteristic curve. If the grid potential fluctuates too much so as to reach a value below or above the bends in the characteristic curve inaccurate speech reproduction results.

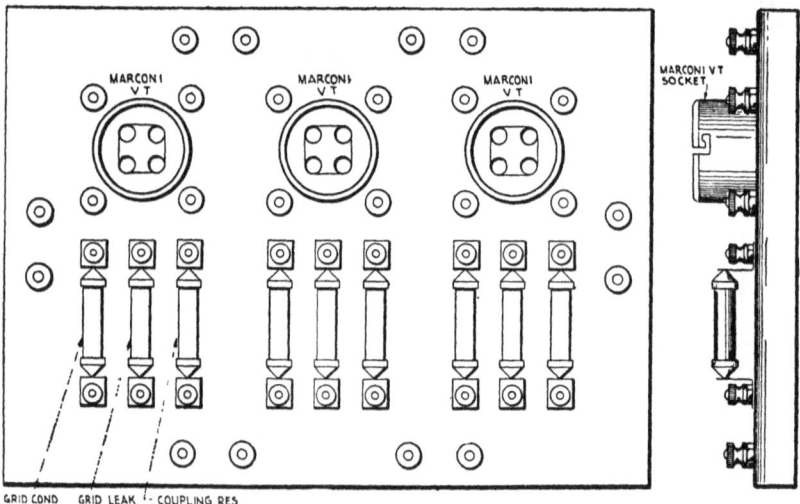

Fig. 221c. Plan and side views of the valve adapter manufactured by the Marconi Company for use with the Marconi V T

It is desirable to speak into the transmitter in such a way that the breath does not strike the diaphragm. The sound waves only are necessary to vibrate the diaphragm.

Several transmitters may be connected in parallel for higher powers provided they are all of the same type. If one should happen to carry a little more than its share of current it will become heated and its resistance will be lowered. It will then take even a greater portion of the current and will soon become inoperative. The microphones should be arranged on a stand so that the voice strikes each equally.

MARCONI V. T. ADAPTER, GRID CONDENSER AND GRID LEAKS.—In many instances in this volume the circuit diagrams for the Marconi V. T. call for the use of several valves. It is a genuine convenience to possess a valve adapter having three bulbs and the requisite grid condensers, grid leaks and coupling resistances mounted with appropriate binding posts for immediate connection.

Such an arrangement permits the use of any kind of a circuit to which the Marconi V. T.'s are adapted.

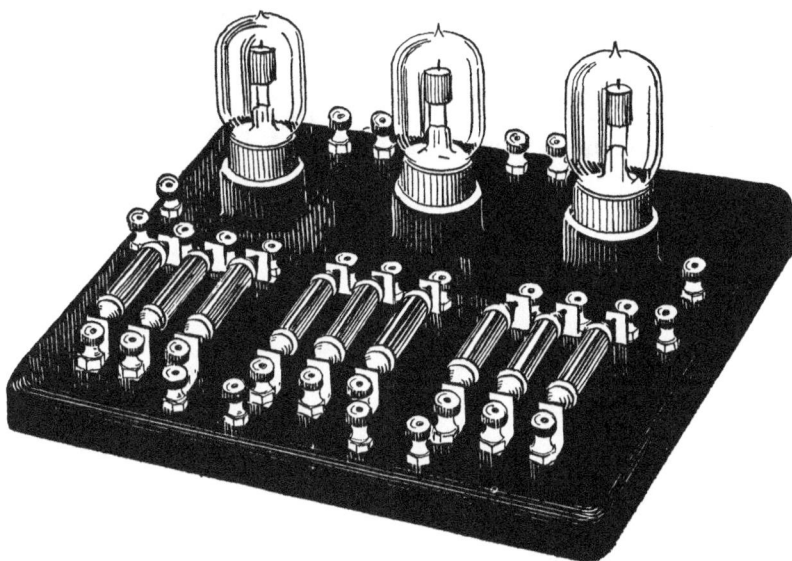

FIG. 221d. Marconi three-valve adapter in perspective.

Fig. 221c is a plan view of the three-valve mounting manufactured by the Marconi Company. Four binding posts for connection to the grid plate and filament terminals are mounted around the base of each valve. A *grid condenser* (which has the same dimensions and the same appearance as a grid leak), a *grid leak* and *coupling resistance* are mounted immediately in front of each valve. A *special filament rheostat* is mounted in the base of each socket for variation of the filament temperature. The facility with which the connections can be changed with an adapter of this type needs no further emphasis.

Figs. 184, 185, 186, 212, 214, 221, 222, 256 are circuits and apparatus to which the Marconi valve mountings are admirably adapted. For long distance reception with the loop circuits of Fig. 256, two of these three valve mountings should be used providing a six-stage amplifier. With the grid leaks, condensers and coupling resistances already in place it is but a matter of a few moments to connect them up in any way desired.

The three-valve base may, for example, be mounted on the top of a cabinet receiver. The first valve may be used with a regenerative coupler as a detector, and the second and third bulbs as audio frequency amplifiers.

It is to be noted that the grid condensers in Figs. 221c and d have the same dimensions as the grid leaks. These condensers may be purchased ready for use in capacities from 0.0001 to 0.005 mfd. The grid leaks are supplied in resistances from 200,000 to 5,000,000 ohms (5 megohms).

In most of the resistance coupled amplifier circuits heretofore shown, the coupling resistances were of 2 megohms but as will be noted from the curve of Fig. 178 good amplifications can be secured with other valves of resistance. Take as an illustration an 8-stage resistance coupled amplifier in which the first three valves are used as radio frequency amplifiers, the fourth valve as a detector and the remaining four as audio frequency amplifiers; the following values of resistance have been found suitable for maximum magnification. Valves 1 and 2, 2 and 3, and 3 and 4, should be coupled through Marconi resistances of 500,000 ohms each, but valves 4 and 5 and all remaining valves should be coupled by resistances of 2 megohms each. In a circuit of this type the Marconi valve adapters will be found extremely useful.*

---

*The three-valve adapter is not yet on the market, but a single valve adapter is now available and is of course more flexible for complex circuits than that described.

CHAPTER X

CABINET RECEIVERS AND ACCESSORIES

GENERAL.—A few designs given by amateur experimenters will be shown. The arrangement of the apparatus of a receiving set is unimportant as long as the internal connections are not bunched. The experimenter must bear in mind that two parallel leads from a tuning coil may add a very considerable capacity to the circuit, and the mutual effects between parallel wires may in some instances reduce the signals. It is well to keep the leads as far apart as possible. Regenerative coupling may be obtained in a receiver cabinet with the wires of the grid and plate circuits separated a couple of feet.

A panel or cabinet receiver adds nothing to the electrical efficiency of a set, except that it is manipulated with less difficulty than the isolated instrument type. The saving of space may be an argument in favor of the cabinet receiver.

CABINET RECEIVER FOR THE MARCONI V. T.'s.—The design in Fig. 222 is offered as a suggestion. The cabinet includes the component parts of the wiring diagram Fig. 223 which indicates a short wave (200 meter) receiver and *three Marconi V. T.'s in cascade.* The first bulb is used as a *detector bulb* and the next two as *amplifiers.* The detector circuits are the same as those of Fig. 183, except for the addition of the 20,000 ohm graphite potentiometer *P-1.*

V-1, Fig. 223, is the antenna variometer and *V*-2 the secondary variometer. They are identical in construction. The outer cylinders of the variometers are $4\frac{1}{2}''$ in diameter, $1\frac{1}{2}''$ long, and wound full with No. 26 s.c.c. The inner cylinders are $4''$ in diameter, $1\frac{1}{2}''$ long and wound full with No. 26 s.c.c. *C*-1 is a standard 21 plate variable condenser or any other variable condenser of about the same range of capacity. The capacity of *C*-3 is 0.0001 mfd.; that of *C*-4, 0.005 mfd. *R, R, R,* are 2 megohm grid leaks and *R*-1 is a coupling resistance. The potential of *B* is 80 volts. Class II Marconi V. T.'s should be used with this circuit.

The three Marconi V. T.'s in Fig. 222 are mounted on the top of the cabinet. The two variometers are located inside of the cabinet, and are controlled by knobs on the outside. The 10-ohm filament rheostat is mounted below the variable coupling condenser *C*-1. The plate battery potentiometer is placed on the right side. If the *A* battery has a potential of 4 volts and the *B* battery of 80 volts the potentiometer and the rheostat may be omitted. The grid condensers *C*-4 are mounted at convenient points inside the cabinet, as shown. The grid leaks are preferably placed on the front of the panel where they can be readily removed. Appropriate scales for the variable condenser and the variometers are mounted on the front of the panel. The cabinet is approximately 13½″ long, 7″ high and 7″ deep.   The B battery is placed behind the right hand variometer.

This detector set will be found particularly useful in connection with the short wave *wireless telephone tube transmitter* of Figs. 219 and 220. The complete set will permit conversation up to 50 miles.

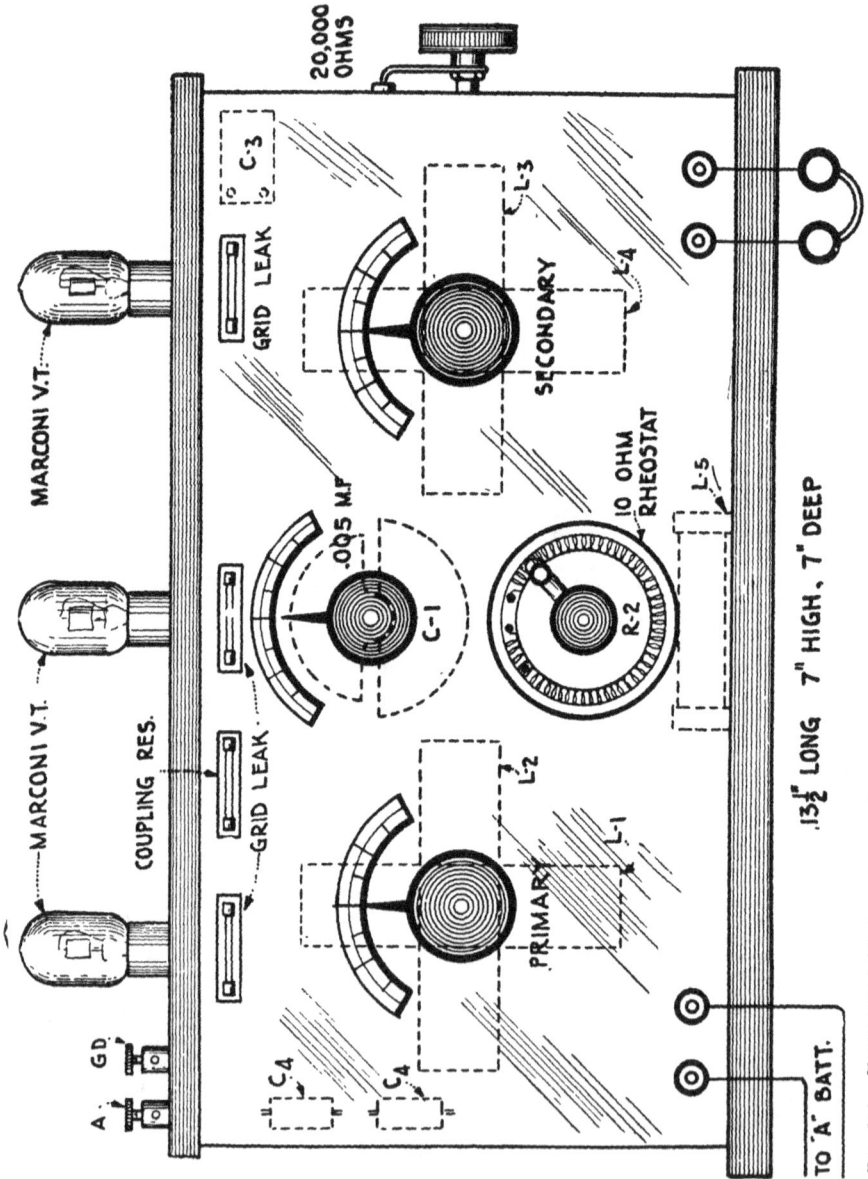

Fig. 222. Short wave cabinet receiver for use with the Marconi V. T.'s. Variometers are used as the tuning elements. The circuit utilizes one V. T. as a detector and two additional V. T.'s as amplifiers.

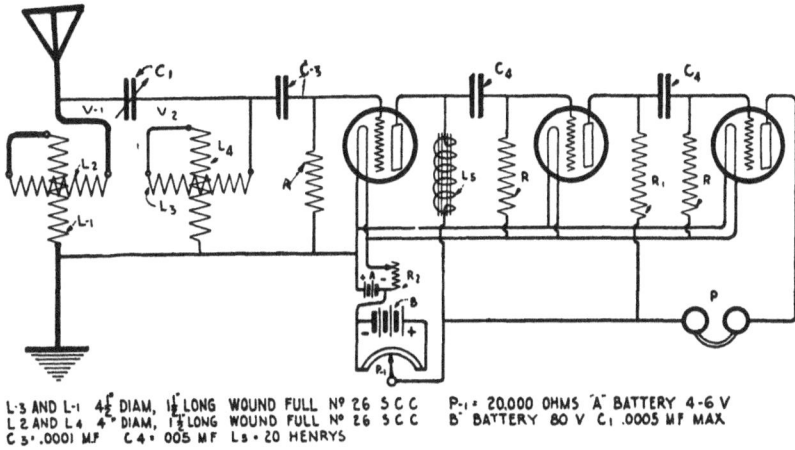

L·3 AND L·1 4½ DIAM, 1½ LONG WOUND FULL № 26 S C C   P·₁ ₌ 20,000 OHMS ‾A‾ BATTERY 4-6 V
L 2 AND L 4 4″ DIAM, 1½ LONG WOUND FULL № 26 S C C   B‾ BATTERY 80 V C₁ .0005 MF MAX
C 3 ₌ .0001 MF   C 4 ₌ 005 MF   L s ₌ 20 HENRYS

Fɪɢ. 223.   Wiring diagram of the cabinet set in Fig. 222.

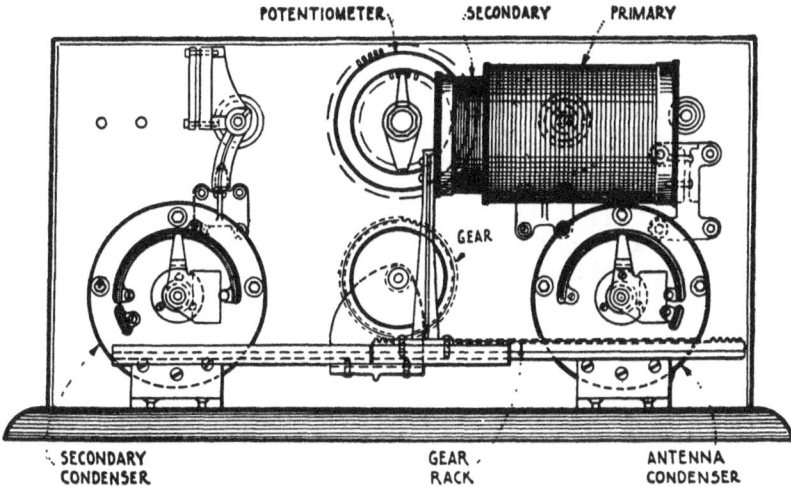

Fɪɢ. 224.   Rear view of a commercial type of receiving tuner showing the internal construction and particularly the mounting of the tuning transformer and the primary and secondary condensers.

COMMERCIAL PRACTICE.—Fig. 224 is in reality a rear view of the Marconi type 106 tuner, published to give the beginner an idea of the inner cabinet construction of commercial tuners.   The secondary coil of the coupler is mounted on a gear rack moving in a metal guide.   A large gear controlled by a knob on the front of the panel operates the coupling mechanism.   The "tens" and "units" switches are placed on the panel in front of the transformer primary.   Since the rotary plates of the variable condensers lie in a horizontal position they must be balanced by weights in order to hold their position at any particular setting.

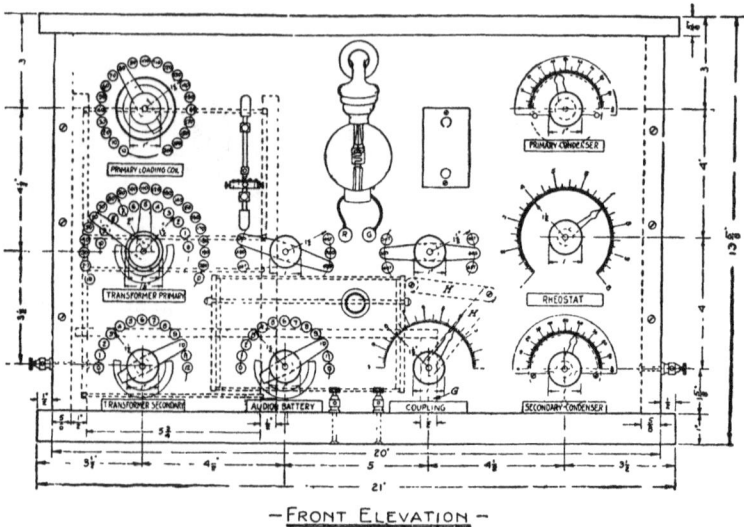

FIG. 225. Plan view of a cabinet designed by an amateur.

AMATEUR PRACTICE.—A panel set designed by A. C. Burroway of Ohio shows the excellent workmanship of some amateur designs. Fig. 225 is a plan view, Fig. 226 a side view, and Fig. 227 a partial circuit diagram, which can be worked out to completion by studying the diagrams already given. The valve mounting is for an old time audion but a Marconi V. T. mounting may be substituted. A perikon detector is mounted alongside of the valve which is to be used in event the valve goes out of commission.

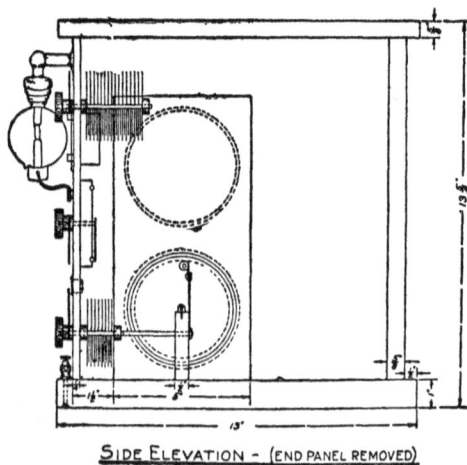

FIG. 226. Side view of the cabinet set of Fig. 225.

The function of the various switches and knobs are clearly marked on the diagram. The method of varying the coupling by a bell-crank and lever deserves particular attention. The voltage of the B battery is controlled by a multi-point switch.

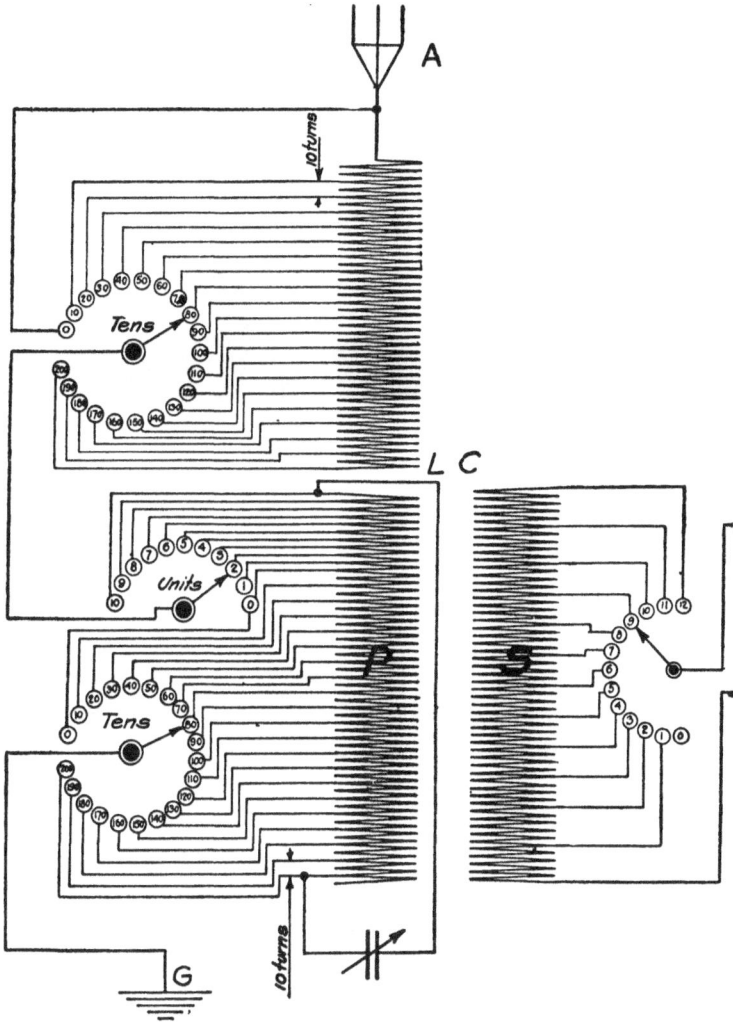

Fig 227. Circuit diagram for the cabinet set of Fig. 225.

The aerial tuning inductance is $4\frac{1}{8}$" x 6" wound with 200 turns of No. 26 B. & S. The primary coil has the same dimensions wound with 210 turns of No. 26 B. & S. The secondary coil is $3\frac{3}{4}$" x 6" wound with No. 30 B. & S. The tuner will be found suitable for the use of crystal rectifiers at wave lengths up to 3500 meters and with valves up to 2500 meters.

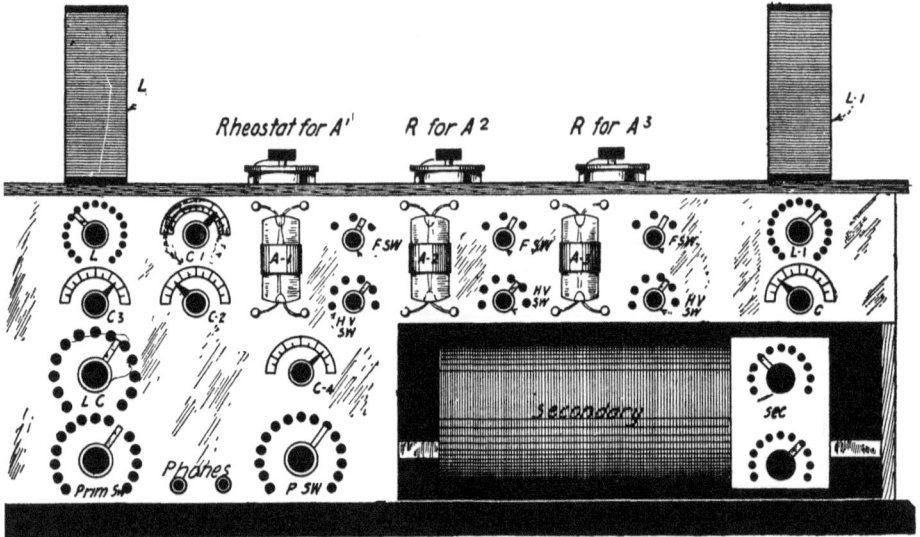

FIG. 228. Cabinet receiver for use with a valve detector and two stage amplifier.

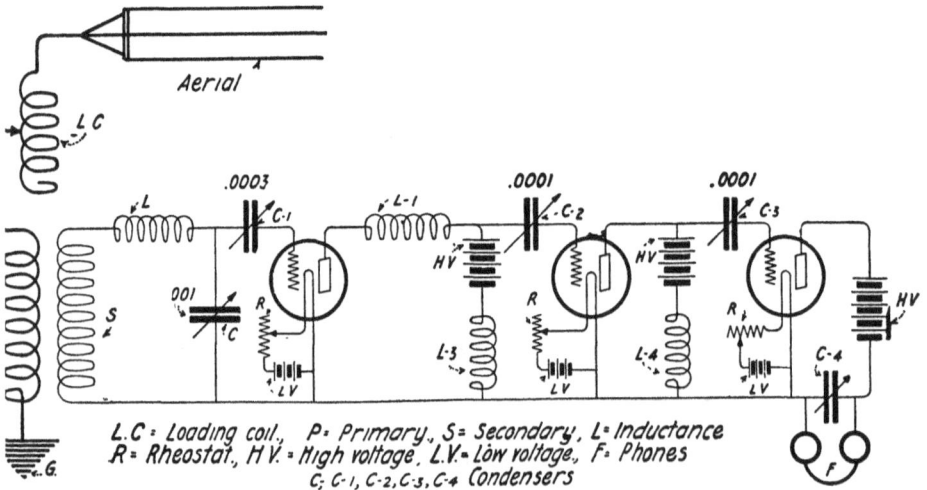

L.C = Loading coil., P = Primary., S = Secondary, L = Inductance
R = Rheostat., H.V. = High voltage, L.V. = Low voltage, F = Phones
C; C-1, C-2, C-3, C-4 Condensers

FIG. 229. Circuit for the cabinet set of Fig. 228.

Chas. Doty of New York has shown the design of Fig. 228 which uses three valves connected in cascade. The circuit and the cabinet can be adapted to the Marconi V. T. The primary of the coupler is stationary with the secondary sliding in and out as shown. L and L-1 are secondary loading coils mounted on the top.

The circuit is shown in Fig. 229. P is 7″ x 12″ wound with No. 28 s.s.c. S is 6″ x 12″ wound with No. 30 s.s.c. L and L-1 are 4″ x 30″

wound with No. 30 s.s.c. L-3 and L-4 are the secondaries of $\frac{1}{2}''$ spark coils. Mr. Doty used 3 high voltage batteries but as the preceding diagrams demonstrate, one battery may be used as well. The inductance L-1 may be omitted.

R. Hoaglund is responsible for the set in Fig. 230 and the construction is self-explanatory. The tuner was designed for crystal rectifiers and has inductances of the correct dimensions to respond to 2500 meters. The

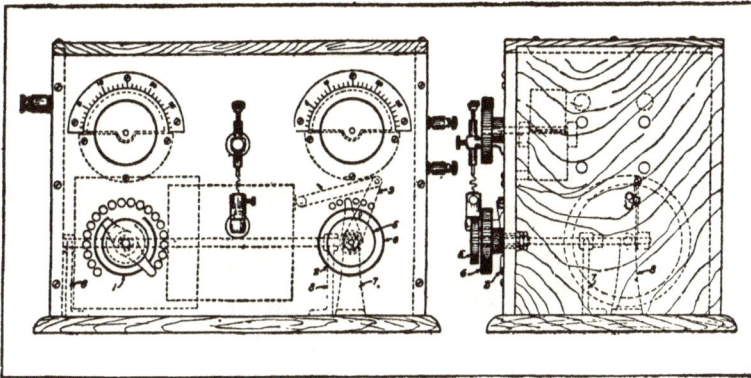

FIG. 230. Cabinet receiver for reception at wave lengths up to 2500 meters.

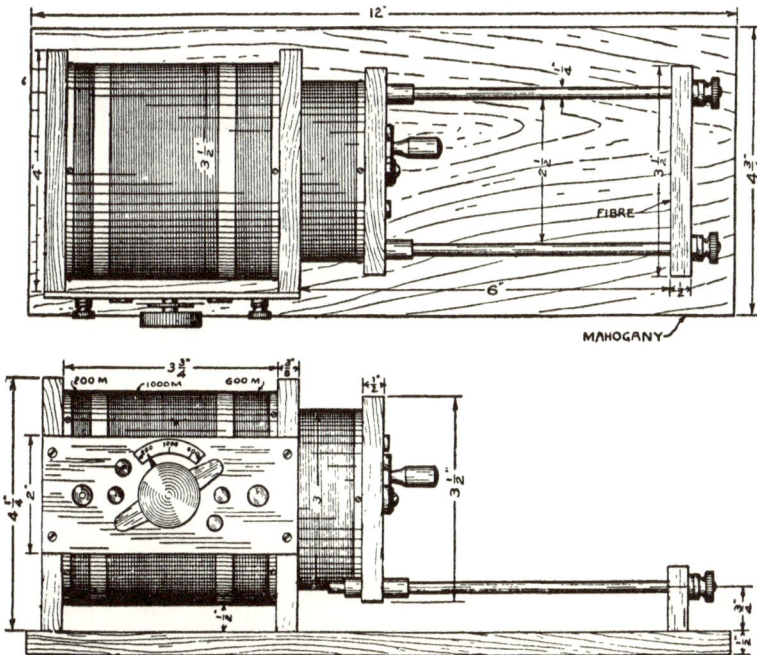

FIG 231. Design of a simple "loose coupler" for short wave reception.

primary and secondary variable condensers are mounted to the top of the cabinet and the coupler immediately underneath.

Fig. 231 shows a good design for a *simple "loose"* coupler. The primary is split into three sections to cover 200, 600 and 1000 meters. The secondary has a 3-point switch to cover the same range of wave lengths. The secondary has 266 turns of No. 30 s.s.c., with taps taken at 60, 180, and 266 turns respectively. The requisite turns for the primary will have to be calculated for each station for the number varies with different aerials. The method of computation is explained on pages 169 to 177.

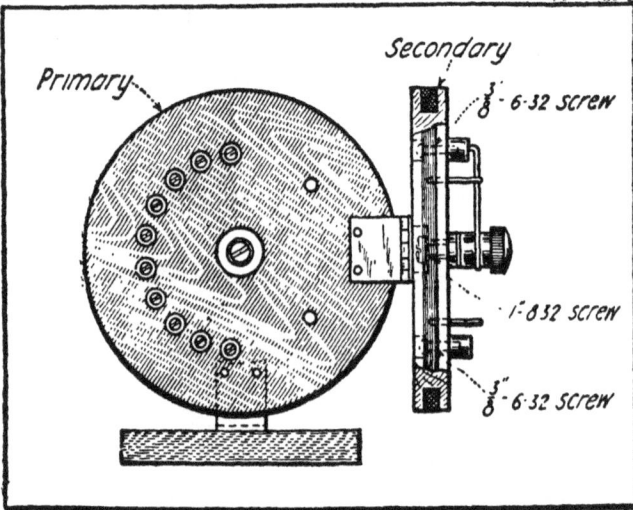

FIG. 232. "Hinged" coupler suitable for amateur wave lengths.

The hinged coupler in Fig. 232 is suitable for reception at moderate wave lengths. The primary coil is wound with 100 turns of No. 28 s.s.c. and the secondary coil with 200 turns of No. 32 s.c.c. The primary is tapped every 10 turns, the secondary every 20 turns. The discs for support of the coils are $3\frac{1}{4}$″ in diameter, and the groove for the winding is $\frac{1}{8}$″ wide, $\frac{5}{16}$″ deep. The switches for inductance variation are mounted directly on the discs.

CHAPTER XI

DESIGN OF WAVEMETERS—TUNING THE AMATEUR'S TRANSMITTER—MEASUREMENTS OF INDUCTANCE, CAPACITANCE AND HIGH FREQUENCY RESISTANCE—GENERAL RADIO MEASUREMENTS

GENERAL.—The frequency of an oscillation circuit may be calculated from the relation $n = \dfrac{1}{2\pi\sqrt{LC}}$, where L is the inductance in henries, and C the capacitance in farads. Hence, when we have a coil L shunted by a variable condenser C, and a calibration curve for the condenser is provided, we may determine the frequency of the circuit for each setting of the condenser.

Knowing the frequency the *equivalent wave length* is equal to $\dfrac{300,000,000}{\text{frequency}}$. The scale of the condenser may then be marked in wave lengths rather than in the corresponding frequencies.

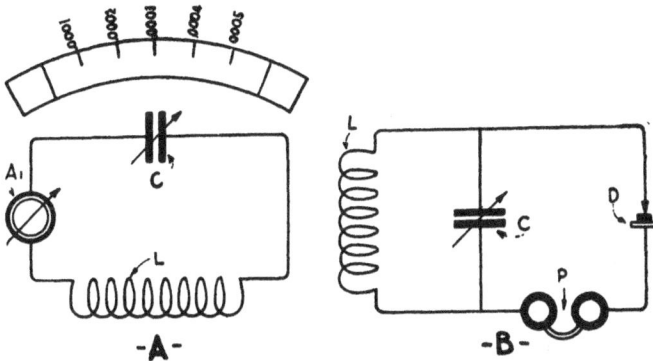

Fig. 233. *A*, simple oscillation circuit which when calibrated may be used as a wavemeter. *B*, wavemeter circuit using a crystal rectifier and head telephone for locating resonance.

Any oscillation circuit like that of Fig. 233 may be converted into a *wavemeter*. It is only necessary to determine the *equivalent wave length* of the circuit for various settings of the variable condenser. The results

289

may be put in a table or plotted as a curve for future reference.   Assume that as in Fig. 233 the capacitance of the condenser C at various settings varies from 0.0001 to 0.0005 mfd., and the inductance of the coil L is 90,000 cms.   The wave length for 0.0001 mfd. in shunt may be found from the following:

$$\lambda = 59.6 \ \sqrt{L\,C}$$

or

$$\lambda = 59.6 \ \sqrt{90,000 \times 0.0001} = 178 \text{ meters.}$$

The wave lengths for the other settings are as follows:

```
0.0002 = 252 meters
0.0003 = 309    "
0.0004 = 357    "
0.0005 = 400    "
```

If the amateur can purchase a *calibrated variable condenser* with the capacity given for several settings of the scale he may compute the dimensions of a suitable inductance from formula (**40**) page 54 and use the formula above for computing the wave length.

Such a wavemeter will prove to be fairly accurate if the necessary computations are carefully carried out.   A wavemeter of the range computed above will be satisfactory for tuning the amateur transmitter but one of considerably greater range will be required for calibration in receiving work.

USES OF THE WAVEMETER.—The wavemeter is not only useful for tuning a transmitting set but for carrying out many other radio frequency measurements.   They are enumerated below.

```
(1) Measurement of wave length
(2)      "      "  frequency
(3)      "      "  decrement
(4)      "      "  inductance
(5)      "      "  capacitance
(6)      "      "  mutual inductance
(7)      "      "  coefficient of coupling
(8)      "      "  distributed capacitance of coils
```

These various applications of the wavemeter will be treated in the following paragraphs.

RESONANCE INDICATORS.—We may expect from the discussion on reactance and its effects in Chapter III that the current in a wavemeter circuit reaches its maximum amplitude when resonance is established with the circuit under measurement.   Similarly, the potential difference across the terminals of the condenser C, Fig. 253a, attains a maximum at resonance.   The fact that resonance has been obtained may be determined by either a "*current*" or "*voltage*" indicator connected in the wavemeter circuit.   For current indications, a milliammeter—0 to 200 milliamperes— is placed in series with the wavemeter circuit at A-1 in Fig. 233.   For some measurements a *"current square"* meter, that is, one in which the scale divisions are proportional to I^2, is preferred.

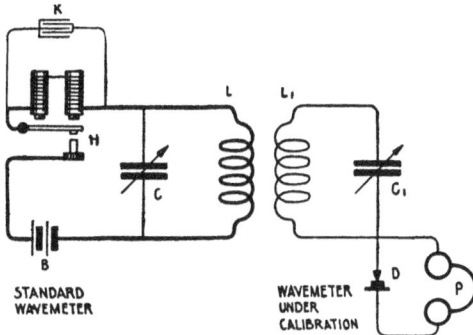

To indicate a voltage maximum across the condenser C, in circuit B Fig. 233, it is shunted by a carborundum rectifier D with a head telephone in series. If the decrement of the circuit under test is relatively small the current maximum as read by observation of A-1 or heard in the telephone P, will be sharply defined for any small movement of the variable condenser. The detector D is used when the source of oscillations is weak.

FIG. 234. Simple method of calibrating a wavemeter from a standard.

With strong oscillations, the "current square" meter may be used.*

CALIBRATION OF A WAVEMETER FROM A STANDARD.—

A home-made meter can be calibrated from a standard wavemeter as in Fig. 234. $L\,C$ is the standard wavemeter, excited by a buzzer and a battery; L-1, C-1 the wavemeter to be calibrated. The detector D, of the wavemeter, is connected *unilaterally* although it may be shunted to the condenser as in Fig. 233 at B.

The meter is calibrated as follows: Set $L\,C$ at some wave length and vary the capacitance of C-1 until a sharp maximum is heard in the head telephone P. The standard and the meter under calibration are then set at the same wave length. Continue to do this throughout the scale of C-1 and keep an accurate record of the results. For accuracy use low voltages at B, and, to prevent sparking, shunt the magnets of the buzzer by a 1 mfd. condenser.†

A more accurate method of calibration is shown in Fig. 235. In the method of Fig. 234, the added capacity of the buzzer leads will affect slightly the calibration of the standard wavemeter. This error may be eliminated by the circuits and method of Fig. 235. In that figure B is a *standard wavemeter*, C the *wavemeter to be calibrated*, and A represents a condenser and coil of a magnitude permitting resonance with B.

To carry out the calibration, first set B at some wave length, then put the buzzer in operation and vary the capacitance of condenser C until $L\,C$ is in resonance with L', C' as noted in the telephone. Then put wavemeter C in inductive relation to A and vary condenser C-2 until wavemeter C resonates with A. The wave length of C is now the same as B. The same procedure should be carried out over the entire scale of condenser C-2.

---

*An enumeration of all the various resonance indicators is given on page 189 of the author's "Practical Wireless Telegraphy."

†In all radio measurements where the wave meter or a resonating circuit is inductively coupled to the source, the coupling should be as loose as is consistent with the proper operation of the current or voltage indicator. Only in this way can accuracy be obtained.

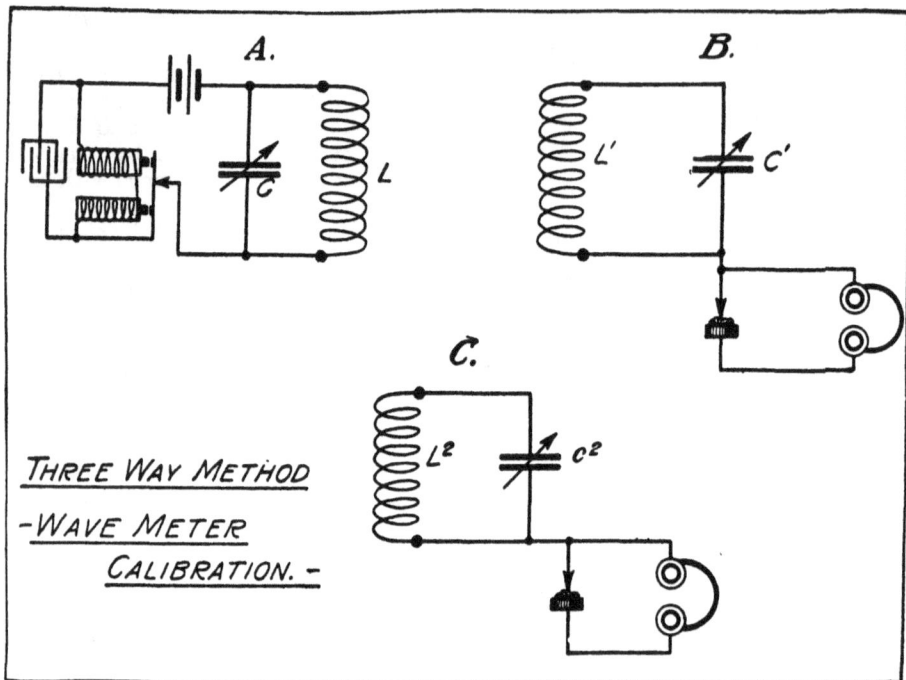

THREE WAY METHOD
-WAVE METER
CALIBRATION. -

FIG. 235 Accurate method of calibrating a wavemeter from a standard.

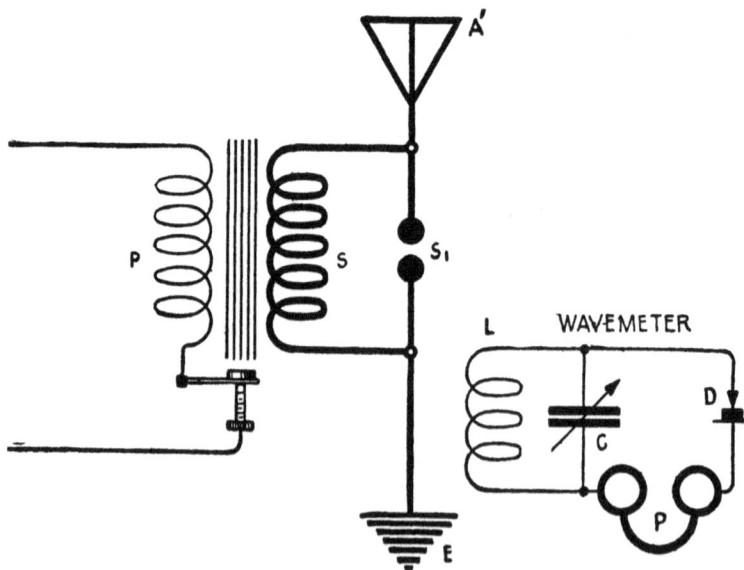

FIG. 236 Showing how the natural wave length of an antenna is measured by a wavemeter.

TUNING THE AMATEUR'S TRANSMITTER.—The first measurement is that of the *natural wave length of the antenna.* To carry out the measurement as in Fig. 236 proceed as follows:

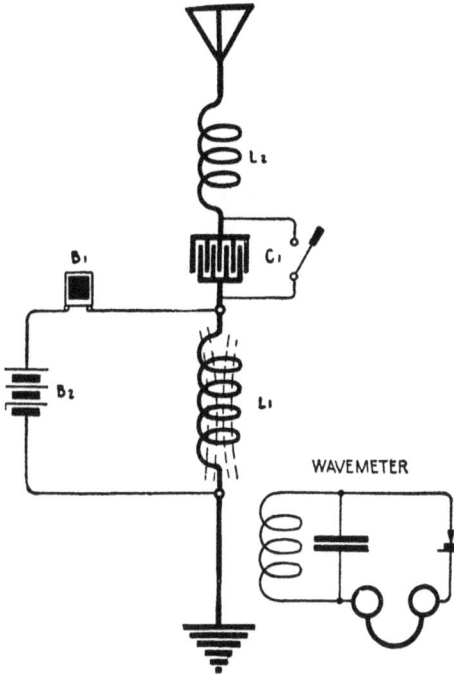

Put a spark gap *S*-1 in the antenna circuit and connect its terminals to the secondary of an *induction coil* or *high voltage transformer.* First be sure that the detector is adjusted and then with the spark discharging, vary the capacitance of the wavemeter condenser *C* until response is heard in the telephone. The wave length of the wavemeter at the setting for resonance is the wave length of the antenna circuit.

Assume that the reading turns out to be 175 meters, and that 200 meters is desired. Then put the secondary coil of the transmitting oscillation transformer in the antenna circuit, and add turns until the wavemeter indicates 200 meters.

The antenna may be *excited by a buzzer* as in Fig. 237. The coil *L*-1 in that diagram is the secondary of the transmitting oscillation transformer. *C*-1 is a short wave condenser (if used), and *L*-2 an aerial tuning coil.

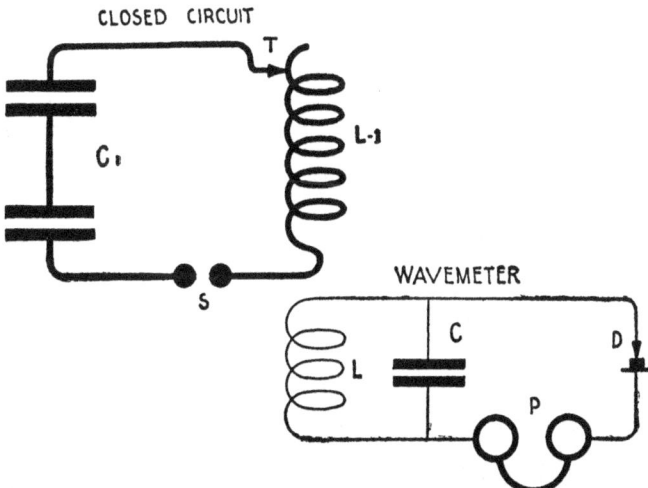

FIG. 237 Showing how an aerial may be set into oscillation by a simple vibrating buzzer.

FIG. 238. Measurement of the wave length of the closed oscillation or spark gap circuit.

MEASUREMENT OF THE CLOSED CIRCUIT.—The equivalent wave length is obtained as in Fig. 238. L-1, C-1, S is the spark gap circuit. The coil L of the wavemeter is placed in inductive relation to L-1, the spark gap energized by the high voltage transformer, and the capacitance of condenser C varied until resonance is secured. The wave length is read from the wavemeter scale.

To *reduce the wave length*, cut out turns at L-1 by the tap T or reduce the capacitance of C-1. To *increase the wave length* reverse the process.

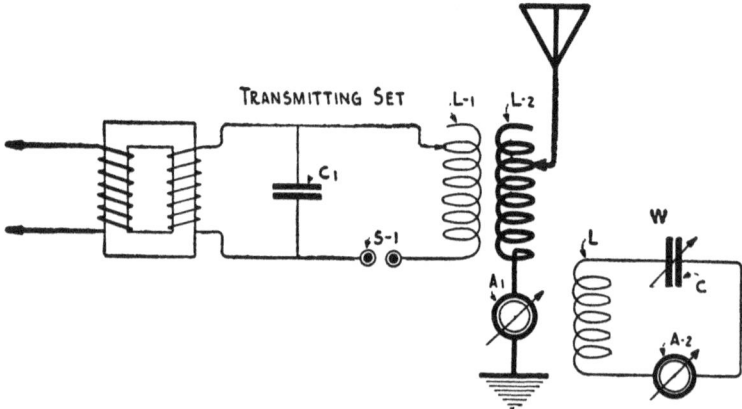

FIG. 239. Showing the use of the wavemeter for measuring the length of the radiated wave. This circuit and arrangement of apparatus is also suitable for measuring the decrement of the radiated wave.

MEASUREMENT OF THE RADIATED WAVE.—After tuning the open and closed circuits individually couple them together and vary the coupling until the antenna ammeter A-1, Fig. 239, reads a maximum. Then place the wavemeter in inductive relation to the earth lead and *tune* it to resonance with the sending set. If the spark gap quenches well only one sharp maximum will be observed at the "current square" meter A-2. (The radiated wave may be read by a head telephone and crystal detector as well.) If the spark gap does not quench properly, *two maximums* will be observed at A-2. This indicates that two waves are radiated, and this the *law will not permit*. *The secondary must then be drawn away from the primary until the two waves merge into one.* The circuits of Fig. 239 are applicable to *decrement measurements* as explained on page 301.

RESONANCE CURVES.—These curves depict the relation between the antenna current at the resonant wave length and at wave lengths off resonance. They show in a general way the over-all distribution of energy, and permit direct calculation of the decrement. The dot and dash curve in Fig. 240 shows the results of *close coupling* at the oscillation transformer and indicates *imperfect quenching* of the primary oscillations. The dotted curve indicates the result of the *coupling* between the closed and open circuits being *reduced* until better quenching is secured. The two waves are closer. The solid-line curve indicates *good quenching*, as the coupling at the oscillation transformer has been reduced until *single wave emission* is obtained.

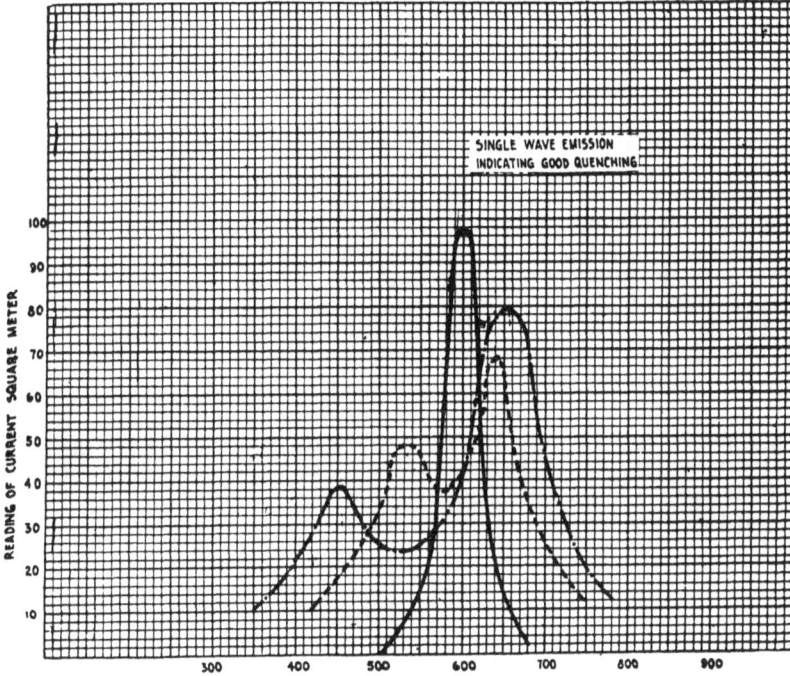

FIG. 240   Resonance curves of the antenna oscillations in a radio transmitter showing the effects of various degrees of quenching. With relatively close couplings double wave emission occurs.

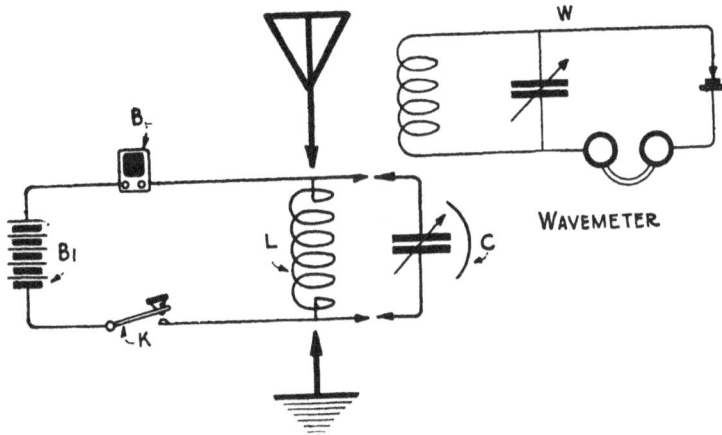

FIG. 241.   Circuit and apparatus for measuring the capacitance of a wireless telegraph aerial.

The data for these curves may be secured with the circuit arrangements in Fig. 239. With the spark discharging at S-1, vary the wave length of wavemeter LC at C, around the resonance point, and note the reading of A-2 for each setting of condenser C. The wave lengths obtained in this way are the ordinates of the curve. The abscissas are the readings of the "current square" meter. The curve is made by locating points on cross-section paper, corresponding to each set of readings, after which a line is drawn common to them all.

SIMPLE MEASUREMENT OF ANTENNA CAPACITANCE.—An approximate measurement (Austin) is outlined in Fig. 241. L is a coil that will increase the natural wave length of the antenna 4 *or* 5 *times*. C is a calibrated variable condenser and W a standard wave meter.

First connect L in the antenna circuit, excite the antenna with the buzzer, and measure the wave length at W. Then take coil L out of the antenna circuit and shunt it by a standard variable condenser C. Vary C until the wave length of L C is the same as before. The capacitance of C is now that of the antenna neglecting the distributed inductance of the antenna system.

MEASUREMENT OF ANTENNA INDUCTANCE AND CAPACITANCE SIMULTANEOUSLY.—Two standard coils L-1 and L-2 of different values are inserted successively in the antenna circuit as in Fig. 242 and the wave length read for each coil at W. In the case of L-1,

$$\lambda_1 = 59.6 \ \sqrt{(L_1 + L_a) \ C_a} \tag{66}$$

In the case of L-2,

$$\lambda_2 = 59.6 \ \sqrt{(L_2 + L_a) \ C_a}$$

Eliminating C_a,

$$L_a = \frac{L_2 \ \lambda^2{}_1 - L_1 \ \lambda^2{}_2}{\lambda^2{}_2 - \lambda^2{}_1} \tag{67}$$

Where

$\lambda_1 = $ wave length with L_1
$\lambda_2 = $ wave length with L_2
$L_a = $ antenna inductance
$C_a = $ antenna capacitance.

Since λ_2, L_2 and L_a are now known the antenna capacitance,

$$C_a = \frac{\lambda^2{}_2}{3552 \ (L_2 + L_a)} \tag{68}$$

The measurement is perhaps more accurate if an undamped source of oscillations such as an oscillating tube is coupled to the antenna, and a low-reading ammeter is placed in series with the antenna circuit. The undamped source is then tuned to the antenna circuit as observed by the aerial ammeter. The wavemeter is then tuned to the undamped source with the antenna disconnected. The same procedure is carried out with the loadings L-1 and L-2. The remainder of the process follows the outline above. If the oscillating tube has 1 ampere in its oscillating circuit the antenna ammeter should read from 0 to 100 milliamperes. This permits fairly loose coupling between the source and the antenna circuit.

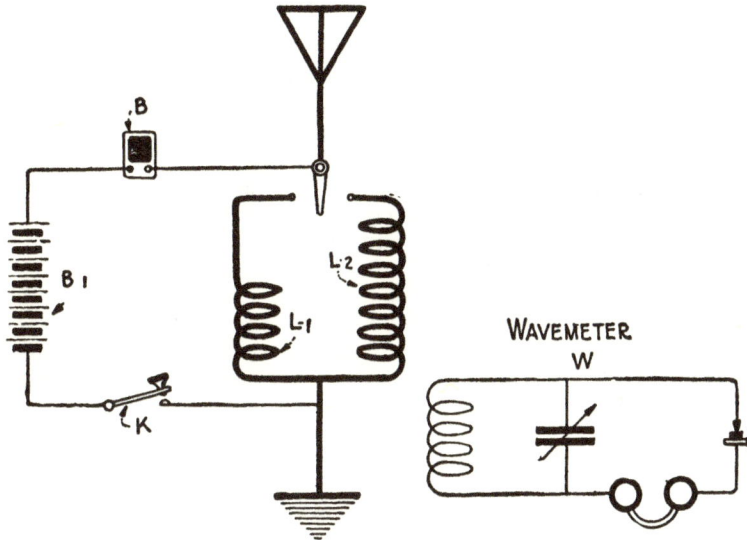

Fig. 242. Circuit and apparatus for the simultaneous determination of the inductance and capacitance of a wireless telegraph aerial.

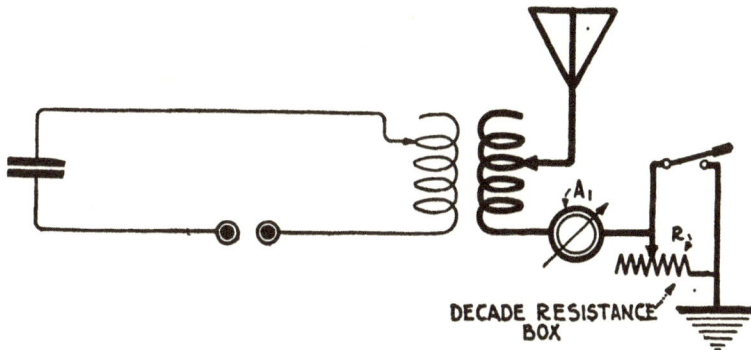

Fig. 243. Method of measuring antenna resistance.

THE MEASUREMENT OF ANTENNA RESISTANCE AND POWER.—This may be done as in Fig. 243. R is a continuously variable resistance (decade box built especially for this work by the General Radio Company) with a range from 0 to 20 ohms. A-1 is the antenna ammeter With R out of the circuit the antenna current is measured at A-1 in the usual manner. R is then inserted and varied until A-1 reads one-half the former reading; that is $\frac{1}{2}I$. Then R_a, the antenna resistance is equal to R. If $R = 10$ ohms and $I = 3$ amperes, then, the antenna power $= I^2 R = 3^2 \times 10 = 90$ watts.

In explanation of this measurement of antenna resistance it is obvious that an added resistance equal to the antenna resistance will reduce the antenna current to one-half its natural value, provided the resistance does not influence the inductance and capacitance of the antenna system.

MEASUREMENT OF THE EFFECTIVE INDUCTANCE CAPACITANCE AND RESISTANCE OF AN AERIAL.—The effective values are

those which, when lumped in a series circuit to which a given e.m.f. is applied, will give the same natural frequency of oscillation and the same current as a given antenna circuit, whether the applied e.m.f. is damped or undamped. If the effective values can be measured, an artificial aerial can be constructed which will duplicate exactly the actual antenna.

The method of measurement now to be described was suggested by J. R. Miller in B. of S. circular No. 326 (Oct. 23rd, 1918). The circuit is shown in Fig. 244. L-1, C-1 and R-1 are respectively a variable inductance, a variable capacity, and a variable resistance of values somewhere in the region of those to be expected from the practical antenna. Switch S-1 places the antenna coil L-3 either in series with the antenna or in series with L-1, C-1, R-1. U is a source of undamped oscillations —a Marconi V. T. connected up for the production of radio frequencies. D is a damped oscillation spark set, the spark gap having *magnesium* electrodes. Switch S-2 connects L-2 to either source.

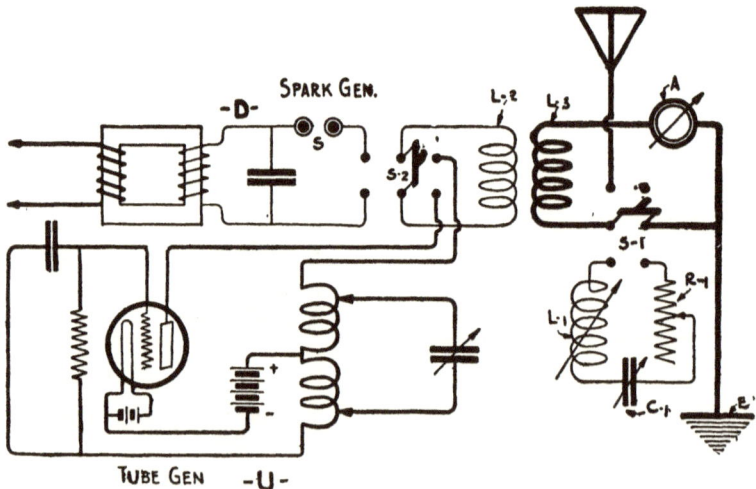

FIG. 244. Miller's method of measuring the effective inductance, capacitance and resistance of an antenna circuit.

The theory on which this measurement is based will first be given. When a source of undamped oscillations in a primary circuit induces current in a secondary tuned circuit the current in the secondary for a given e.m.f. depends only upon the resistance of the secondary circuit. In the case of damped oscillations in the primary, the current in the secondary for a given e.m.f. and primary decrement depends upon the decrement of the secondary; that is upon the resistance and the ratio $\frac{C}{L}$. The higher the decrement of the primary circuit relative to the decrement of the secondary the greater is the dependence of the current in the secondary upon its own decrement.

Advantage is taken of these facts in determining the effective inductance L_e, the effective capacitance C_e, and the effective resistance R_e of the antenna system as in Fig. 244. The procedure is as follows:

The frequency of the undamped source—the tube generator—is tuned to the antenna, and the circuit L-1, C-1 then tuned to the source. The resistance R-1 is then varied until the ammeter A reads the same in both positions of the antenna switch S-1. The resistance of the circuit L-1, C-1 is then equal to the effective resistance of the antenna and L-1, $C\text{-}1 = L_e\,C_e$ (and $R\text{-}1 = R_e$).

The damped source—the spark gap circuit—is then coupled and tuned to the antenna by varying its frequency and noting the reading of the ammeter A. The coil L-3 is then thrown to the L-1, C-1 circuit by switch S-1. If the current increases, the value of C-1 is greater than C_e and vice versa. L-1 and C-1 are then varied, meanwhile maintaining resonance (while R-1 is left unchanged) until the ammeter A reads the same in both positions of the switch S-1. C-1 then $= C_e$ and L-1 $= L_e$.

The coil L-1 is preferably a variometer. If the setting of the variometer has to be changed considerably in order to obtain a balance with the switch S-1 in the two positions, its resistance must be taken into account and eliminated. The resistance of the variometer may be eliminated by shorting it and varying R-1 until equal readings are obtained at A with switch S-1 in the two positions, using undamped oscillations. Then $R_e = R\text{-}1$. Now that L, C and R are known, the decrement may be calculated from .

$$\delta = \frac{\pi\,R\,\sqrt{C}}{\sqrt{L}}$$

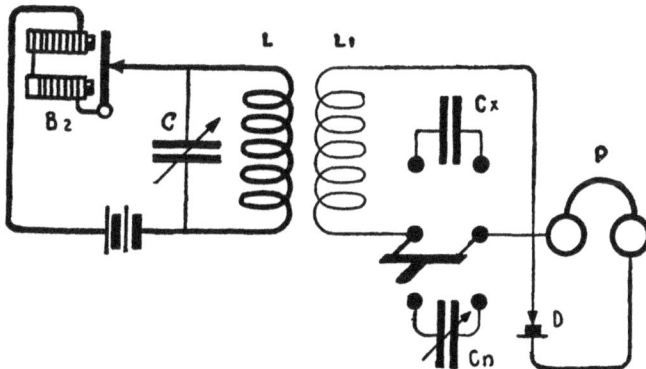

Fig. 245. Method of measuring capacities at radio frequencies.

MEASUREMENT OF CAPACITY AT RADIO FREQUENCIES.—

The circuit is shown in Fig. 245. $L\,C$ is a wavemeter excited by a buzzer

B-2. *L*-1 is a variable inductance which may be shunted either by a calibrated condenser *C-n* or by the condenser of unknown capacity *C-x*. *D* is a crystal rectifier, and *P* a head telephone.

To carry out the measurement proceed as follows: With the switch thrown to *C-x* vary the wave length of the wavemeter at *C* until resonance is obtained. Then throw the switch to *C-n* and vary its capacity until resonance is again secured. Then *C-x = C-n*.

If the effective capacity of *C-x* is desired for some other particular frequency, then set *C L* at that frequency and vary *L*-1 until *L*-1, *C-x* is in resonance. Then proceed as above.

MEASUREMENT OF INDUCTANCE AT RADIO FREQUENCIES. —The circuit of Fig. 245 applies also to this measurement. To find the inductance of *L*-1, for example, throw the switch to *C-n* and tune *L C* to resonance. Note the wave length of the wavemeter and the capacity of *C-n*. Then

$$L\text{-}1 = \frac{\lambda^2}{3552 \times C\text{-}n} \tag{69}$$

MEASUREMENT OF MUTUAL INDUCTANCE AT RADIO FREQUENCIES.—Assume that the coils *L*-2 and *L*-3 in Fig. 246 are the primary and secondary of a transmitting oscillation transformer. Their mutual inductance may be determined approximately as follows: Put jumpers across the posts *A* and *B*, and *C* and *D*. Then tune the

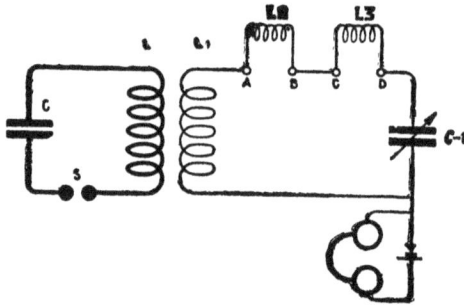

FIG. 246. Measurement of mutual inductance at radio frequencies.

wavemeter *L*-1, *C*-1 to the spark gap circuit *L C S* or any source of oscillations. Designate the capacity of the wavemeter condenser *C*-1 at resonance as *C*-2. Then connect *L*-2 and *L*-3 in series and find a new value of *C*-1 for resonance and designate it as *C*-3. Then, if the magnetic fields of the two coils assist, a value of inductance will be obtained equal to *L*-2+*L*-3+2 *M*. And in terms of the wavemeter settings,

$$L\text{-}2 + L\text{-}3 + 2\,M = \left(\frac{C\text{-}2}{C\text{-}3} - 1 \right) L\text{-}1$$

Next reverse the connections at either C, D, or A, B, and re-tune the wavemeter. Let the new reading of the condenser C-1 be C-4, then

$$L\text{-}2 + L\text{-}3 - 2\,M = \left(\frac{C\text{-}2}{C\text{-}4} - 1 \right) L\text{-}1$$

If $L\text{-}5 = L\text{-}2 + L\text{-}3 + 2\,M$, and $L\text{-}6 = L\text{-}2 + L\text{-}3 - 2\,M$, then

$$M = \frac{L\text{-}5 - L\text{-}6}{4} \tag{70}$$

where $M =$ the mutual inductance in microhenries or centimeters, according to the way L-1 is expressed.

The coefficient of coupling $K = \dfrac{M}{\sqrt{L_2\,L_3}}$. L-2 and L-3 may be measured as in the preceding paragraph.

MEASUREMENT OF THE LOGARITHMIC DECREMENT.— The decrement according to U. S. statute, must not exceed 0.2 per complete cycle. The number of oscillations for a given decrement,

$$M = \frac{4.6}{\delta} \tag{71}$$

where $4.6 = \log_e 100$.
$\delta = $ decrement.

Hence, a decrement of 0.2 corresponds to $\dfrac{4.6}{0.2} = 23$ complete oscillations.

The decrement of the radiated wave may be measured by the circuit of Fig. 239. The transmitter is assumed to be tuned for single wave emission. If two waves widely separated exist, the measurement can be applied to each wave.

The procedure is as follows: Connect a "current square" meter in series with the wavemeter circuit and place the wavemeter in inductive relation to the ground lead of the antenna system. Locate the point of resonance on the condenser C and then change the coupling between the wavemeter and the antenna circuit until a fair deflection of the meter A-2 is obtained. Then increase the capacitance until A-2 reads one-half the value obtained at resonance and note the capacitance of the condenser C. Then decrease the capacitance of C to some value below resonance where A-2 again reads one-half the value obtained at resonance and again note the capacitance of C. The sum of the decrement of the circuit under measurement and the decrement of the wavemeter is then found from the following formula:

$$\delta_1 + \delta_2 = 3.1416 \frac{C_2 - C_1}{C_2 + C_1} \tag{72}$$

where

δ_1 = decrement of the radiated wave

δ_2 = decrement of the wavemeter

C_2 = capacitance of wavemeter condenser above resonance where $I^2 = \frac{1}{2} I^2_r$

C_1 = capacitance of wavemeter condenser below resonance where $I^2 = \frac{1}{2} I^2_r$

Suppose, for example, $C_2 = 0.003$ mfd., and $C_1 = 0.0028$ mfd., then,

$$\delta_1 + \delta_2 = \frac{0\ 003 - 0\ 0028}{0\ 003 + 0.0028} \times 3.1416 = 0.108$$

This is the sum of the decrements, and δ_2 the decrement of the wavemeter, must be subtracted from $\delta_1 + \delta_2$ to obtain δ_1 the decrement of the circuit under measurement.

In terms of wave length, the decrement formula may be written

$$\delta_1 + \delta_2 = \frac{(\lambda^2_r - \lambda^2)}{\lambda_r \ \lambda} \tag{73}$$

where λ_r = wave length at resonance

λ = wave length below resonance where the current square meter reads ½ the deflection at resonance.

DETERMINING THE DECREMENT OF THE WAVEMETER.— There are several methods of determining the decrement of the wavemeter, but the most accurate is that shown in Fig. 247. The wavemeter L, C, A-2 is excited by a *tube generator* like that of Fig. 218. The frequency of the generating tube is varied principally at C-1. The generating circuit should be designed to cover the range of frequencies of the wavemeter. The procedure is then the same as outlined in connection with formula (72), the capacity of C being set above and below resonance at points where $I^2 = \frac{1}{2} I^2_r$, and where I^2_r is the current at resonance.

Since the *source is undamped*, the *resulting decrement* is δ_2, that of the *wavemeter*. The decrement of the wavemeter should be thus determined at 5 or 10 points on the condenser scale and the results plotted in the form of a curve for future reference. The decrement values may be marked directly on the condenser scale.

To summarize the foregoing, first measure the decrement according to (72) and then subtract the decrement of the meter as determined above. The result is the *decrement of the radiated wave.*

The decrement of the amateur's transmitter is minimized by a good earth connection, by using antenna wires of low resistance, and by providing a spark gap which gives *good quenching.* If the gap does not quench well the coupling at the oscillation transformer must be reduced until good quenching is secured.

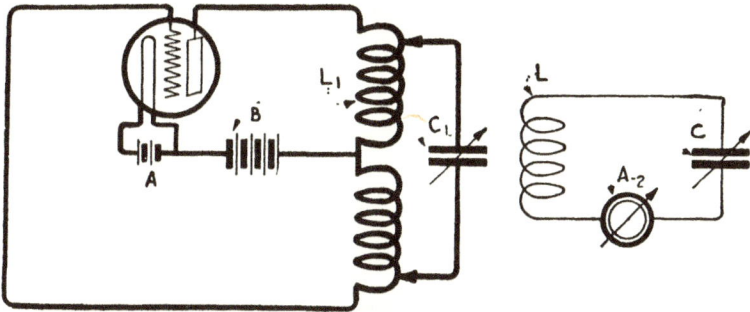

FIG. 247. Measurement of the decrement of a wavemeter through the use of a vacuum tube generator.

DIRECT READING DECREMETER.—Direct determinations of the decrement can be made by the Kolster decremeter, the theory of which has been given in the book "Radio Instruments and Measurements." Any wave meter can be made a direct reading decremeter. Consider, for example, the decrement formula (a formula which does not take into account the average of the decrements on both sides of the point of resonance in a distorted curve),

$$\delta_1 + \delta_2 = \pi \ \frac{C_r - C}{C} \qquad (74)$$

Using the "half deflection" method, it is clear that the decrement for any set of conditions corresponds to the displacement of the condenser's moving plates which varies the capacity by the amount $C_r - C$. The displacement of the plates for a given decrement will be different for various values of C.* Therefore at successive points on the condenser scale any

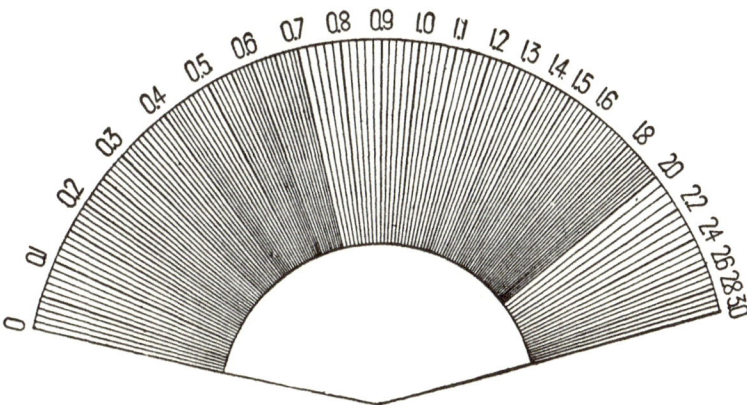

FIG. 248. Direct reading decrement scale prepared by the Bureau of Standards which is applicable to any wavemeter having a condenser of semi-circular plates.

―――――――――

*The total capacity in the circuit

displacement of the moving plates which changes the square of the resonance current to $\frac{1}{2} I^2{}_r$, means a certain value of $\delta_1 + \delta_2$.   A special scale

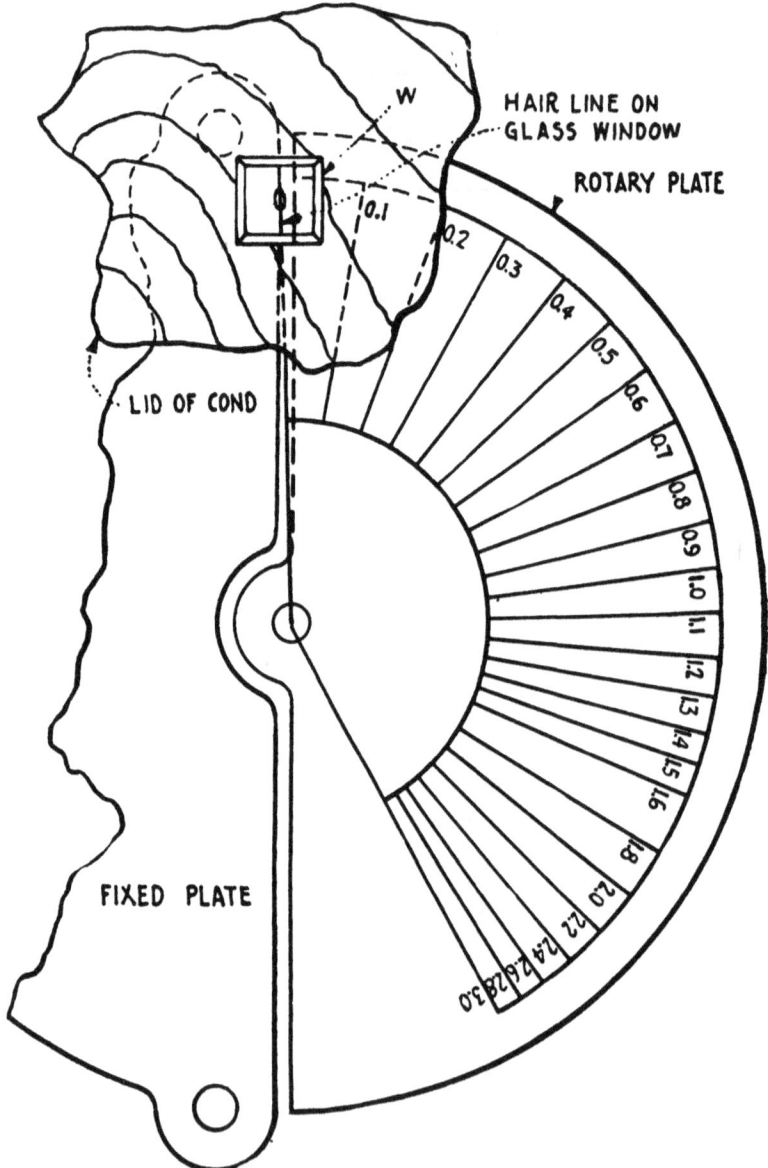

FIG. 249.   Showing the decrement scale attached to the movable plates of a variable condenser.*

---

*The hair line should be placed on the opposite side of the condenser scale and in coincidence with the zero graduation on the capacity scale.

may, therefore, be computed and attached to any variable condenser provided the relation between the capacity and the displacement of the moving plates is known. Such a scale has been computed by the Bureau of Standards for any variable condenser with *semi-circular plates* (the more common type). This is shown in Fig. 248. The Bureau has shown that the graduations of the decrement scale of such a condenser vary as the *logarithm of the angle of rotation*. The scale may be placed on the unused half of the condenser top (opposite the condenser scale) or it may be attached to the moving plates of the variable condenser. The latter method is shown in Fig. 249. A small glass window with a *hair line* in the center is placed on top through which observations of the decrement are made. The *zero mark* on the decrement scale must coincide with the scale reading for "zero" capacity when the movable plates are positioned for maximum capacity.

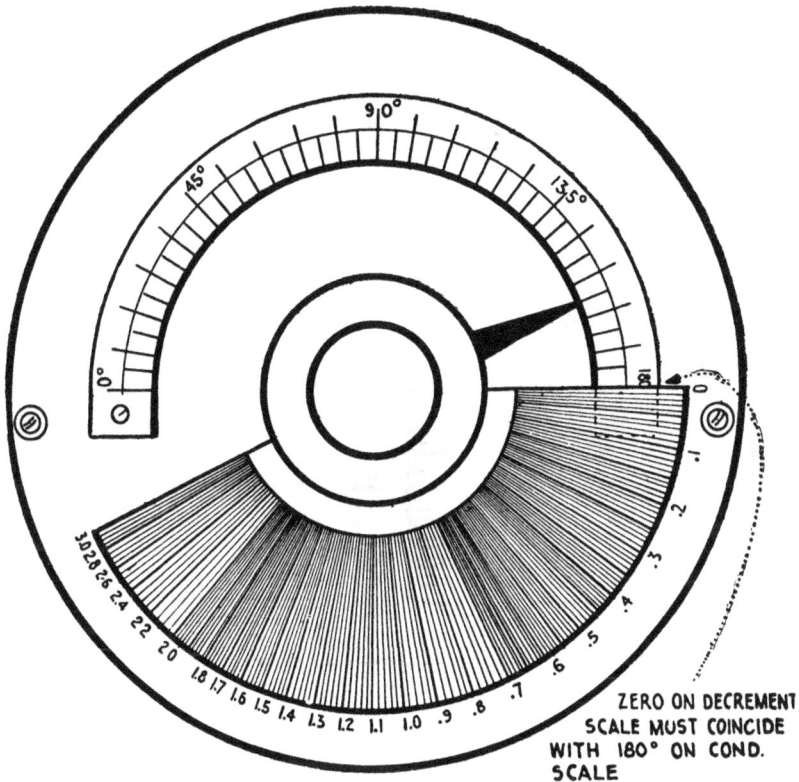

FIG. 250. Showing the decrement scale attached to the cover of the variable condenser of a wavemeter.

The decrement scale mounted on the unused half of the cover is shown in Fig. 250, where, the zero mark on the decrement scale coincides with the graduation on the condenser scale for maximum capacitance.

To measure the decrement with this scale proceed as follows: Insert a current square meter in series with the decremeter circuit. Locate resonance and note the reading of the current square meter. Then note the tabulation on the decrement scale on either side of resonance where the meter reads $\frac{1}{2}I^2_r$.

The difference between the two readings is then $\delta_1 + \delta_2$. The decrement δ_2 of the wavemeter is determined as in Fig. 247. If, for example, the readings of the decrement scale on both sides of resonance are 0.6 and 0.16, then,

$$\delta_1 + \delta_2 = 0.16 - 0.6 = 0.1$$

and, if the decrement of the decremeter at the resonance capacity = 0.033, then

$$\delta_1 = 0.10 - .033 = 0.067$$

Although this decrement scale gives accurate readings, for large decrements (from 0.2 to 2.0), it is rather difficult to obtain precision at decrements below 0.15. The author finds it wise to use a large hand glass in order to magnify the scale at the lower divisions.

Every amateur transmitting station should be provided with a decremeter as with this instrument it is possible to determine the *sharpness of the radiated wave*, and whether or not it comes within the restrictions of the law.

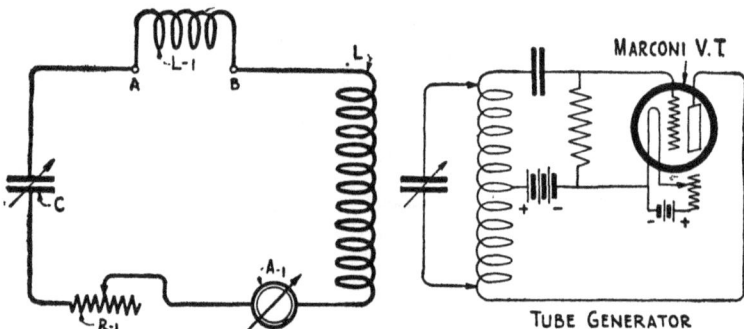

FIG. 251. Circuits for determining the radio frequency resistance of a wavemeter or a coil.

If a calibrated variable condenser is available, the experimenter may predetermine the dimensions of an inductance for a given range of wave lengths by use of formula (72). He will then possess a wavemeter which if a current measuring instrument be inserted in the circuit will permit tuning and decrement measurements to be carried out. With the decrement scale given here it is not essential to know the wave length of the resonating circuit or the capacitance of the variable condenser. As long as the condenser has semi-circular plates and the inductance has the correct dimensions to establish resonance with the transmitter, the decrement can be determined directly as outlined above.

MEASUREMENT OF THE RESISTANCE OF A WAVEMETER AND A COIL.—The circuit is shown in Fig. 251. L, C, A-1 is a wavemeter and L-1 a coil whose resistance is to be measured. Coil L is coupled to a vacuum tube generator of variable frequency. A-1 is a small ammeter and R-1 is a standard high frequency resistance box. The procedure follows: Cut L-1 out of the circuit and connect together the posts A and B. Tune the source until a maximum I is indicated at A-1,* with R-1 set at zero. Then insert some resistance at R-1, and observe the corresponding current I_1. Then the resistance of the wavemeter circuit,

$$R_e = \frac{R\text{-}1}{\dfrac{I}{I_1} - I} \qquad (75)$$

To find the resistance of L-1, cut it in the circuit, retune to resonance and measure the resistance again. Then subtract the resistance of the wavemeter (determined as above) from the value now obtained and the result will be the resistance of the coil.

If a current square meter is used at A-1, add resistance at R-1 until the deflection is one-quarter of that obtained in the first observation. Then $R_e = R$.

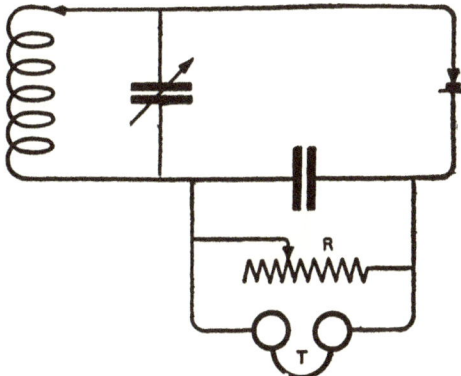

FIG. 252. Measurement of the audibility of incoming signals.

AUDIBILITY OF INCOMING SIGNALS.—A comparative measurement of the strength of signals may be made as in Fig. 252 where the telephones T are shunted by a variable resistance or impedance R. If s is the impedance of the shunt, I_t, the least audible current in the telephone, I, the total current in the shunt and telephones, and t, the impedance of the telephone, then

$$\frac{I}{I_t} = \frac{s+t}{s} \qquad (76)$$

The ratio $\dfrac{s+t}{s}$ is called the *audibility factor*.

---

*A thermocouple and galvanometer. See pages 181 and 182 "Radio Instruments and Measurements."

Since the impedance varies with the frequency, wave-form and other constants of the circuit, it is necessary to calibrate the apparatus for each setting under actual conditions of use.

The ordinary way is to ignore the impedance and call s and t, the resistance of the shunt and telephone respectively. The resistance R in Fig. 252 is then varied until the signals are just audible. Then if

$s = 50$ ohms and $t = 2000$ ohms, $I = \dfrac{50 + 2000}{50} \times I_t = 40 I_t$, that is, the signal

is 40 times audibility.

Although the latter method has many disadvantages it is still widely used. Calibrated audibility meters may be purchased.

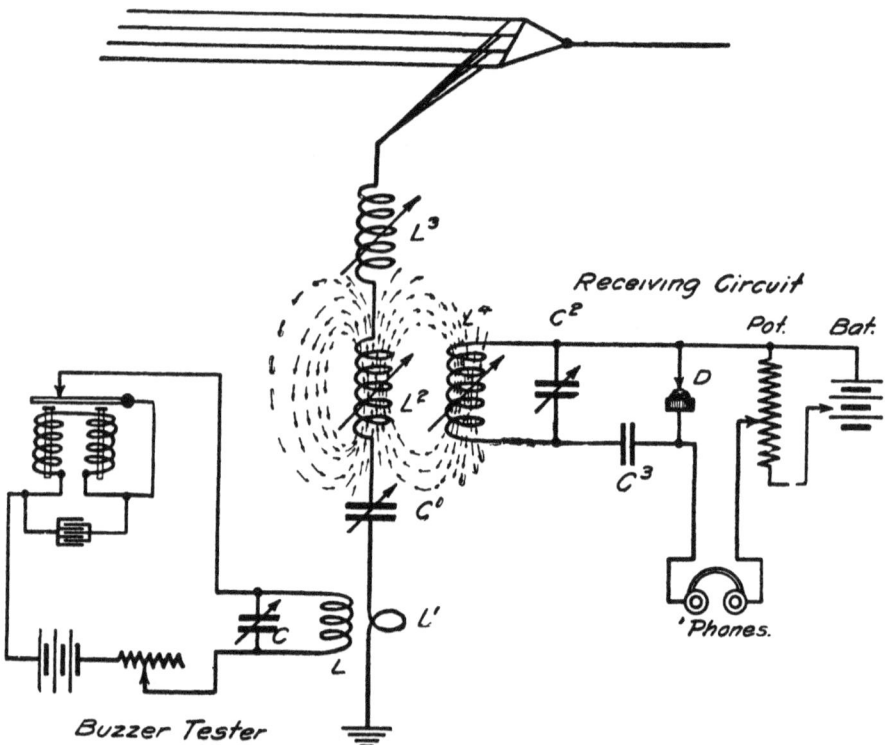

Fig. 253. The calibration of a receiving set by a wavemeter. With this apparatus the complete receiving system may be tuned to any wave length in advance of reception.

DISTRIBUTED CAPACITANCE OF A COIL.—Using the circuit of Fig. 245, the distributed capacitance of the coil L-1 may be determined as follows: Determine the apparent inductance of L-1 at two settings of C-n, tuning the circuit L-1, C-n to the wavemeter for each setting. Then use formula (69).

Letting

$L_{a1} =$ the apparent inductance for C-1 (one value of capacitance of C-n)

and

L_{a2} = the apparent inductance for C-2 (the second value of C-n)

Then

$$C_0 = \left(\frac{L_{a1}}{L_{a2}} - 1 \right) \frac{C_1 C_2}{C_2 - C_1} \tag{77}$$

Where

C_0 = distributed capacitance in mfds.

CALIBRATION OF A RECEIVING SET.—The circuit appears in Fig. 253. L C is a wavemeter excited by a buzzer, which is placed in inductive relation to the antenna system at L'. The wave length of the complete receiving set is then determined as follows: Start the buzzer and set the wavemeter at some wave length. *Vary the inductance and capacitance in the open and closed circuits and change the coupling until a sharp maximum is heard in the head telephones.* The wave length of the receiving set is then the same as the wavemeter. Using loose coupling at L,L', the oscillation detector can be adjusted for maximum sensitiveness by trying different points on the crystal and adjusting the potentiometer *Pot.*

A wavemeter used in this way has a good deal of value around the experimental station as it assures the operator that his apparatus is set at the correct wave length for reception from a given transmitter; otherwise much experimental tuning will be necessary in order to pick up a station.

HOME-MADE WAVEMETER FOR THE AMATEUR STATION. —The wave length curve of Fig. 253a has been prepared to show the amateur who does not possess the facilities for calibration how to construct a wavemeter with which he may tune a 200-meter transmitter.

This curve fits the General Radio Company's type 124-A No. 501 variable air condenser. The inductance for the wavemeter consists of 30 turns of No. 18 cotton covered annunciator wire, wound on a tube 5″ outside diameter. At 20° on the variable condenser, the wave length is 175 meters and at 100 degrees it is 560 meters. This will cover the amateur's transmitter.

Thus, in order to obtain a wavemeter for tuning purposes, the experimenter need only purchase the above mentioned condenser and construct a coil according to the dimensions given above. The detector circuit, as shown in the drawing next to the curve, is connected unilaterally to the wavemeter. A Marconi V. T. should be used for sharp readings and to permit the wavemeter to be loosely coupled to circuit under measurement. The *grid of the valve only* is connected to one terminal of the variable condenser C.

For the curve given, the leads from the coil to the condenser are 6″ long.

MEASUREMENT OF HIGH VOLTAGES.—The simplest method is the "sphere gap voltmeter" mentioned in Chapter IV. Two spark balls 2 centimeters in diameter are mounted on an insulating base and connected in shunt to the high voltage source. The balls are separated until sparking just ceases. The following table of voltages, indicates the maximum e.m.f. for different gap lengths.

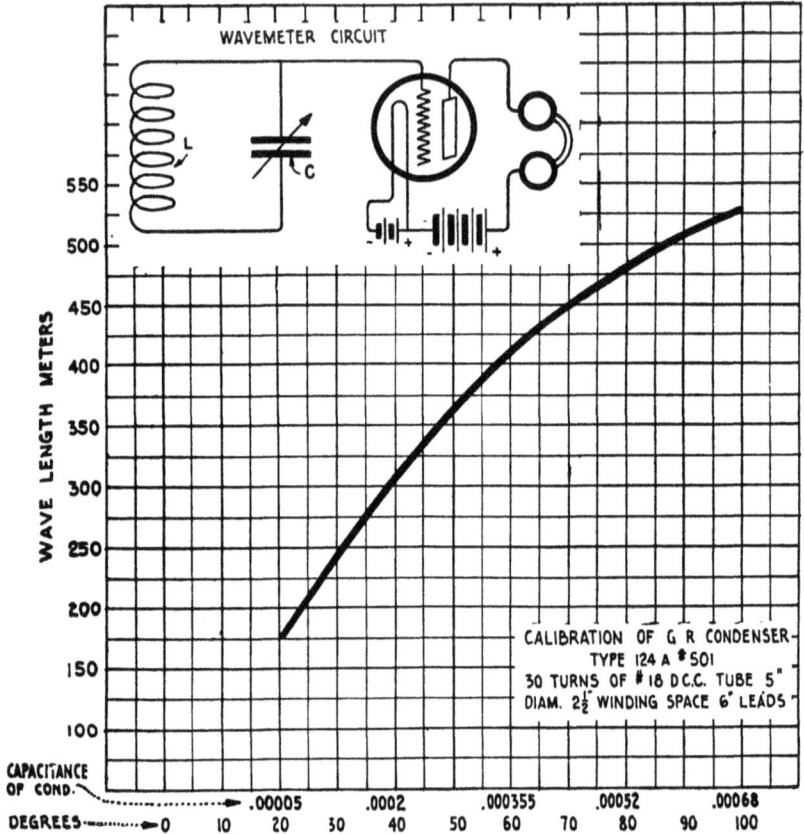

FIG. 253a. Wave length curve for the General Radio Company's condenser type 124–No 501 when shunted by a coil of 30 turns of No. 18 annunciator wire wound on a bakelite tube 5″ in diameter. The range of this wavemeter is correct for tuning the amateur transmitter

TABLE XVI

Spark Voltage	Length of gap in cm.
4700	0.1
8100	0.2
11400	0.3
14500	0.4
17500	0.5
20400	0.6
31300	1.0
47400	2.0

FIG. 253b. Table showing the voltages corresponding to various spark gap lengths using spark balls 2 cm in diameter.

CHAPTER XII

CLOSED COIL AERIALS—DIRECTIVE TRANSMITTERS AND RECEIVERS—CONSTRUCTION OF A DIRECTION FINDER

GENERAL.—For many years closed coil aerials have been employed by the Marconi Companies in their direction finding apparatus. Weagant has been particularly active in using these coils for long distance reception. It is well known that a closed circuit loop exhibits directional characteristics both in transmission and reception. It is useful not only for locating the direction of a station but as an in-door aerial as well.

It is doubtful whether the amateur experimenter will ever utilize the loop aerial for radio transmission, but it is interesting to know that a low power, portable transmitting set using a small loop aerial and permitting communication over 20 miles was developed during the war. The *radiating* properties of the loop are notoriously low, but the loss in radiation is made up by the use of super-sensitive vacuum tube amplifiers at the receiving station.

Any closed coil aerial connected to an amplifying receiver may be used as a *direction finder*. The operating characteristics of a loop may be explained by the sketches in Fig. 254a and b. When the plane of the loop L points in the direction of the advancing wave as at Fig. 254a, the e.m.f. induced therein is a maximum, but if the plane of the loop is at right angles to the advancing wave as in Fig. 254b, substantially no current is set up in it. If then, the loop be turned on its axis X, as at Fig. 254c, a position can be found for either maximum or minimum signals from any given transmitter.

The curve of reception for the closed coil or loop aerial is shown in Fig. 254d. Maximum signals are obtained when the plane of the coil lies in the direction $A\ B$; and minimum signals in the direction $A_2\ B_2$. In any other direction, such as $A_1\ B_1$ the strength of the signal varies as the cosine of the angle which the plane of the loop makes with the plane of the advancing wave motion.

The *position* of the loop for *minimum signals* is generally more sharply defined than that for *maximum signals*. Hence, the pointer on the direction finding scale is mounted at a right angle to the plane of the loop.

Returning now to Fig. 254d, if the small circle R represents the smallest current that will give an audible signal, then for any position of the coil aerial within the angles $R\ M_1\ M_2$ and $R\ M_3\ M_4$, no signals are obtained.

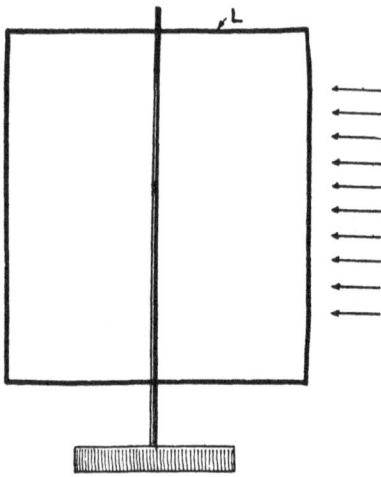

FIGURE 254 A

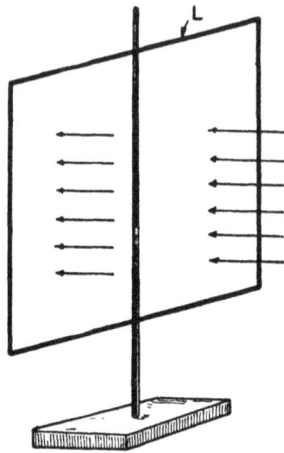

FIGURE 254-B

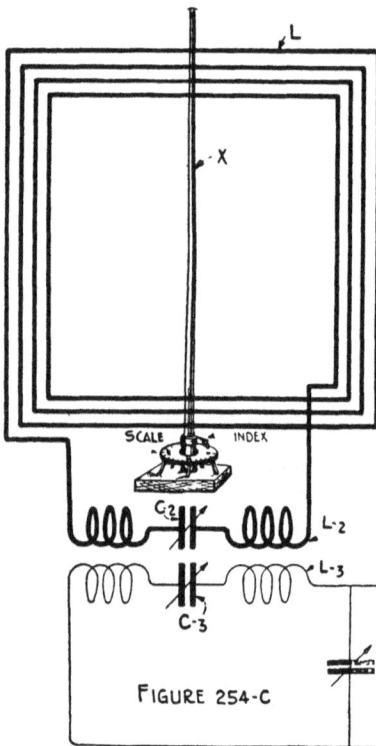

FIGURE 254-C

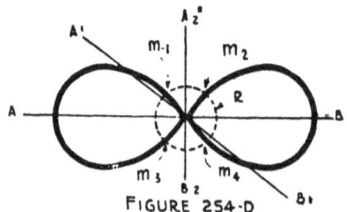

FIGURE 254-D

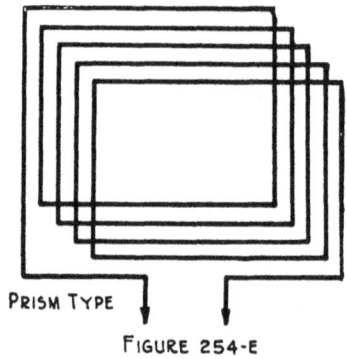

PRISM TYPE

FIGURE 254-E

FIG. 254.   A, showing the position of a closed loop aerial for maximum induction from an advancing wave
B, the position of the loop for minimum induction   C, loop aerial mounted on an axis for determining
the direction of a sending station   D, curve of reception for the loop aerial   E, loop aerial in the
form of a square prism.

To determine the direction of a transmitter the positions M_1 M_2 are noted on the scale where the signals just become inaudible. The loop is then turned 180° and the positions M_3 M_4 for minimum signals observed. By taking an average of the scale readings at M_1 M_2, and at M_3 M_4 and placing the loop along the line of the resultant the pointer P will lie in the direction of the station sending.

In practice the loop L Fig. 254c is mounted on a pole or stand so that it can be turned through an arc of 180°. The scale is stationary and its zero index should be placed in line with either the geographical north pole or the north magnetic pole.

The loop L may be made a flat spiral as in Fig. 254c, or a square prism as in Fig. 254e, or it may be wound in regular or irregular multilayers. Litzendraht wire is generally used but solid copper wire will do nearly as well.

In addition to their directional characteristics it will be evident that these loops will eliminate a certain amount of interference from stations whose signals are not desired, for if the *plane of the coil points in the direction of the station desired, a signal of the same wave length coming at right angles to the plane will not be heard*. Signals at any other angle will be heard with less intensity than with a non-directional antenna.

DIMENSIONS OF LOOPS.—For strongest signals the natural wave length of the loop should be near the wave length of the signal to be received. These loops do not respond well when loaded to wave lengths two or three times the fundamental unless a cascade amplifier is employed.

The turns should not be too close or the distributed capacitance will be excessive. The turns are usually spaced. An increase of the number of turns up to a certain point increases the signals, but beyond this point no increase is obtained.

For long wave lengths, 20 to 70 turns embracing a comparatively small area are used. Loops of greater area with 1 to 4 turns are employed for short waves.

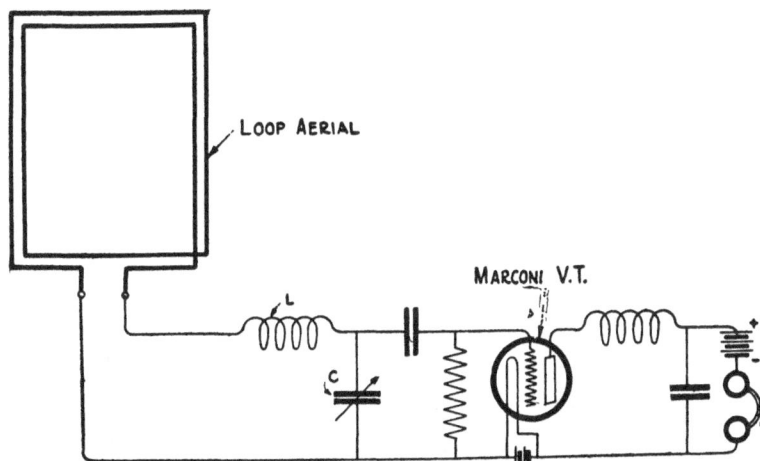

FIG 255. Loop aerial that has been found suitable for the reception of wave lengths between 2500 and 7000 meters The frame is 9 ft square wound with 22 turns of 3x16x38 Litzendraht or No. 14 r.c. wire. The turns are spaced 0 1 inch.

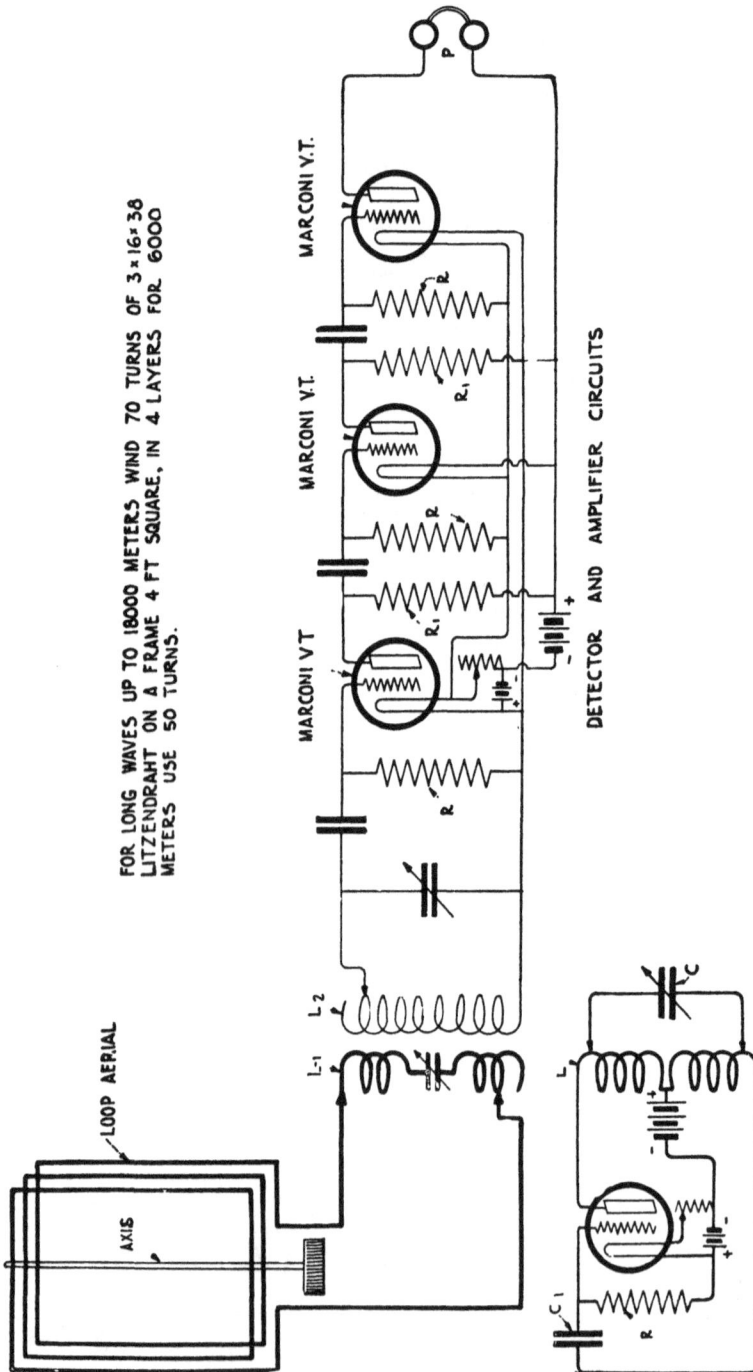

FOR LONG WAVES UP TO 18000 METERS WIND 70 TURNS OF 3 × 16 × 38 LITZENDRAHT ON A FRAME 4 FT SQUARE, IN 4 LAYERS FOR 6000 METERS USE 50 TURNS.

LOOP AERIAL

AXIS

MARCONI V T   MARCONI V.T.   MARCONI V.T.

DETECTOR AND AMPLIFIER CIRCUITS

TUBE GENERATOR

FIG 256. Loop aerial and associated detector circuits suitable for wave lengths between 7500 and 18,000 meters. The frame is 4 ft square wound with 70 turns of 3x16x38 standard Litzendraht in 4 layers. A resistance-coupled cascade amplifier of 3 to 8 stages is coupled inductively to the loop. For heterodyne reception a vacuum tube generator with a single valve is coupled inductively to the receiver circuits. This system will permit the reception of undamped oscillations from high power stations operating at long wave lengths, over distances of several thousand miles. The loop may be erected indoors in wooden buildings with good results. For reception at 6000 meters 50 turns wound on the same frame will be found suitable

The Signal Corps has obtained good results at wave lengths up to 600 meters with a square loop 8' on a side, wound with three turns of 3 x 16 x 38 Litzendraht cable spaced ½". The natural wave length of such a loop is near 160 meters and with the series condenser C-2 and the inductance L-2, Fig. 254c, waves up to 600 meters can be tuned in. A loop 4' square with 4 turns spaced ¼" is also suitable.

For *waves* of moderate length up to 8000 meters a loop of 22 turns spaced 0.1 of an inch, on a frame 9' square, has been found suitable. A circuit that has been used with a loop of these dimensions is shown in Fig. 255. For long waves the loading coil L is used to increase the wave length of the system.

For direction finding, the primary inductance L-2 Fig. 254c is preferably split in the center so that the variable condenser C-2 may be inserted.

The Research Department of the Marconi Company has achieved noteworthy results with the *loop* and circuits of Fig. 256 with which signals have been received at long wave lengths over distances of several thousand miles. For wave lengths up to 18,000 meters the frame of the loop should be 4' square wound with 70 turns of 3 x 16 x 38 Litzendraht in 4 layers. The loop is coupled inductively to a 3- to 6-stage resistance coupled amplifier using Marconi V. T.'s. The detection and amplifier circuit is the circuit of Fig. 184 and data for this are given in connection with that figure.

For heterodyne or beat reception the local radio frequency current is supplied by another Marconi V. T. which is connected up as an oscillation generator. The coil L of the generating circuit is coupled inductively either to L-1 or L-2. The *tube generating circuit* is substantially that of Fig. 218. For long wave reception, the coil L and the condenser C must be of sufficient magnitude to generate oscillations between 18,000 and 50,000 cycles.

Throughout Fig. 256 grid leaks of 2 megohms are indicated at R. The coupling resistances R-1 have resistance of 2 megohms each. The dimensions of the tuning elements for the circuit can be taken from Fig. 134.

A suitable loop for 6000 meters is one four feet square wound with 50 turns of 3 x 16 x 38 standard Litzendraht. The turns may be wound in 3½ layers and the circuit Fig. 256 should be used.

Good results will be obtained when the loop aerial is erected inside of wooden buildings. The signals are reduced considerably when mounted inside of buildings with metal frames.

The experimenter must bear in mind that the signals are maximum when the plane of the loop points in the direction of the sending station. Hence, for reception from all points of the compass the loop must be mounted so that it can be turned through an arc of 180°. The loop may be suspended from the limb of a tree or any convenient structure.

The use of these loops opens to the experimenter a most interesting field of research. For one thing the fact that signals may be received over great distances with the antenna structure mounted indoors is in itself a good argument to warrant their adoption.

TWO-LOOP DIRECTION FINDER.—The circuits of a direction finding system used during the war are shown in Fig. 257. Two loops, known as the *main* and *auxiliary* loop, are mounted at a right angle on a common axis. The loops should be 4′ square for moderate wave lengths, the main loop being wound with 12 turns of No. 18 annunciator wire, and the auxiliary loop with 33 turns of the same size wire. To reduce the distributed capacitance of the loop the turns are preferably slightly spaced. Both loops are connected to a common detector circuit primary *P*. The connections from the auxiliary loop can be reversed by the switch *S*-2 and the switch *S*-1 disconnects the primary from the main loop and connects it to a coil *L*-1 which has the same inductance as the main loop. *C*-1 is a variable condenser of 0.001 mfd.

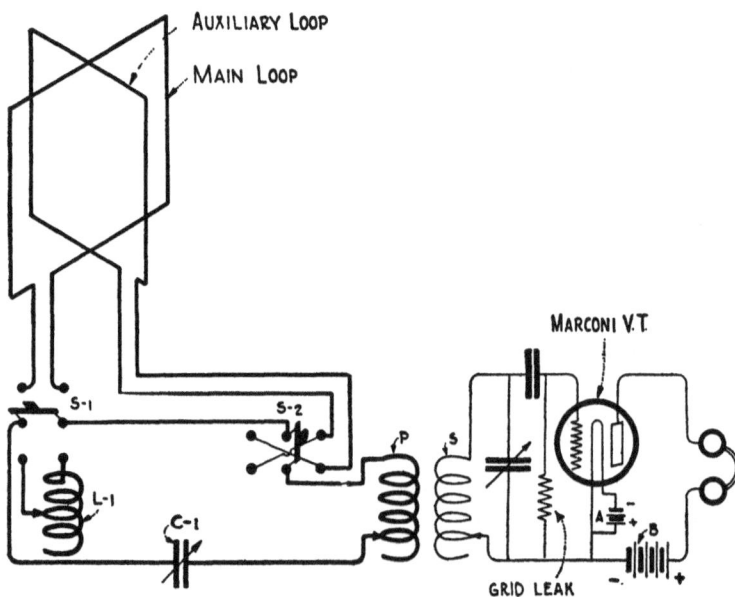

Fig. 257. Two-loop direction finding system used during the recent war. The direction of the sending station is determined by turning the loops until the signals are a maximum. The auxiliary loop acts as a check on the position of the main loop for maximum induction, for when the plane of the main loop does not lie in the general direction of the sending station the auxiliary loop becomes active and indicates by its effect upon the signals that the direction of the sending station has not been accurately located. When the correct position of the main loop is found the auxiliary loop is inactive and it will then neither strengthen or weaken the signals when the reversing switch is thrown from one position to the other.

The method of locating a transmitting station is as follows: Throw switch *S*-2 to the left, close *S*-1 to the top, and turn the loops for maximum signals. The main loop is now somewhat in line with the transmitter. Then throw *S*-1 to the inductance *L*-1 and if signals are still being received, the auxiliary antenna evidently is receiving signals.

Then revolve the loops still further (up to 90°) and change *S*-2 from one position to the other with the switch *S*-2 closed to the main loop. As long as current is flowing *in the auxiliary antenna* reversing *S*-2 will either increase or decrease the signals.

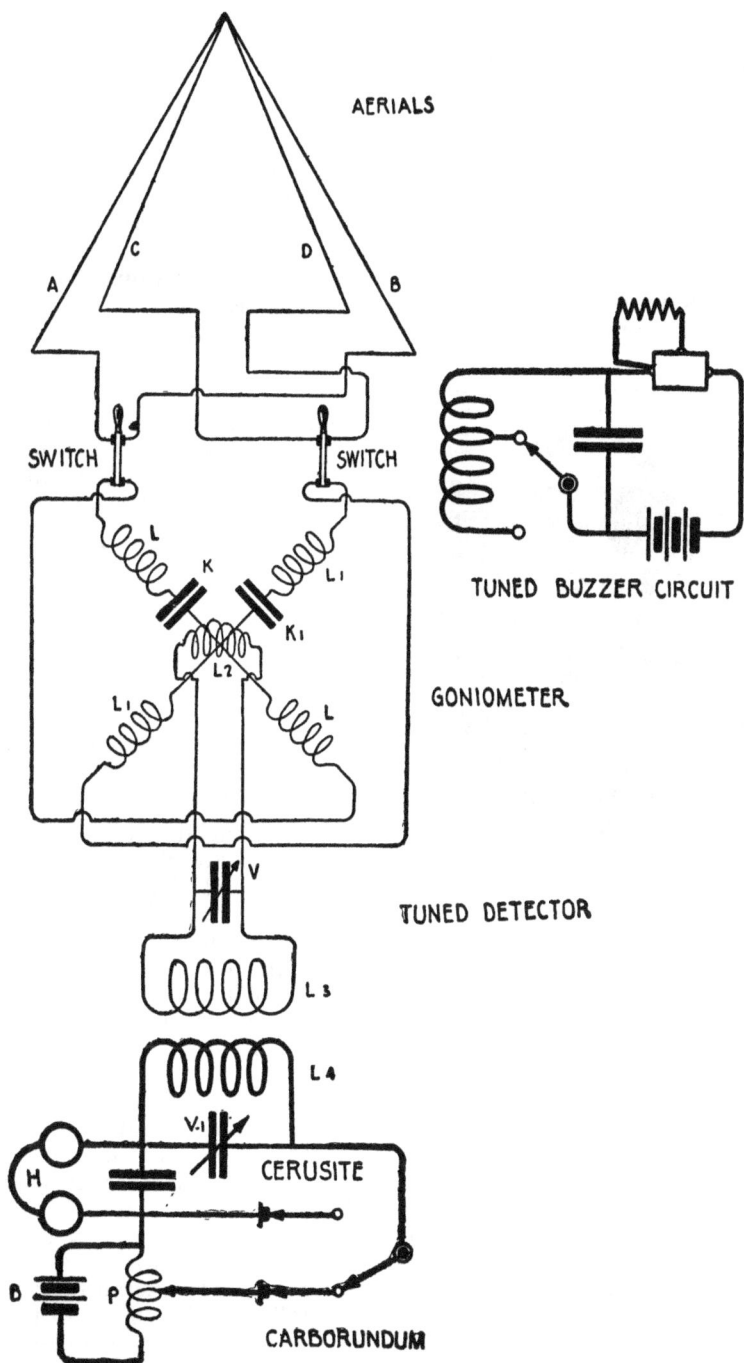

AERIALS

C   D

A   B

SWITCH   SWITCH

L   K   L₁

K₁

L₂

L₁   L

GONIOMETER

V

TUNED DETECTOR

L₃

L₄

V₁

H   CERUSITE

B   P

CARBORUNDUM

TUNED BUZZER CIRCUIT

FIG. 258 Circuits of the Marconi-Bellini-Tosi direction finder which has been in commercial use for a number of years. Energy is collected from the passing wave by two triangular loop aerials mounted at right angles. The direction of the sending station is determined by an "exploring" coil mounted inside an instrument known as a goniometer

By turning the loops farther a position will be found where, with the switch S-2 in either position, the signals will not be affected. The plane of the main loop now points along the line of the direction of the sending station. It then receives the maximum induction from the passing wave, whereas the auxiliary loop is not affected.

As pointed out before, the position of a loop aerial for maximum signals is not sharply defined, but by means of the balancing system provided here the position of the main loop for maximum signals can be very accurately located. The auxiliary loop evidently serves as a good check on the position of the main loop for maximum signals, for if the main loop is not in the position for maximum induction, the auxiliary loop will become active and show by its effect upon the signals from the main loop that the direction of the sending station has not been accurately located.

MARCONI TWO-LOOP DIRECTION FINDER. — This system has been used commercially for a number of years. Two *single turn* or *multi-turn loop* aerials are supported by any convenient structure at a right angle and their terminals are connected to the primaries of a device called a *goniometer*. These primaries combine to produce a *resultant magnetic field* which acts upon a common detector circuit through the agency of a *rotating exploring coil*.

The fundamental circuit is shown in Fig. 258. Loop *A B* is erected at a right angle to loop *C D*. The former is connected to the goniometer primary *L* and the latter to the goniometer primary *L*-1.

FIG. 259. The Marconi radiogoniometer—a radio instrument for locating the direction of a sending station.

Condensers *K* and *K*-1 are inserted in the middle of each primary. They are mounted on a common shaft so that their capacity may be varied simultaneously.

The exploring coil *L*-2 rotates in the resulting magnetic field and its terminals are connected to the detecting circuit through the intermediate circuit *L*-2, *V*, *L*-3. The intermediate circuit may be omitted and the coil *L*-2 connected directly to the terminals of the variable condenser *C*-1. A Marconi V. T. may be substituted for the crystal detectors shown.

The construction of the goniometer is shown in Fig. 259. It will be noted that the goniometer coils (marked *L* and *L*-1 in Fig. 258) are crossed at a right angle. One suitable for amateur use will be described later.

The operation of the direction finder is as follows: When one of the loops points accurately in the direction of the transmitter, it alone is active, and the other is inactive. If the exploring coil *L*-2 is then turned

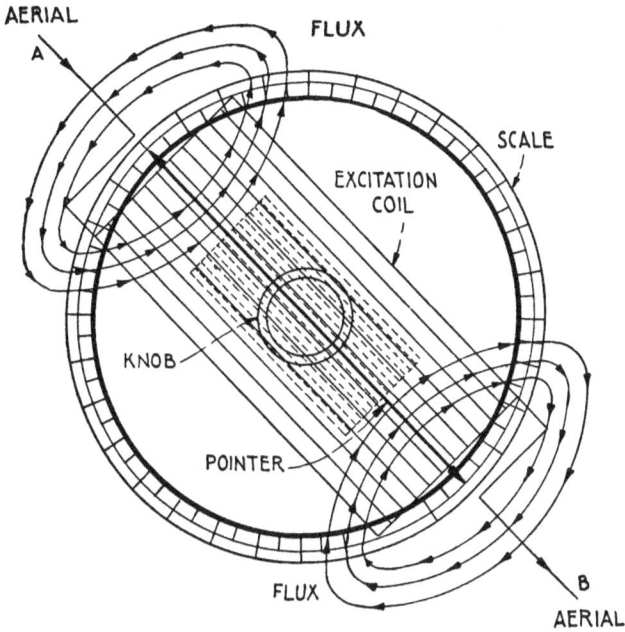

FIG. 260.   Showing the magnetic field within the Marconi goniometer when one loop alone is acted upon by the advancing wave.

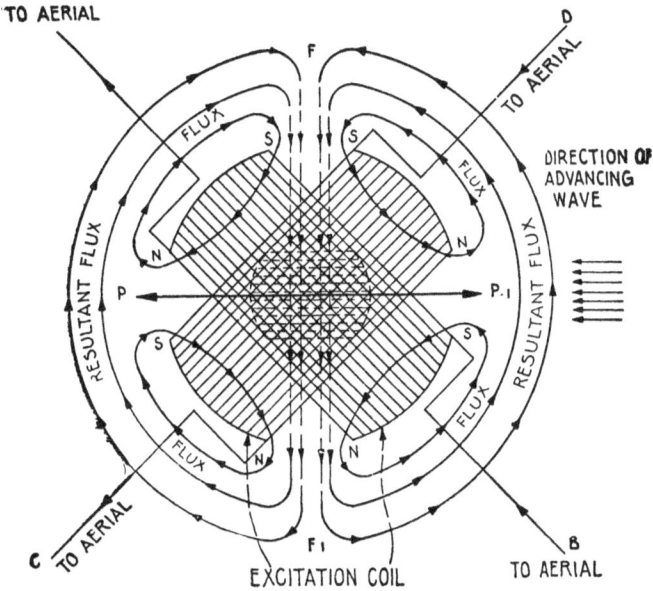

FIG. 261   Showing the magnetic field within the goniometer when both loops are acted upon equally by the advancing wave.

to receive the maximum energy from the corresponding goniometer primary, the pointer mounted on its shaft will point in the direction of the plane of the loop. The same results are obtained from the other loop when it alone is acted upon.

If the advancing wave strikes the two loops at any angle, a resultant field will be set up within the goniometer bearing a direction at right angles to the direction of the advancing waves. Then, when the exploring coil L-2 is turned to receive the maximum induction from the resulting

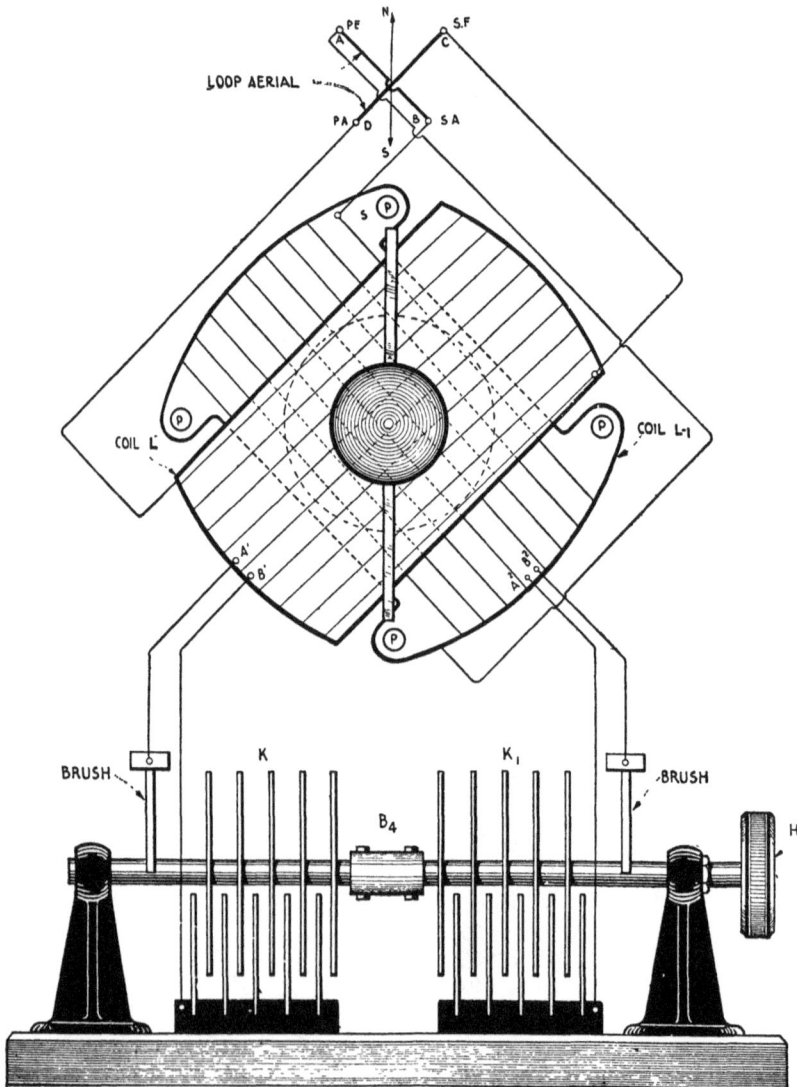

FIG. 262 Plan view of a goniometer suitable for the amateur station The connections between the triangular aerials and the goniometer primaries are indicated Tuning is effected by two variable condensers, one connected at the center of each goniometer primary.

field, the *pointer will be along the* direction of the transmitting station. The resultant fields for two cases are shown in Figs. 260 and 261. Fig. 260 shows the field around the goniometer primary when one loop is being acted upon alone. Fig. 261 shows the resultant field when the advancing wave acts upon both loops at equal angles. For any other angle, the resultant field will shift accordingly.

A suitable goniometer will now be described for reception at amateur wave lengths. The dimensions and the constructional details appear in Figs. 262 to 265. The chief part of the equipment is the *goniometer*, which utilizes the energy of both loop aerials to affect the detecting circuit. Briefly, it will be noted that the goniometer consists of two identical rectangular coils mounted co-axially at an angle of 90° with an exploring coil located in the center. Each coil is wound with 8 turns of No. 26 wire and is interrupted at the exact center so that a series tuning condenser may be inserted.

The exploring coil mounted on a vertical axis at the center of the rectangles is constructed like the rotating coil of a variometer and is wound full with No. 28 s.s.c. wire. The terminals of the ball winding are connected to the detecting circuit as in Fig. 258. The ends of the goniometer primaries are connected to the two triangular loops in the manner shown in Fig. 262.

The chief requirements of an operative direction finder set are satisfied if the general design outlined above is adhered to. It makes little difference, from a mechanical standpoint, in what manner the experimenter mounts the rectangles and the exploring coil, for if the two rectangles are permanently fixed at a right angle, and the exploring coil is accurately centered inside, a workable apparatus will result. The advantage of the Marconi radiogoniometer over some other systems is that the loop aerials need not be turned, as the direction of a sending station is determined simply by rotating the exploring coil.

Fig. 262 is a top view of the two goniometer primaries, and shows the connections from these to the loop antennae. The primaries L and L-1 cross each other at a right angle, the primary L being mounted slightly above L-1. Coil L-1 is interrupted at A^2, B^2 for connection with the variable condenser K-1; similarly coil L at A', B' to include the variable condenser K. The condensers are identical in construction, and in capacity, and both are mounted on a common shaft rotated by the control handle H. The condensers are insulated from each other by the hard rubber sleeve B-4. A pointer which moves over a 180° scale should be attached to the shaft

Each goniometer primary has exactly 8 turns plus the length of the top pieces (Fig. 262), spaced $\frac{1}{4}''$, slots being sawed in the top and bottom frames to take the wire.

Fig. 263 affords an idea of the goniometer in elevation. Each rectangular coil is $6\frac{1}{2}''$ high, $5''$ long (at its maximum diameter) and $2\frac{1}{2}''$ wide. A $\frac{3}{16}''$ brass rod passes through the top, through the support for the coils L and L-1, and finally terminates in a bearing in the base. The variometer ball is mounted at the exact middle of the goniometer. The dimensions of the ball are given in Fig. 264. The shaft also carries the control handle and a pointer which moves over the scale shown in Fig. 265.

Fig. 264a shows the dimensions of the top and bottom supports for the rectangular coils. The support for one coil A is spread out at the ends to take machine screws for holding the supporting hard rubber rods R-1 in Fig. 263. The support B for the other coil is screwed to A, but a slight air gap is left to permit the wire to be fished through during

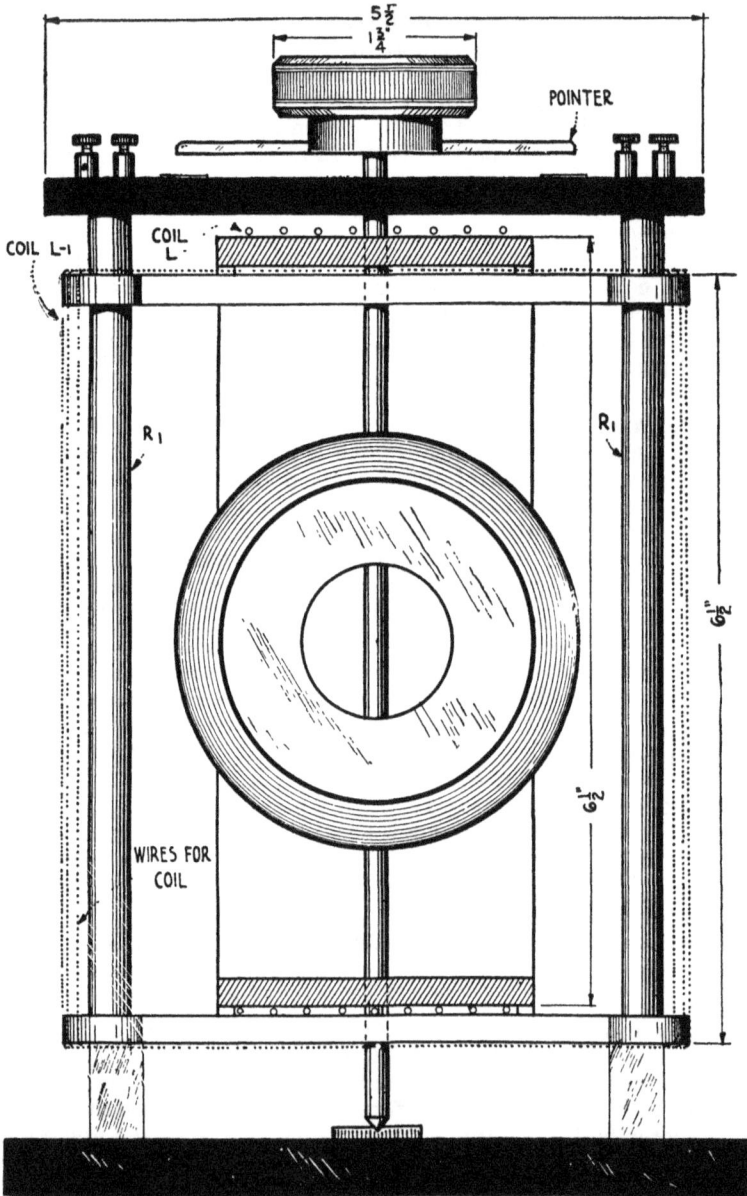

FIG. 263. Amateur type of goniometer in elevation.

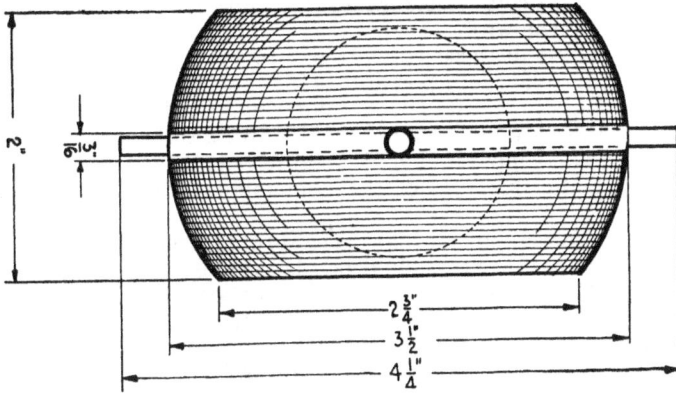

Fig. 264. Dimensions of the exploring coil.

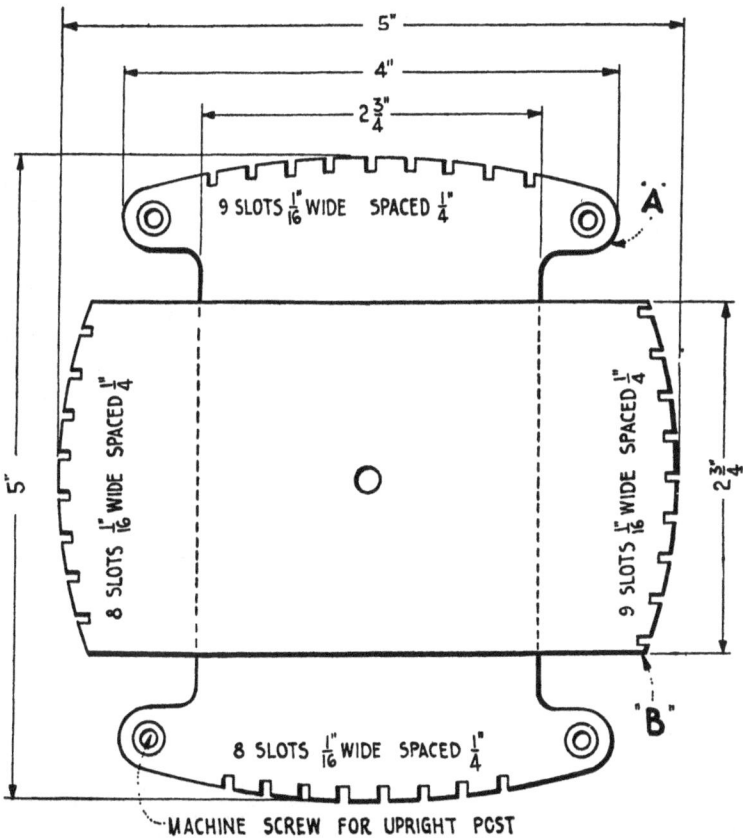

9 SLOTS $\frac{1}{16}$" WIDE SPACED $\frac{1}{4}$"

8 SLOTS $\frac{1}{16}$" WIDE SPACED $\frac{1}{4}$"

9 SLOTS $\frac{1}{16}$" WIDE SPACED $\frac{1}{4}$"

8 SLOTS $\frac{1}{16}$" WIDE SPACED $\frac{1}{4}$"

MACHINE SCREW FOR UPRIGHT POST

Fig. 264a. Dimensions of the top and bottom supports of the goniometer.

the winding of the coils (see Fig. 263). Slots $\frac{1}{16}''$ wide are cut in both A and B to take the wire.

Careful scrutiny of these sketches will reveal that the variometer ball winding must be placed in position before the turns are wound on the frames. Placing these turns would indeed be tedious were it not for the fact that they are few in number. It will also be clear that the circumference of the supports A and B represents a circle $5''$ in diameter. That is to say the diameter of the goniometer looking down from the top is $5''$.

Note that the scale in Fig. 265 is calibrated from $0°$ to $180°$ in both directions. The zero index should be placed along the line marked $N\,S$ in Fig. 262. If the line $N\,S$ is in the direction of the geographical or magnetic north pole the direction of a sending station is readily located with the aid of a map. The possible towns or cities from which such signals emanate may be recorded for future reference.*

Suitable loop aerials for moderate wave lengths are shown in Fig. 266, where each has the form of an isosceles triangle. Each loop has but one turn for short waves up to 600 meters, for longer waves two or three turns may be used. As long as they are symmetrical, the loops may have the shape of a square, a rectangle, or a circle. The aerial in Fig. 266 is 40 feet high and 30 feet long at the base. The lead-in wires W_1 and W_2 should be placed at least 5 feet above the earth. The two loops must be symmetrical in every respect.

CALIBRATION OF THE GONIOMETER.—The terminals of the two-loop aerials in Fig. 262 are marked $S\,F$, $P\,A$, $S\,A$, $P\,F$. These letters refer to their positions aboard ship, and mean respectively, *starboard forward, port aft, starboard aft* and *port forward*. Thus the line $N\,S$ in Fig. 262 would be the bow and stern line of the vessel. Whether or not the terminals of the loops are connected to the proper terminals of the goniometer primaries can be checked up by a wavemeter set into excitation by a buzzer. The method of calibration follows: Place the coil of the wavemeter (with buzzer active) in abutment with the side of the loop $P\,F$. Rotate the exploring coil (after resonance has been established by turning the variable condensers) until a signal maximum is obtained. The plane of the exploring coil should then coincide with the plane of the corresponding goniometer primary, in this case the coil L-1 Fig. 262; that is, the pointer should lie on the goniometer scale in the position $45°$–$135°$.

Then place the wavemeter coil in abutment with the other loop at the side $S\,F$ and follow the same process. If the second loop requires re-tuning of the detector circuit for maximum signals it is an indication that the loops are unsymmetrical. This should be corrected immediately and it will probably be found that the wires of one of the loops are not pulled taut.

Now place the wavemeter exactly midway between the sides of the loops corresponding to $P\,F$ and $S\,F$, and turn the direction finding

---

*This apparatus is a bilateral direction finder or radio compass That is, it gives the general line of direction, but does not indicate on which side of the loop the signal is received A unilateral apparatus is in use in the U S Navy which gives the absolute direction with a fair degree of accuracy In this system a vertical antenna is balanced against a closed circuit loop, advantage being taken of the relative phases of the currents in the loop and in the vertical antenna, and of the directional characteristics of the loop to obtain a reading

pointer for a signal maximum. The pointer should then lie on the position $0°-180°$ and the signal should disappear in the position $90°-90°$. If the maximum is obtained in the position $90°-90°$, reverse the connections to one of the goniometer primaries.

In the actual Marconi set, a small wavemeter, giving $\lambda = 300$ and $\lambda = 600$, is placed immediately above the goniometer box with the four leads from the two loops through the corners of the box. Thus since both loops are excited symmetrically, the instrument can be tuned to any given wave length in advance of the reception of signals. The complete goniometer can at the same time be checked up for accuracy and correct connections.

In order that the signals from the buzzer will be heard when the wavemeter is a small distance from the loops a Marconi V. T. should be used as the oscillation detector. During calibration the wavemeter should be set at various wave lengths and the positions of the variable condensers K and K-1, and the secondary variable V-1 (Fig. 258) noted.

The goniometer and aerials just described will respond up to 600 meters, but by the use of suitable loadings or multiturn loops, much

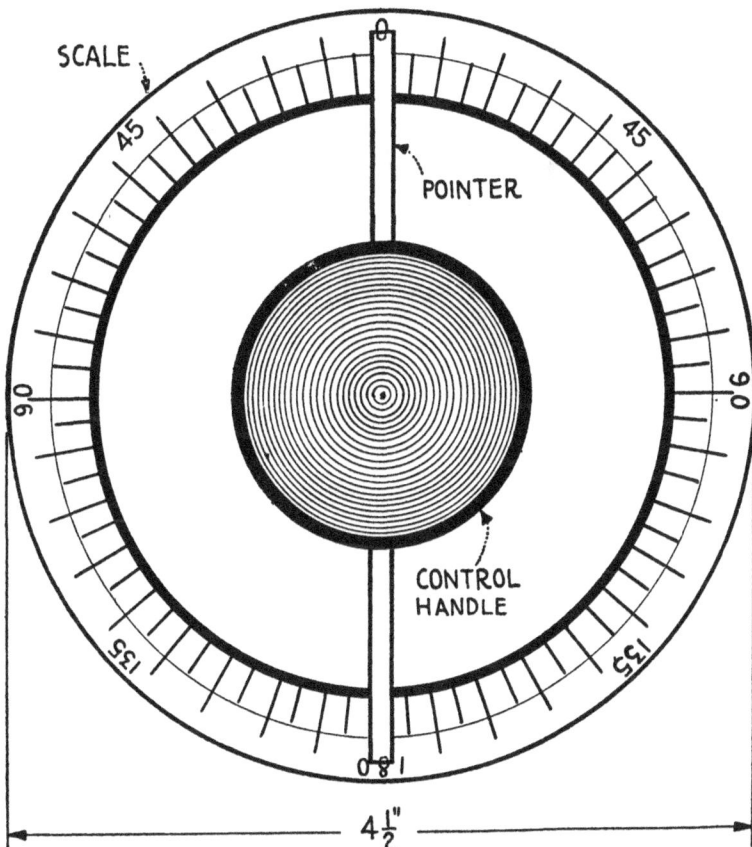

FIG 265   Details of a suitable goniometer scale.

longer waves can be received. The possible wave length range is of course governed largely by the dimensions of the loops.

To find the direction of a sending station proceed as follows: Tune the loops to the desired wave length (if known) by a wavemeter. Turn the handle of the exploring coil until a signal maximum is obtained. The pointer should then lie along the general direction of the sending station. Now that the general line of direction of the sending station has been determined, other means must be employed to find out on what side of the receiving loop the signals are transmitted. The observer, if he be at a land station and the land station is not on an island, obviously can locate the absolute direction of the ship sending, but if he is aboard a ship or on an island other means must be employed to indicate on which side of the receiving station, the station sending is located.

If the position of the pointer for a maximum is not sharply defined, the operator should observe the scale readings on both sides of maximum where the signals just disappear. A mean of these readings is

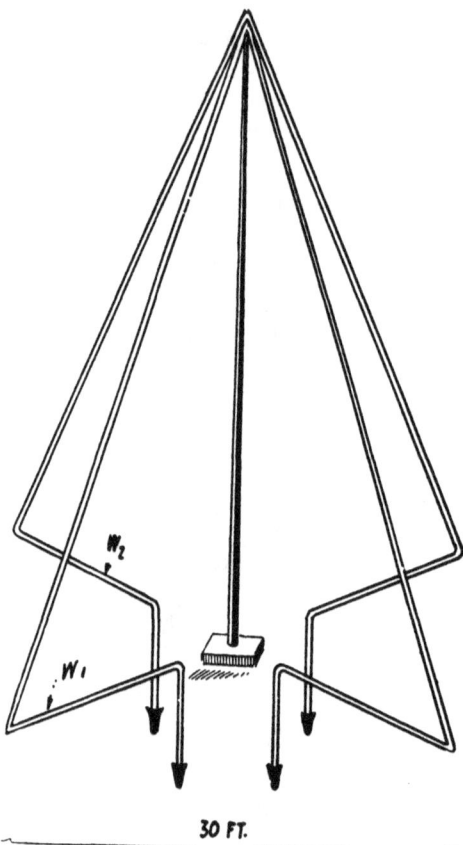

FIG. 266. Triangular loop aerials for use with the Marconi direction finder For long wave length reception multi-turn loops are employed.

then taken for the maximum. The three-bulb Marconi V. T. set of Fig. 183 should be used for long distance reception at short wave lengths. The goniometer described here will be used in connection with Weagant's static eliminator explained briefly in the following chapter.

CHAPTER XIII

WEAGANT STATIC ELIMINATOR—UNDERGROUND AERIALS

GENERAL.—Until Weagant's epoch-making discovery, no satisfactory solution of the static problem had been presented. His researches seem to indicate that the predominant types of *static waves* may be classified under two headings, viz., "grinders" and "clicks." The former are the kind that cause a continual grinding sound in the telephones and are especially strong in the temperate zone during the summer months. The latter are more like the detonations of cannon fire taking place at irregular intervals.

Like wireless waves, *clicks* appear to be propagated *horizontally;* while *grinders* are propagated *vertically* and they affect simultaneously aerials separated by great distances. This latter observation served as a basis for the development of practical apparatus and the results obtained in the earliest experiments justified the hypothesis to a large degree.

The interference from grinders is most prevalent in the warm season from noon to the sunrise of the following morning. Clicks are most noticeable during the cooler periods of the year and day. They do not interfere with reception as extensively as grinders, unless the incoming signals happen to be very weak.

The interference of grinders may be eliminated satisfactorily by Weagant's two-aerial system, but to do away with the "jamming" of both grinders and clicks, he devised a three-aerial arrangement which has proved very satisfactory.

It may be well to mention here that one reason why the former attempts to eliminate static interference have proved unsuccessful was due to the belief generally held that the frequency of the static currents in receiving antennae differed from the frequency of the signal. Weagant has proven conclusively that the static currents partake of the frequency of the antenna system whatever it may happen to be, and hence any attempts to utilize simple balancing circuits outside of those used in the system devised by Weagant prove fruitless.

Obviously it is impossible to separate two currents of the same frequency in the same circuit in any ordinary radio receiving system. However, by employing the principle discovered by Weagant an interfering current of the same frequency but of different wave form than the desired current can be balanced out most effectively. As will be presently understood, he took advantage of the apparent vertical propagation of static "waves" as contrasted with the horizontal propagation of wireless waves to devise a system whereby the signal currents can be retained while the static currents are annulled.

WEAGANT'S TWO-AERIAL SYSTEM.—The fundamental principle may be explained by the aid of Fig. 267. Let loops A' and B' be separated one-half wave length from center to center and assume that the arrows A indicate passing signal waves. Then let the downward pointing vertical arrows represent vertically propagated static waves. Evidently loops A' and B' are acted upon simultaneously by the static waves, whereas any half cycle of the signal wave acts first on one loop and then on the other. That is when the positive half of the signal wave acts upon loop B', the negative half acts upon A'.

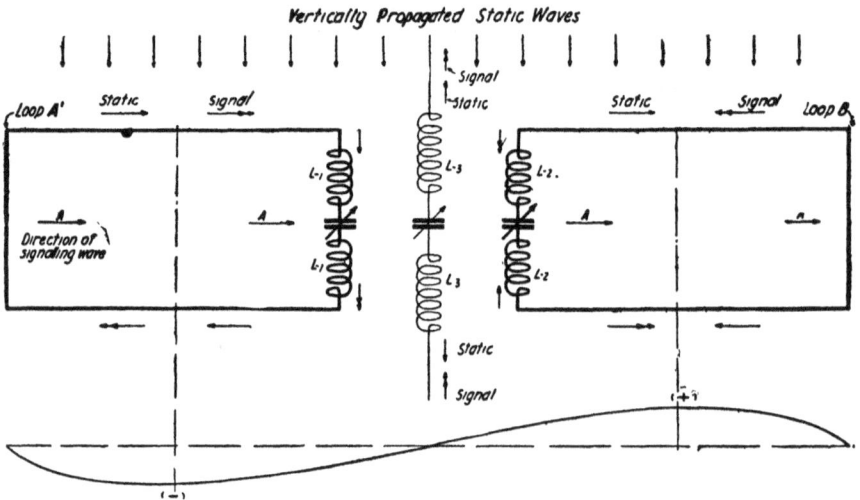

FIG. 267. Showing the fundamental principles of Weagant's two-aerial system for the elimination of the interference of "grinders"—a type of static which apparently is propagated vertically. For amateur wave lengths each loop should be 350 feet in length.

If now both loops are coupled to a common detector circuit L-3, the effect will be as follows: The static currents will flow in the same direction in both loops (as indicated by the single pointed arrows) and in the particular illustration under consideration, will go up through the coil L-2 and down through the coil L-1. The magnetic fields of L-1 and L-2 are thus in opposition and substantially no static currents will flow in the detecting circuit L-3.

The signaling currents in the two loops, however, flow in opposite directions (as indicated by the two-pointed arrows) and as a result, the signal current goes down the coil L-1 and down the coil L-2. Their e.m.f.'s add and if the loops are spaced ½ wave length the signals will be twice as strong as with one loop.

In practice L-1 and L-2 are the primaries of a goniometer such as described in Figs. 258 to 266, while L-3 is the inner or rotating coil.

To balance out static it is then only necessary to turn the rotating coil until a signal maximum is obtained, the balancing being done by the goniometer itself. In one position of the goniometer the signals will be a maximum and the static interference a minimum. If the exploring coil is now turned 90°, the static interference will be a maximum and the signals a minimum.

One-half wave length separation of the loops will give the best results but a general order of the result will be obtained with any spacing. For one-half wave length separation at 10,000 meters, the distance from center to center of the loops Fig. 267 should be 5000 meters = 16,250 feet (approx.) = 3 miles. The entire antenna system will then be 6 miles long. Such an installation is now in operation at Lakewood, N. J., and the results secured are most satisfactory.

THE PRACTICAL CIRCUIT FOR THE TWO-AERIAL SYSTEM.
—This is shown diagrammatically in Fig. 268. Long single turn loops spaced from 5 to 15 feet between wires are employed. To improve the tuning suitable loadings L-9 and L-10 are inserted in the middle of each loop.

The balancing is done in a goniometer L-1, L-2, L-3 like that of Fig. 258. The switches S-3 and S-4 are provided so that the primary condensers may be connected in series or in parallel with the loops. Switches S-1 and S-2 are employed to reverse the connections between the goniometer and the incoming leads. This is essential in order to obtain a satisfactory balance.

Another switch S-7 is placed between the rotating coil L-3 of the goniometer and the detector circuit. C-3 is a variable condenser connected in series with L-3. L-9 and L-10 are loading coils of 30 millihenries each for $\lambda = 12,000$ meters and 5 millihenries for $\lambda = 6000$ meters. Damping resistances of 1000 ohms are sometimes connected in the leads.

The procedure for balancing out static currents is as follows: First tune one loop to the desired signal and then the other. Then rotate L-3 until the *signals are a maximum and static a minimum*. Then to ascertain if a better balance can be obtained try series or parallel goniometer tuning and reversing the leads.

THE TWO-AERIAL SYSTEM APPLIED TO AMATEUR USE.—
The amateur with the space to erect these loop aerials should be able to utilize the Weagant system to good advantage. The upper and lower wires of the loops can be mounted on fence posts or perhaps permission can be obtained to erect them on local telephone poles or between trees. The upper and lower wires of a barbed wire fence may be employed if the fence lies in the proper direction for the station which it is desired to receive.

For $\lambda = 200$ meters, the centers of the Weagant loops should be spaced $\frac{200}{2} = 100$ meters = 325'; and the whole antenna system will be $2 \times 325 = 650'$ long. A simple goniometer may be used for balancing purposes and no loadings will be required. *Series goniometer* tuning will perhaps provide a balance and hence the circuit of Fig. 268 can be considerably

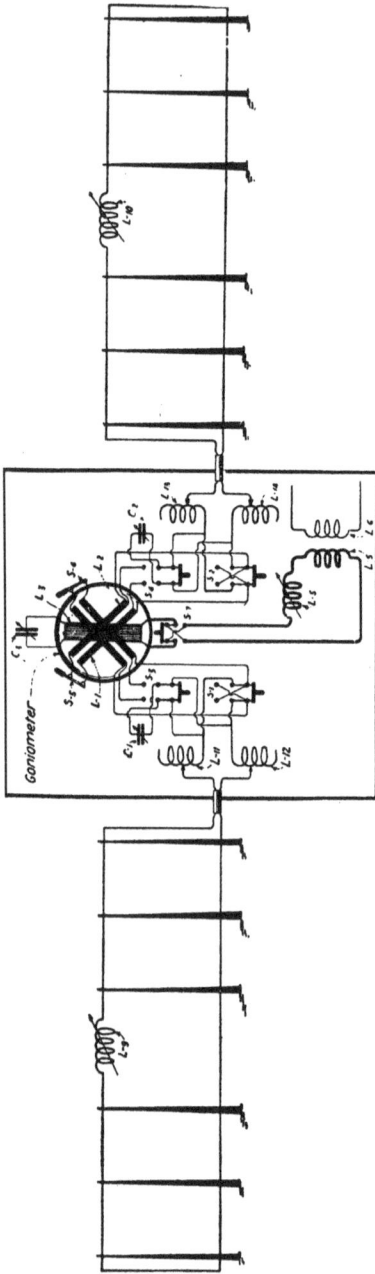

Fig. 268. The practical circuit for the Weagant two-aerial system. The static currents are balanced out in a goniometer, the primaries of which are coupled to the loops

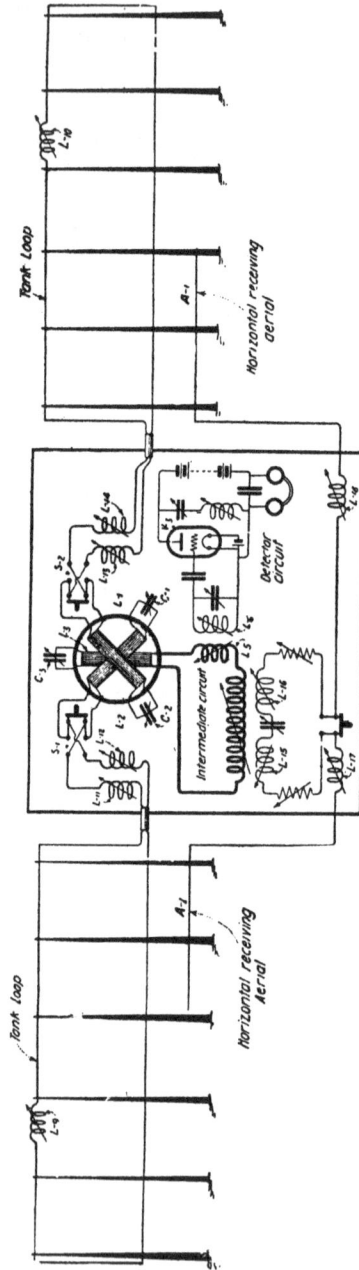

Fig. 269. Weagant's three-aerial system for the elimination of static interference. This is the most successful system of static elimination yet devised. It effectually annuls the interferences of both 'grinders' and 'clicks' permitting 24-hour communication over great distances. In this circuit the loop aerials are a source of static currents only whereas the long low horizontal aerial immediately underneath is the main receiving aerial which picks up both static and signal currents. The static currents generated in the loops are balanced against those in the receiving aerial resulting in substantially complete annulment while the signal currents are retained An installation of this kind is now in daily operation at Lakewood, N. J. The two loops are 6 miles in length and the horizontal receiving antenna is 6,000 feet in length. In practice, means are provided to reverse the phase relations of the currents in L-15 and L-16 in respect to the intermediate circuit

simplified, by eliminating the switches *S*-3 and *S*-4. The variable condensers *C*-1 and *C*-2 should be connected in place of the knife blade switches *S*-5 and *S*-6. The reversing switches *S*-1 should remain.

The experimenter must remember that these loops possess directional characteristics and that for maximum signals the planes of loops must point in the direction of the sending station. Fair signals can, however, be obtained at a considerable angle to the loops. But any signal propagated perpendicularly to the planes of the loops will not be heard even when the loops are tuned to resonance with it. If the relative phases of the currents in two loops are adjusted by the series condensers, the system can be made uni-directional receiving with maximum intensity off either end desired.* For such work the loops should have one-quarter wave length separation.

The necessary value for the loadings *L*-9 and *L*-10 for $\lambda = 200$ meters are not known, but it is recommended that coils of 0.5 millihenry be tried. If the tuning is not satisfactory other values may be tried until sharp resonance is secured The coils should be of the air core type baked in a waterproof compound.

The working circuits of the *Weagant three-aerial system* are shown in Fig. 269. In this method the two loops are used in such a way that they have been called a *"static tank "* The main receiving aerial is a *long low horizontal wire A*-1, erected immediately underneath the loops. The two loops and the receiving aerial are coupled to a common detector circuit through suitable goniometers. The goniometer which is connected between the two loops is adjusted to give *static currents* but no *signals*. The horizontal receiving aerials on the other hand, *picks up both static and signals*. The *static currents generated in the static tank are then balanced against those in the receiving aerial, leaving the signal currents which are heard in the head telephone*. Experiment proves that the static tank is a source of static currents generated by both *grinders* and *clicks*. Hence, both the grinder and click currents induced in the receiving aerial may be balanced out and the signal currents retained. It is due to the fact that the static tank picks up some of the energy of both grinders and clicks that the three-aerial system is superior to the two-aerial arrangement. The latter is only effective in reducing the interference of grinders. The three-aerial system is the *most perfect static eliminator* yet devised as it permits reception at all times except during local lightning storms.

UNDERGROUND AERIALS.—The long low horizontal aerials used extensively by the U. S. Navy were first employed by G. Marconi. Signals were received over great distances. Zenneck's "Wireless Telegraphy" reports that similar antennae have been used by Braun, Kieblitz and others. Weagant used these aerials at New Orleans early in 1914.

At present the method is to bury these aerials two or three feet below the earth's surface. But the mode of operation is the same if they are buried below or placed slightly above the earth's surface.

The fundamental idea of the underground aerial system is disclosed in Fig. 270 where two long horizontal aerials extending in opposite

---

*For a more comprehensive description of the Weagant system, see the author's report in the April and July, 1919. issues of The Wireless Age

directions are laid below the surface of the earth. The receiving station is located midway between them. It is apparent that the electrostatic capacity of the underground wires in respect to the true conducting earth plays an important part in their operation, so much so that these aerials probably *function as loops*. This is substantiated by the observation that there is an *optimum length* which can be employed for a given range of wave lengths and that even the material of the insulation covering the wire governs the optimum length. Experimenters unacquainted with the facts have been given the impression that underground aerials give better signals than other types, but this is not true as will soon be found out by actual test. The signals are generally weaker than those obtained with antennae above the earth and the signal to static ratio is not improved except through the balancing out effects obtained by the wave length separation of the two aerials. A small frame aerial will give just as good results with much less labor in installation.

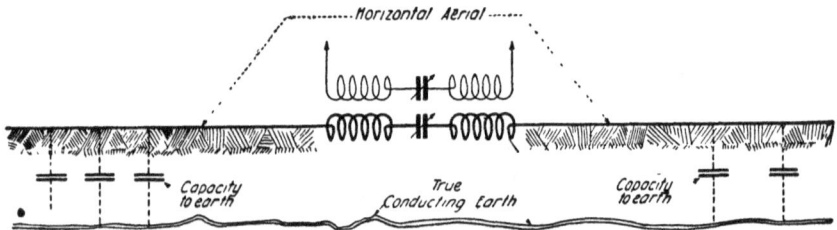

Fig. 270 Showing the disposition of underground aerials and illustrating their electrostatic capacity to earth The system usually employs two aerials extending in opposite directions, with the receiving apparatus installed at the center.

Experiment indicates that if the system of Fig. 270 is employed for reception at wave lengths between 450 and 1000 meters, and No. 12 d.r.c. is used, each wire should be about 150 feet in length. If standard 20,000-volt high tension cable is employed the wires should be about 250' in length. In the first case, the complete antenna system will be 300' long, in the second case 500' long. Whether or not the average experimenter will be inclined to dig a 500-foot ditch is a matter on which the author would rather not pass an opinion.

Although the receiving set is generally coupled to the center of the underground antenna system as in Fig. 270, a single wire grounded at one end may be used as shown in Fig. 271. The series antenna condenser of Fig. 270 is usually necessary for maximum signals.

For short wave lengths the finding of the "optimum" length of the underground aerials—the length for best signals—is a tedious job and must be determined by experiment. The procedure is to bury a 300-foot high tension cable say 3', and bare the cable every 20' for the last 100' toward the free end. The cable is brought to the surface at each break and different lengths tried until the best signals are secured. The optimum length varies markedly with the nature of the soil and the size and insulation of the wire. If tests indicate that the strongest signals apparently are to be obtained between two breaks the intervening 20-foot length must be tapped at three or four places and further trials made for maximum signals.

When the optimum length is thus found, the splices are taped and then vulcanized, or insulated in any serviceable way. The free ends must be particularly well insulated. The last two feet of the wire at the free end may be inserted in a rubber tube and filled with sulphur, or the ends may be vulcanized or coated with a good insulating compound. All "grounds" and "leaks" must be carefully avoided.

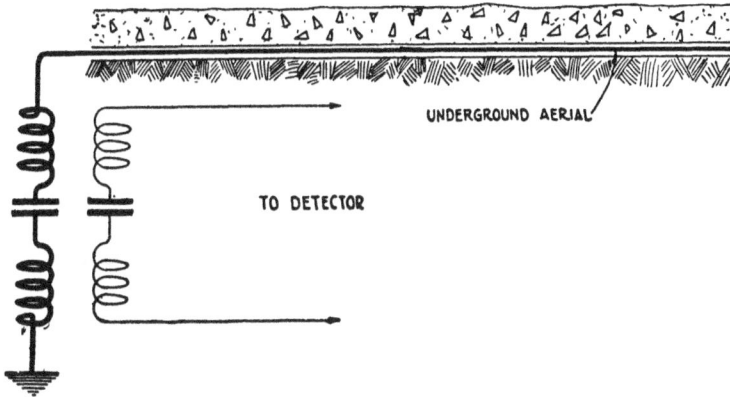

UNDERGROUND AERIAL

TO DETECTOR

Fig. 271. Generally in the underground system two aerials are laid down in opposite directions, but in some cases a single wire connected to earth at one end is used.

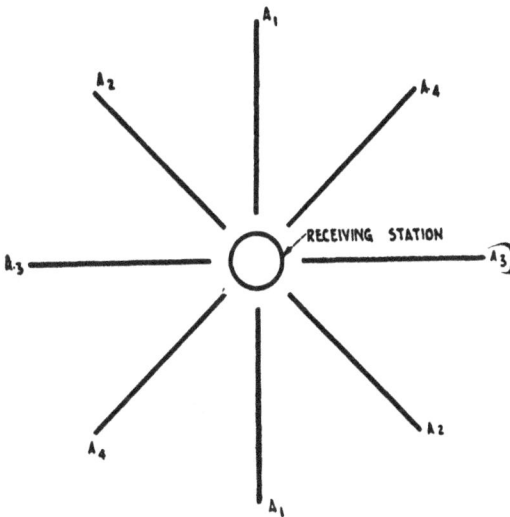

RECEIVING STATION

Fig. 272. For reception from all points of the compass several aerials covering six or eight directions should be laid down

As in the complete metallic loop system the plane of the underground aerial must point in the direction of the sending station. For universal reception, 8 points of the compass should be covered as in Fig. 272 and switches provided for each pair of aerials. Or if the majority of stations lie in a definite direction one pair of aerials may be laid down in the plane of these stations. They should be accurately surveyed for any given transmitting station.

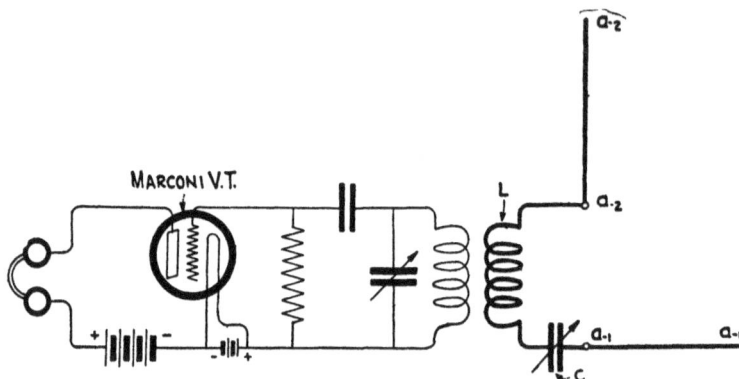

FIG. 273. For "stand by" operation with underground aerials, two aerials are laid down at a right angle If one runs north and south, the other should extend east and west.

A "stand-by" circuit is obtained by having one aerial at a right angle to the other as in Fig. 273. Thus, if A_1 runs north and south, A_2 should run east and west.

For long waves between 4000 and 15,000 meters, two 1200-foot aerials made of No. 10 or 12 d.r.c. wire are used. Better signals are often obtained with greater lengths up to 2400', which is usually the optimum length that can be employed. The probable reason for this is that at such lengths, the capacity to earth is excessive, providing several paths to earth. Rogers, who has conducted numerous experiments with these aerials, claims that the optimum length may be considerably increased by covering the conductors with metallic casings at equally spaced intervals. These casings are not in contact but they are connected to earth. They are well insulated from the aerial conductor.

In comparing the practicability of underground aerials with common types, it becomes evident that only the amateur who lives on a farm or in some isolated community will be able to use an underground aerial. Extensive ditch digging in congested areas is liable to induce litigation, particularly if the ditch crosses two or three private properties! The question to be settled is, which of the two undertakings involves the least labor for reception at short wave lengths, the digging of a 600-foot ditch or the erection of a 100-foot flat top aerial?

Without doubt, the small frame aerials erected indoors will appeal most to the ambitious experimenter who desires to receive long distance signals by an up-to-date method.

CHAPTER XIV

LONG DISTANCE RELAYS BY RADIO—WAVE LENGTH OF HIGH POWER STATIONS— GENERAL INFORMATION RE- GARDING INSTALLATION

GENERAL.—One of the chief pleasures of the advanced experi- menter is the transmission of wireless telegrams over great distances. The despatch of wireless signals with low-power sets and small aerials over distances up to 1500 miles indicates the progressive age in which we live. Imagine the effect of such a suggestion 25 years ago! Only the most ardent dreamer would have dared to cherish such a theory. Yet, to-day mere boys are doing this work with apparatus which they have generally constructed themselves.

The beginner must realize, however, that such long distance com- munication can only be effected in the temperate zone at night, during the favorable months of the year. During the day much greater power inputs than those used by amateur stations are required to establish continuous communication over such distances. The daylight range of a 200-meter set is limited to something like 200 miles, but at night the above mentioned distance of 1500 miles has been exceeded.

SELECTION OF STATIONS AND PERSONNEL FOR LONG DISTANCE RELAYING.—Usually, stations which are known to have done consistent long distance service are selected for this work. One organization has divided the U. S. into six sections, known as the At- lantic, Gulf, West Gulf, Central, Rocky Mountain, and Pacific divisions. Each section is under a division manager.

Division managers appoint the assistant division managers and the district superintendents. Applications for appointment to the trunk line of the organization are made to the district superintendents, who have charge of all traffic arrangements. The superintendents report to the assistant division managers, the assistant division managers report to the division managers; and they, in turn, report to the traffic manager, who is responsible to and reports to the Board of Directors.

The superintendents form the points of contact with the general membership of the league. In addition to this they see that local clubs are formed and affiliated with the league in every town of their district with a population large enough to warrant an organization. These local clubs aim to handle local traffic and to educate the radio beginner of the locality, that he may be of greater value in the radio field and understand the proper operation of his station. The affiliation of these numberless radio clubs throughout the country will build up an organization large and powerful enough to offer determined and effective resistance against any future measures which interfere unduly with amateur radio work.

Needless to say the owner of a *relay station* should be an *expert operator;* capable of taking down at least 25 words per minute. He should be thoroughly familiar with the methods of handling traffic at Government and commercial stations. He should know the *U. S. laws* and the *International regulations*; and should enforce them in his immediate vicinity. A campaign of education conducted by a local radio club is the most effective means of impressing the beginner with the requirements of a law-abiding station. The relay station should conform to the wave length restrictions and make sure that the power input does not exceed that required by law, viz., $\frac{1}{2}$ kw. near Government stations and 1 kw. at greater distances. The station should be neatly wired and installed in accordance with the underwriters' rules.

A three-stage, vacuum tube set should be used for long distance reception. A regenerative set will serve if the distance to be covered is not too great.

The station should be equipped with a *quick operating change-over switch* so that no unnecessary time is lost in changing from transmission to reception.

Prolonged calling of stations should be avoided in the despatch of traffic. Brief calls at short intervals will be just as effective. Needless repetition of words should be avoided except when "static" interference is strong. Sending schedules should be arranged in advance to be sure that the receiving operator is at his post.

Neither a relay station or a "cross-town" station should "hog" the air. All good operators have equal rights but a beginner should "keep out" until he becomes an efficient operator. A "ten-word-a-minute" operator is a nuisance and should not be allowed to operate his set when located in the vicinity of a great number of stations although the law at present permits him to do so.

A station when communicating no more than 5 or 10 miles should never use full power but should employ the lowest power consistent with readable signals The key should never be "held down" for testing while other stations are working Attention to these suggestions will go toward maintaining peace and harmony. Always "listen in" before you send in order to see what is going on and whether or not your sending will cause interference.

REGARDING INSTALLATION.—All power circuits should be placed in metal conduit or metal moulding and the outside casing grounded to earth. Protective condensers should be placed on the power lines. Transmitting aerials should never be erected over power, lighting, or telephone wires, but if possible, should be swung at a right angle to them.

The power circuits of the radio transmitter should be connected up with No. 10 or No. 12 r.c. wire. the secondary circuit with No. 16 or 18 wire. The primary circuits should be equipped with appropriate fuses. The antenna lead-in wires should equal in conductivity the flat top portion of the aerial.

Receiving sets may be wired up with No. 18 annunciator wire. There is no advantage in having more than two wires in the receiving aerial. The transmitting aerial should have at least four wires. Two-foot hard rubber or bakelite insulators should be used at either end of a transmitting aerial.

WAVE LENGTH OF HIGH POWER STATIONS.—At the present writing the wave length name and call letters of active high power stations are as follows:

Call	Wave Lengths	Location	Characteristics
XDA	4,000	Mexico City, Mexico	Damped
GB	7,500	Glace Bay, Nova Scotia	Damped
NWW	9,800	Tuckerton, N. J.	Undamped
OUI	10,500–15,000	Eilvese, Germany	Undamped
IDO	11,000	Rome, Italy	Undamped
LCM	11,500	Stavenger, Norway	Undamped
NPM	8,100–11,000	Pearl Harbor, Hawaii	Undamped
UA	11,500	Nantes, France	Undamped
NDD	9,200–13,600	Sayville, L. I.	Undamped
POZ	12,600	Nauen, Germany	Undamped
NPL	9,800–13,360	San Diego, Cal.	Undamped
NFF	13,600	New Brunswick, N. J.	Undamped
MVV	14,000	Carnarvon, Wales	Undamped
YN	15,500	Lyons, France	Undamped
NSS	16,900	Annapolis, Md.	Undamped
NAA	6,000	Arlington, Va.	Undamped
NBA	7,000	Balboa, C. Z.	Undamped
NAB	5,700	Boston, Mass.	Undamped
NPO	12,000	Cavite, P. I.	Undamped
NAO	4,700	Charleston, S. C.	Undamped
NPA	7,600	Cordova, Alaska	Undamped
NAJ	5,700	Great Lakes, Ill	Undamped
NPN	5,000	Guam, Mareanna Islands	Undamped
NAW	4,500	Guantanamo, Cuba	Undamped
NAR	6,500	Key West, Fla.	Undamped
NAT	5,500	New Orleans, La.	Undamped
NPC	5,250	Puget Sound, Wash	Undamped
NPG	8,600–4,800	San Francisco, Cal	Undamped
NAV	5,250	San Juan, Porto Rico	Undamped
NPV	6,000–3,000	Tutuila, Samoa	Undamped
BZM	5,000	St. Johns, Newfoundland	Undamped
BZR	5,000	Bermuda, W. I	Undamped
BWP	2,000	Punta Delgada, Azores	Undamped
BXY	5,000	Hong Kong, China	Undamped
BYC	4,500	Horsea, England	Undamped
FL	10,000	Eiffel Tower, Paris	Undamped
MFT	6,000	Clifden, Ireland	Damped
FCI	6,500	Coltano, Italy	Damped
TSR	7,000–5,000	Petrograd, Russia	Damped
VKT	2,200	Naura, Pacific Ocean	Damped
NAA	2,500	Arlington, Va., U. S. A. (Time Signals)	Damped
JJC	12,000	Funabashi, Japan	Damped
PMM	6,100	Java, Dutch East Indies	Undamped

These data are subject to change.

Fig. 274. Table of wave lengths covering the more important high power stations now in operation.

TOOLS.—The first essential tools are a pair of 6-inch pliers, two or three screw drivers of different sizes, a soldering torch, an electrician's knife, a soldering iron, a small wrench and a pair of "nippers." A hand drill with an assortment of small drills is very useful. Other tools will be added from time to time.

INDEX

APPENDIX

USEFUL INFORMATION CONCERNING
MARCONI V. T.'s

The following summary may be of benefit to the amateur experimenter who is doubtful regarding the exact values of the coupling resistances to be used in cascade amplifiers with the Marconi V. T., and who desires to know under what conditions the resistance coupled amplifier is superior to transformer coupling, etc. It is believed that the following data will clear the matter in all its details. The experimenter should bear in mind that Marconi tubes are sold in two grades, known as Class I and Class II.

(1) For *extremely high amplifications* using five valves or more, successive valves should be coupled through *resistances*.

(2) For *radio frequency amplification* with resistance coupling (which is practical for wave lengths above 3000 meters, but it is of no value for 200 meters) a coupling resistance of ½ *megohm* should be employed.

(3) For *audio frequency amplification* the coupling resistances should be of 2 megohms. These will provide a voltage amplification factor of 7 for each tube, and the method is of practical value up to 7 stages of amplification. A total amplification of (7^7) or 823,543 will be obtained.

(4) For *two-stage amplification*, transformer coupling is desirable, particularly on short wave lengths.

(5) Using two stages of audio frequency amplification with transformer couplings, an amplification of 400 will be obtained. Three stages of amplification are practical with transformer couplings, provided the filament currents for each tube are regulated by rheostats to prevent "howling."

(6) Using Class II Marconi tubes in connection with one of the Federal Company's audio frequency transformers, a voltage amplification of 20 is obtainable.

(7) When using Class II Marconi V. T.'s, grid leaks of 2 megohms each, should be connected to the grids and the *positive side* of the filaments.

(8) Using class I Marconi V. T.'s, grid leaks of 2 megohms should be connected between the grid and either the positive or negative side of the filament, depending upon the particular tube in use. The correct connection is readily found by experiment.

(9) Class II Marconi V. T.'s require a plate battery of 60-80 volts for amplification and 45 volts for detection.

USEFUL TABLE FOR DETERMINING THE WAVE LENGTH, FREQUENCY AND OSCILLATION CONSTANT OF RADIO FREQUENCY CIRCUITS

λ = wave-length in meters.

λ^2 = wave-length squared.

n = number of oscillations per second.

O = $\sqrt{LC}$ and is called the oscillation constant.

C = capacity in microfarads.

L = inductance in centimeters (1,000 cms. = 1 microhenry).

D = difference of L C for 1 meter.

λ	λ^2	n	O or $\sqrt{LC}$	L C	D
100	10,000	3000000	1.68	2.82	.042
110	12,100	2727272	1.80	3.24	.084
120	14,400	2500000	2.02	4.08	.067
130	16,900	2307600	2.18	4.75	.077
140	19,600	2142600	2.35	5.52	.083
150	22,500	2000000	2.52	6.35	.081
160	25,600	1874800	2.68	7.16	.096
170	28,900	1764600	2.85	8.12	.10
180	32,400	1666600	3.02	9.12	.105
190	36,100	1578800	3.19	10.17	.112
200	40,000	1500000	3.36	11.29	.11
210	44,100	1428400	3.52	12.39	.123
220	48,400	1363500	3.69	13.62	.128
230	52,900	1304200	3.86	14.90	.134
240	57,600	1250000	4.03	16.24	.131
250	62,500	1200000	4.19	17.55	.146
260	67,600	1153800	4.36	19.01	.15
270	72,900	1111000	4.53	20.52	.157
280	78,400	1071300	4.70	22.09	.163
290	84,100	1034300	4.87	23.72	.158
300	90,000	1000000	5.03	25.30	.174
310	96,100	967700	5.20	27.04	.18
320	102,400	937400	5.37	28.84	.185
330	108,900	909100	5.54	30.69	.18
340	115,600	882300	5.70	32.49	.197
350	122,500	857100	5.87	34.46	.20
360	129,600	833300	6.04	36.48	.208
370	136,900	810800	6.21	38.56	.215
380	144,400	789400	6.38	40.71	.206
390	152,100	769200	6.54	42.77	.226

λ	λ²	n	O or $\sqrt{L.C}$	L C	D
400	160,000	750000	6.71	45.03	.23
410	168,100	731700	6.88	47.33	.237
420	176,400	714300	7.05	49.70	.228
430	184,900	697700	7.21	51.98	.248
440	193,600	681800	7.38	54.46	.254
450	202,500	666700	7.55	57.00	.26
460	211,600	652200	7.72	59.60	.265
470	220,900	638300	7.89	62.25	.255
480	230,400	625000	8.05	64.80	.277
490	240,100	612200	8.22	67.57	.282
500	250,000	600000	8.39	70.39	.288
510	260,100	588200	8.56	73.27	.277
520	270,400	576900	8.72	76.04	.299
530	280,900	566000	8.89	79.03	.305
540	291.600	555600	9.06	82.08	.311
550	302,500	545400	9.23	85.19	.317
560	313,600	535700	9.40	88.36	.303
570	324,900	526300	9.56	91.39	.328
580	336,400	517200	9.73	94.67	.334
590	348,100	508500	9.90	98.01	.340
600	360,000	500000	10.07	101.41	.324
610	372,100	491800	10.23	104.65	.35
620	384,400	483900	10.40	108.15	.358
630	396,900	476200	10.57	111.73	.362
640	409,600	486800	10.74	115.35	.346
650	422,500	461500	10.90	118.81	.373
660	435,600	454600	11.07	122.54	.380
670	448,900	447800	11.24	126.34	.385
680	462,400	441200	11.41	130.19	.391
690	476,100	434800	11.58	134.10	.373
700	490,000	428600	11.74	137.83	.403
710	504,100	422500	11.91	141.86	.407
720	518,400	416700	12.08	145.93	.414
730	532,900	411000	12.25	150.07	.394
740	547,600	405400	12.41	154.01	.426
750	562,500	400000	12 58	158.27	.430
760	577,600	394800	12.75	162.57	.426
770	592,900	389600	12.92	166.83	.443
780	608,400	384600	13.09	171.35	.432
790	624,100	379800	13.25	175.57	.453
800	640,000	375000	13.42	180 10	.459
810	656,100	370400	13.59	184.69	.464
820	672,400	365900	13.76	189.33	.472
830	688,900	361400	13.93	194.05	.448
840	705,600	357100	14.09	198.53	.482

λ	λ²	n	O or √L̅C̅	L C	D
850	722,500	352900	14.26	203.35	.489
860	739,600	348800	14.43	208.24	.493
870	756,900	344800	14.60	213.17	.469
880	774,400	340900	14.76	217.86	.504
890	792,100	337100	14.93	222.90	.511
900	810,000	333300	15.10	228.01	.516
910	828,100	329700	15.27	233.17	.492
920	846,400	326100	15.43	238.09	.527
930	864,900	322600	15.60	243.36	.534
940	883,600	319100	15.77	248.70	.538
950	902,500	315800	15.94	254.08	.545
960	921,600	312500	16.11	259.53	.518
970	940,900	309300	16.27	264.71	.567
980	960,400	306100	16.44	270.38	.552
990	980,100	303000	16.61	275.90	.567
1000	1,000,000	300000	16.78	281.57	.543
1010	1020,100	297030	16.94	287.00	.570
1020	1040,400	294120	17.11	292.70	.590
1030	1060,900	291260	17.28	298.60	.590
1040	1081,600	288450	17.45	304.50	.60
1050	1102,550	285710	17.62	310.50	.56
1060	1123,600	283010	17.78	316.10	.61
1070	1144,900	280370	17.95	322.20	.61
1080	1166,400	277780	18.12	328.30	.62
1090	1188,100	275230	18.29	334.50	59
1100	1210,000	272730	18.45	340.40	.63
1110	1232,100	270270	18.62	346.70	.64
1120	1254,400	267850	18.79	353.10	.64
1130	1276,900	265480	18.96	359.50	.65
1140	1299,600	263150	19.13	366.00	.61
1150	1322,500	260860	19.29	372.10	.66
1160	1345,600	258610	19.46	378.70	.66
1170	1368,900	256400	19.63	385.30	.68
1180	1392,400	254230	19.80	392.10	.67
1190	1416,100	252100	19.97	398.80	.64
1200	1224,000	250000	20.13	405.20	.69
1210	1464,100	247930	20.30	412.10	.69
1220	1488,400	245900	20.47	419.00	.70
1230	1512,900	243900	20.64	426.00	.66
1240	1537,600	241930	20.80	432.60	.71
1250	1562,500	240000	20.97	439.70	.72
1260	1587,600	238090	21.14	446.90	.72
1270	1612,900	236220	21.31	454.10	.69
1280	1638,400	234370	21.47	461.00	.73
1290	1664,100	232560	21.64	468.30	.74

λ	λ^2	n	$\dfrac{O}{\text{or}}$ $\sqrt{LC}$	L C	D
1300	1,690,000	230760	21.81	475.70	.74
1310	1,716,000	229010	21.98	483.10	.75
1320	1,742,400	227270	22.15	490.60	.72
1330	1,768,900	225560	22.31	497.80	75
1340	1,795,600	223870	22.48	505.30	.77
1350	1,822,500	222220	22.65	513.00	.78
1360	1,849,600	220590	22.82	520.80	.73
1370	1,876,900	218970	22.98	528.10	.78
1380	1,904,400	217390	23.15	535.90	.79
1390	1,932,100	215830	23.32	543.80	.80
1400	1,960,000	214380	23.49	551.80	.80
1410	1,988,100	212760	23.66	559.80	.76
1420	2,016,400	211260	23.82	567.40	.81
1430	2,044,900	209790	23.99	575.50	.82
1440	2,073,600	208340	24.16	583.70	.82
1450	2,102,500	206900	24.33	591.90	.79
1460	2,131,600	205470	24.49	599.80	.83
1470	2,160,900	204080	24.66	608.10	.84
1480	2,190,400	202700	24.83	616.50	.85
1490	2,220,100	201340	25.00	625.00	.80
1500	2,250,000	200000	25.17	633.50	.81
1510	2,280,100	198680	25.33	641.60	.86
1520	2,310,400	197360	25.50	650.20	.88
1530	2,340,900	196070	25.67	659.00	.87
1540	2,371,600	194800	25.84	667.70	.83
1550	2,402,500	193540	26.00	676.00	.89
1560	2,433,600	192310	26.17	684.90	.89
1570	2,464,900	191060	26.34	693.80	.90
1580	2,496,400	189860	26.51	702.80	.90
1590	2,528,100	188670	26.68	711.80	.86
1600	2,560,000	187500	26.84	720.40	.91
1610	2,592,100	186340	27.01	729.50	.92
1620	2,624,400	185190	27.18	738.70	.93
1630	2,656,900	184050	27.35	748.00	.93
1640	2,689,600	182930	27.52	757.30	.89
1650	2,722,500	181820	27.68	766.20	.94
1660	2,755,600	180730	27.85	775.60	.96
1670	2,788,900	179640	28.02	785.20	.94
1680	2,822,400	178570	28.19	794.60	.91
1690	2,856,100	177510	28.35	803.70	.97
1700	2,890,000	176460	28.52	813.40	.97
1710	2,924,100	175440	28.69	823.10	.98
1720	2,958,400	174420	28.86	832.90	.93
1730	2,992,900	173410	29.02	842.20	.98

λ	λ²	n	$\dfrac{O}{\text{or}}$ $\sqrt{LC}$	L C	D
1740	3,026,600	172410	29.19	852.00	1.00
1750	3,062,500	171430	29.36	862.00	1.00
1760	3,097,600	170450	29.53	872.00	1.00
1770	3,132,900	169490	29.70	882.10	.95
1780	3,168,400	168540	29.86	981.60	1.02
1790	3,204,100	167600	30.03	901.80	.92
1800	3,240,000	166670	30.20	912.00	1.03
1810	3,276,100	165750	30.37	922.30	1.04
1820	3,312,400	164840	30.54	932.70	.98
1830	3,348,900	163940	30.70	942.50	1.05
1840	3,385,600	163040	30.87	953.00	1.04
1850	3,422,500	162160	31.04	963.40	1.06
1860	3,459,600	161290	31.21	974.10	1.00
1870	3,496,900	150430	31.37	984.10	1.07
1880	3,534,400	159370	31.54	994.80	1.08
1890	3,572,100	158730	31.71	1005.60	1.08
1900	3,610,000	157890	31.88	1016.40	1.02
1910	3,648,100	157060	32.04	1026.60	1.09
1920	3,686,400	156240	32.21	1037.50	1.10
1930	3,724,900	155440	32.35	1048.50	1.14
1940	3,763,600	154630	32.55	1059.90	1.07
1950	3,802,500	153840	32.72	1070.60	1.05
1960	3,841,600	153060	32.88	1081.10	1.12
1970	3,880,900	152280	33.05	1092.30	1.12
1980	3,920,400	151510	33.22	1103.50	1.14
1990	3,960,100	150750	33.39	1114.90	1.07
2000	4,000,000	150000	33.55	1125.60	1.15
2010	4,040,100	149250	33.72	1137.10	1.25
2020	4,080,400	148520	33.89	1149.60	1.05
2030	4,120,900	147780	34.06	1160.10	1.16
2040	4,166,600	147060	34.23	1171.70	1.10
2050	4,202,500	146340	34.39	1182.70	1.17
2060	4,243,600	145630	34.56	1194.40	1.18
2070	4,284,900	144930	34.73	1206.20	1.18
2080	4,326,400	144230	34.90	1218.00	1.18
2090	4,368,100	143540	35.07	1229.80	1.14
2100	4,410,000	142850	35.23	1241.20	1.20
2110	4,452,100	142180	35.40	1253.20	1.21
2120	4,494,400	141510	35.57	1265.30	1.21
2130	4,536,900	140840	35.74	1277.40	1.15
2140	4,579,600	140180	35.90	1288.90	1.22
2150	4,622,500	139540	36.07	1301.10	1.23
2160	4,665,600	138880	36.24	1313.40	1.23
2170	4,708,900	138240	36.41	1325.70	1.24
2180	4,752,400	137610	36.58	1338.10	1.17
2190	4,796,100	136980	36.74	1349.80	1.26

λ	λ²	n	O or √LC	L C	D
2200	4,840,000	136360	36.91	1362.40	1.25
2210	4,844,100	135740	37.08	1374.90	1.26
2220	4,928,400	135130	37.25	1387.50	1.19
2230	4,972,900	134530	37.41	1399.40	1.28
2240	5,017,600	133930	37.58	1412.20	1.29
2250	5,062,500	133330	37.75	1425.10	1.29
2260	5,107,600	132740	37.92	1438.00	1.22
2270	5,152,900	132160	38.08	1450.20	1.29
2280	5,198,400	131570	38.25	1463.10	1.31
2290	5,244,100	131000	38.42	1476.20	1.31
2300	5,290,000	130430	38.59	1489.30	1.31
2310	5,336,100	129870	38.76	1502.40	1.32
2320	5,382,400	129310	38.93	1515.60	1.25
2330	5,428,900	128750	39.09	1528.10	1.33
2340	5,475,600	128200	39.26	1541.40	1.33
2350	5,522,500	127660	39.43	1554.70	1.34
2360	5,569,600	127120	39.60	1568.10	1.27
2370	5,616,900	126580	39.76	1580.80	1.37
2380	5,644,400	126050	39.93	1594.50	1.35
2390	5,712,100	125520	40.10	1608.00	1.45
2400	5,760,000	125000	40.27	1621.80	1.45
2410	5,808,100	124480	40.45	1636.30	1.21
2420	5,856,400	123960	40.60	1648.40	1.39
2330	5,904,900	123450	40.77	1662.30	1.38
2440	5,953,600	122950	40.94	1676.10	1.39
2450	6,002,500	122450	41.11	1690.00	1.33
2460	6,051,600	121950	41.27	1703.30	1.40
2470	6,100,900	121450	41.44	1717.30	1.41
2480	6,150,400	120960	41.64	1731.40	1.40
2490	6,200,100	120480	41.78	1745.40	1.43
2500	6,250,000	120000	41.95	1759.70	1.36
2510	6,300,100	119520	42.11	1773.30	1.42
2520	6,350,400	119050	42.28	1787.50	1.45
2530	6,400,900	118580	42.45	1802.00	1.44
2540	6,451,600	118120	42.62	1816.40	1.46
2550	6,502,500	117650	42.79	1831.00	1.38
2560	6,553,600	117190	42.95	1844.80	1.46
2570	6,504,900	116730	43.12	1859.40	1.46
2580	6,656,400	116280	43.29	1874.00	1.47
2590	6,708,100	115830	43.46	1888.70	1.39
2600	6,760,000	115380	43.62	1902.60	1.49
2610	6,812,100	114940	43.79	1917.50	1.48
2620	6,864,400	114510	43.96	1932.30	1.51
2630	6,916,900	114070	44.13	1947.40	1.42

λ	λ^2	n	$\dfrac{O}{\text{or}}$ $\sqrt{LC}$	L C	D
2640	6,969,600	113640	44.29	1961.60	1.50
2650	7,022,500	113210	44.46	1976.60	1.51
2660	7,075,600	112780	44.63	1991.70	1.53
2670	7,128,900	112360	44.80	2007.00	1.53
2680	7,182,400	111940	44.97	2022.30	1.43
2690	7,236,100	111530	45.13	2036.60	1.54
2700	7,290,000	111110	45.30	2052.00	1.54
2710	7,344,100	110700	45.47	2067.40	1.56
2720	7,398,400	110290	45.64	2083.00	1.47
2730	7,452,900	109890	45.80	2097.70	1.54
2740	7,507,600	109490	45.97	2113.10	1.58
2750	7,562,500	109090	46.14	2128.90	1.58
2760	7,617,600	108700	46.31	2144.70	1.48
2770	7,672,900	108300	46.47	2159.50	1.57
2780	7,728,400	107920	46.64	2175.20	1.59
2790	7,784,100	107530	46.81	2191.10	1.59
2800	7,840,000	107140	46.98	2207.00	1.60
2810	7,896,100	106760	47.15	2223.00	1.62
2820	7,952,400	106380	47.32	2239.20	1.52
2830	8,008,900	106010	47.48	2254.40	1.62
2840	8,065,600	105630	47.65	2270.60	1.63
2850	8,122,500	105260	47.82	2286.90	1.62
2860	8,179,600	104890	47.99	2303.10	1.54
2870	8,236,900	104530	48.15	2318.50	1.64
2880	8,294,400	104170	48.32	2334.90	1.64
2890	8,352,100	103810	48.49	2351.30	1.50
2900	8,410,000	103450	48.66	2366.30	1.80
2910	8,468,100	103090	48.83	2384.30	1.47
2920	8,526,400	102740	48.99	2399.00	1.77
2930	8,584,900	102390	49.16	2416.70	1.69
2940	8,643,600	102040	49.33	2433.60	1.67
2950	8,702,500	101700	49.50	2450.30	1.58
2960	8,761,600	101350	49.66	2466.10	1.69
2970	8,820,900	101010	49.83	2483.00	1.70
2980	8,880,400	100660	50.00	2500.00	1.71
2990	8,940,100	100320	50.17	2517.00	1.61
3000	9,000,000	100000	50.33	2533.20	1.69
3025	9,150,625	99170	50.75	2575.60	1.71
3050	9,302,500	98560	51.17	2618.40	1.72
3075	9,455,625	97560	51.59	2661.50	1.74
3100	9,610,000	96770	52.01	2705.10	1.75

λ	λ²	n	O or $\sqrt{LC}$	L C	D
3125	9,765,625	96000	52.43	2748.90	1.76
3150	9,922,500	95230	52.85	2793.10	1.78
3175	10,080,625	94490	53.27	2837.80	1.79
3200	10,240,000	93750	53.69	2882.70	1.80
3225	10,400,625	93020	54.11	2927.90	1.83
3250	10,562,500	92310	54.53	2973.70	1.83
3275	10,725,625	91600	54.95	3019.60	1.84
3300	10,890,000	90910	55.37	3065.80	1.87
3325	11,055,625	90220	55.79	3112.60	1.87
3350	11,222,500	89550	56.21	3159.50	1.90
3375	11,280,625	88890	56.63	3207.10	1.90
3400	11,560,000	88230	57.05	3254.80	1.86
3425	11,730,625	87590	57.46	3301.60	1.93
3450	11,902,500	86960	57.88	3350.00	1.95
3475	12,075,625	86330	58.30	3398.90	1.96
3500	12,250,000	85720	58.72	3448.00	1.98
3525	12,425,625	85100	59.14	3497.50	1.99
3550	12,602,500	84510	59.56	3547.40	2.01
3575	12,780,625	89310	59.98	3597.70	2.01
3600	12,960,000	83330	60.40	3648.10	2.03
3625	13,140,625	82750	60.82	3699.00	2.04
3650	13,322,500	82190	61.24	3750.20	2.07
3675	13,505,625	81630	61.66	3802.00	2.07
3700	13,690,000	81090	62.08	3853.80	2.09
3725	13,875,625	80540	62.50	3906.20	2.10
3750	14,062,500	80000	62.92	3958.80	2.12
3775	14,250,625	79470	63.34	4012.00[t]	2.16
3800	14,440,000	78950	63.76	4065.00	2.16
3825	14,630,625	78430	64.18	4119.00	2.16
3850	14,822,500	77920	64.60	4173.00	2.20
3875	15,015,625	77420	65.02	4228.00	2.12
3900	15,210,000	76930	65.43	4281.00	2.20
3925	15,405,625	76440	65.85	4336.00	2.24
3950	15,602,500	75950	66.27	4392.00	2.24
3975	15,800,625	75470	66.69	4448.00	2.24
4000	16,000,000	75000	67.11	4504.00	2.28
4025	16,200,625	74540	67.53	4561.00	2.24
4050	16,402,500	74080	67.95	4617.00	2.32
4075	16,605,625	73620	68.37	4675.00	2.28
4100	16,810,000	73170	68.79	4732.00	2.32

λ	λ^2	n	$\dfrac{O}{\text{or}}$ $\sqrt{LC}$	L C	D
4125	17,015,625	72730	69.21	4790.00	2.32
4150	17,222,500	72290	69.63	4848.00	2 32
4175	17,430,625	71850	70.05	4907.00	2.36
4200	17,640,000	71430	70.47	4966.00	2.40
4225	17,850,625	71010	70.89	5026.00	2.36
4250	18,062,500	70590	71.31	5085.00	2.40
4275	18,275,625	70180	71.73	5145.00	2.44
4300	18,490,000	69770	72.15	5206.00	2.40
4325	18,705,625	69370	72.57	5266.00	2.48
4350	18,922,500	68970	72.99	5328.00	2.40
4375	19,140,625	68580	73.40	5388.00	2 52
4400	19,360,000	68190	73.83	5451.00	2.40
4425	19,580,625	67800	74.24	5511.00	2.52
4450	19,802,500	67420	74.66	5574.00	2.52
4475	20,025,625	67040	75.08	5637.00	2.52
4500	20,250,000.	66670	75.50	5700.00	2 56
4525	20,475,625	66300	75.92	5764.00	2.52
4550	20,702,500	65940	76.34	5827.00	2.60
4575	20,930,625	65580	76.76	5892.00	2.60
4600	21,160,000	65220	77.18	5957.00	2.52
4625	21,390,625	64870	77.60	6020.00	2.68
4650	21,622,500	64520	78.02	6087.00	2.64
4675	21,855,625	64170	78.44	6153.00	2.64
4700	22,090,000	63830	78.86	6219.00	2.64
4725	22,325,625	62490	79.28	6285.00	2.68
4750	22,562,500	63160	79.70	6352.00	2.68
4775	22,800,625	62830	80.12	6419.00	2.64
4800	23,040,000	62500	80.53	6485.00	2.72
4825	23,280,625	62180	80.95	6553.00	2.72
4850	23,522,500	61860	81.37	6621.00	2.76
4875	23,765,625	61540	81.79	6690.00	2.76
4900	24,010,000	61230	82.21	6759.00	2.76
4925	24,255,625	60910	82.63	6828.00	2.76
4950	24,502,500	60610	83.05	6897.00	2.80
4975	24,750,625	60300	83.47	6967.00	2.84
5000	25,000,000	60000	83.89	7038.00	

www.ingramcontent.com/pod-product-compliance
Lightning Source LLC
Chambersburg PA
CBHW022052210326
41519CB00054B/320